Wireless Digital Communications: Architecture, Design and Applications

Wireless Digital Communications: Architecture, Design and Applications

Edited by
Rafael Barrett

WILLFORD PRESS
www.willfordpress.com

Published by Willford Press,
118-35 Queens Blvd., Suite 400,
Forest Hills, NY 11375, USA

ISBN: 978-1-68285-341-2

Cataloging-in-Publication Data

Wireless digital communications : architecture, design and applications / edited by Rafael Barrett.
    p. cm.
Includes bibliographical references and index.
ISBN 978-1-68285-341-2
1. Wireless communication systems. 2. Digital communications. 3. Spread spectrum communications.
4. Computer networks. I. Barrett, Rafael.
TK5103.2 .W57 2017
621.384--dc23

For information on all Willford Press publications
visit our website at www.willfordpress.com

WILLFORD PRESS

Printed in the United States of America.

Contents

Preface..VII

Chapter 1   **Multipath Bandwidth Scavenging in the Internet of Things**...1
           Isabel Montes, Romel Parmis, Roel Ocampo, Cedric Festin

Chapter 2   **Cooperative MIMO Relaying with Orthogonal Space-Time Block Codes**
           in Wireless Channels with and without Keyholes..11
           Tian Zhang, Wei Chen, Wei Zhang, Zhigang Cao

Chapter 3   **Multiple Access Techniques for Next Generation Wireless: Recent Advances and**
           Future Perspectives...26
           Shree Krishna Sharma, Mohammad Patwary, Symeon Chatzinotas

Chapter 4   **Spectrum Sensing: To Cooperate or Not to Cooperate?**..38
           Dongliang Duan, Liuqing Yang, Shuguang Cui

Chapter 5   **Distributed Spectrum Sharing Games Via Congestion Advertisement**..............................46
           Mahdi Azarafrooz, R. Chandramouli

Chapter 6   **Towards Effective Intra-flow Network Coding in Software Defined Wireless**
           Mesh Networks..56
           Donghai Zhu, Xinyu Yang, Peng Zhao, Wei Yu

Chapter 7   **Popular Content Distribution in CR-VANETs with Joint Spectrum Sensing and**
           Channel Access using Coalitional Games..72
           Tianyu Wang, Lingyang Song, Zhu Han

Chapter 8   **Alleviate Cellular Congestion Through Opportunistic Trough Filling**.............................82
           Yichuan Wang, Xin Liu

Chapter 9   **Incentivize Spectrum Leasing in Cognitive Radio Networks by Exploiting**
           Cooperative Retransmission..91
           Xiaoyan Wang, Yusheng Ji, Hao Zhou, Zhi Liu, Jie Li

Chapter 10   **Spectrum Trading for Efficient Spectrum Utilization**...99
           Cong Xiong, Geoffrey Ye Li, Lu Lu, Daquan Feng, Zhi Ding, Helena Mitchell

Chapter 11   **Service Co-evolution in the Internet of Things**...114
           Huu Tam Tran, Harun Baraki Kurt Geihs

Chapter 12   **Portfolio Optimization in Secondary Spectrum Markets**...124
           Praveen K. Muthuswamy, Koushik Kar, Aparna Gupta, Saswati Sarkar,
           Gaurav Kasbekar

Chapter 13   **Spectrum Sensing and Primary user Localization in Cognitive Radio Networks via Sparsity**...143
Lanchao Liu, Zhu Han, Zhiqiang Wu, Lijun Qian

Chapter 14   **Spectrum Hole Identification in IEEE 802.22 WRAN using Unsupervised Learning**..........................157
V.Balaji, S.Anand, C.R.Hota, G.Raghurama

Chapter 15   **Cell Selection in Wireless Two-Tier Networks: A Context-Aware Matching Game**..........................165
Nima Namvar, Walid Saad, Behrouz Maham

Chapter 16   **Distance Based Method for Outlier Detection of Body Sensor Networks**...176
Haibin Zhang, Jiajia Liu, Cheng Zhao

Chapter 17   **Cognitive Relay Networks: A Comprehensive Survey**..184
Ayesha Naeem, Mubashir Husain Rehmani

Permissions

List of Contributors

Index

Preface

This book on wireless digital communications examines the latest developments that have occurred in the fields of wireless data transmission and communication. Wireless technology is implemented in a variety of devices that operate in particular frequencies. Rapid advancement of technology has increased the need for machines that regulate high rates of data transfers. Topics included in the book strive to provide a fair idea about this discipline and to help develop a better understanding of the latest advances within this field. It presents the complex subject of wireless digital communications in the most comprehensive and easy to understand language. As this field is emerging at a rapid pace, the contents of this book will help the readers understand the modern concepts and applications of the subject. This book will prove to be immensely beneficial to students and researchers in this field.

This book is a comprehensive compilation of works of different researchers from varied parts of the world. It includes valuable experiences of the researchers with the sole objective of providing the readers (learners) with a proper knowledge of the concerned field. This book will be beneficial in evoking inspiration and enhancing the knowledge of the interested readers.

In the end, I would like to extend my heartiest thanks to the authors who worked with great determination on their chapters. I also appreciate the publisher's support in the course of the book. I would also like to deeply acknowledge my family who stood by me as a source of inspiration during the project.

Editor

Multipath Bandwidth Scavenging in the Internet of Things[*]

Isabel Montes[1], Romel Parmis[1], Roel Ocampo [1], Cedric Festin [2]

[1]Computer Networks Laboratory, Electrical and Electronics Engineering Institute, University of the Philipines Diliman
[2]Networks and Distributed Systems Group, Department of Computer Science, University of the Philippines Diliman

Abstract

To meet the infrastructure coverage and capacity needed by future IoT applications, service providers may engage in mutually-beneficial modes of collaboration such as cooperative packet forwarding and gatewaying through fixed backhauls and Internet uplinks. In an effort to enable these modes of resource pooling while minimizing negative impact on collaborating providers, we developed a transport-layer approach that would enable IoT nodes to opportunistically scavenge for idle bandwidth across multiple paths. Our approach combines multipath techniques with less-than-best effort (LBE) congestion control methods. Initial tests using the TCP-LP and LEDBAT LBE algorithms on scavenging secondary flows show that this desired functionality can be achieved. To ensure however that IoT nodes are guaranteed at least one flow that fairly competes for fair share of network capacity, one flow called the primary flow uses standard TCP congestion control.

Keywords: Internet of Things, bandwidth scavenging, less-than-best-effort, congestion control, multipath flows

1. Introduction

As the Internet of Things (IoT) continues to evolve, service provides will face new challenges in the provisioning of connectivity requirements, including fixed gateways and backhauls to cloud services for the aggregation, processing, storage and distribution of data obtained from smart objects and devices. Such gateways and backhauls must be engineered to guarantee acceptable service levels given aggregate traffic volumes from a large number of sources of data. These data sources may be spread over large geographic areas, and may even be mobile. These challenges are further exacerbated by the need to strategically locate access points and gateways in a manner that would minimize energy-consuming packet forwarding within the wireless network of objects.

These design challenges will impose significant capital and operating costs to future IoT service providers. To complement long term efforts to engineer for maximum geographic coverage and peak traffic loads, IoT providers servicing overlapping areas may consider mutually-beneficial bilateral commercial agreements enabling transit and cooperative access through their peers' infrastructure and nodes. Such inter-IoT provider cooperation would face several design and implementation challenges, as discussed in the next section

1.1. IoT Service Provider Cooperation: Models and Challenges

Consider two IoT service providers A and B, each with respective member-nodes (smart objects, mobile devices and others) and infrastructure (fixed gateways and backhaul links). Figure 1 depicts three possible models of cooperation along with the default 'no-cooperation' scenario. These are:

- **No cooperation** [Fig. 1a]. Packets from A's nodes may only be forwarded through peer nodes from A, and gatewayed to A's infrastructure

---

[*]This paper is an extended version of [1]. We have added a new section (1.1) that discusses models of IoT service provider cooperation, included additional results, and have revised the rest of the paper to further incorporate useful feedback obtained from reviewers and from the conference

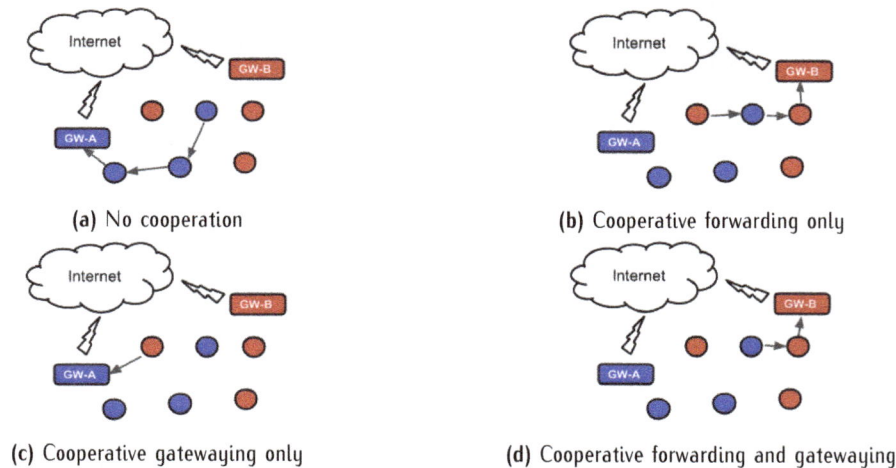

(a) No cooperation

(b) Cooperative forwarding only

(c) Cooperative gatewaying only

(d) Cooperative forwarding and gatewaying

Figure 1. Various IoT service provider cooperative schemes are illustrated in (b)-(d), while the default 'no-cooperation' scenario is depicted in (a) above.

- **Cooperative forwarding only** [Fig. 1b]. Packets from B may be forwarded by nodes of both A and B, but may only be gatewayed via B's infrastructure. This effectively extends the footprint of provider B without having to deploy additional gateways. Provided that A and B have agreed to settlement-free mutual packet forwarding, or otherwise have appropriate (though possibly resource-intensive) accounting mechanisms in place within wireless nodes, there will be no need for Provider A's gateways to account for B's traffic at the gateway.

- **Cooperative gatewaying only** [Fig. 1c]. Packets from B may only be forwarded by peer nodes from B, but may be gatewayed to either A or B's infrastructure. This scheme prevents resources and energy in A's nodes from being consumed by the task of forwarding traffic from B. However, Provider B's coverage is still enhanced by the availability of additional gateways and backhauls from Provider A. Provider B can optimally balance the savings from potentially shortened forwarding path lengths within its wireless network by shunting them to A's gateways with the financial cost of using A's infrastructure in the process. Additionally, it may be more practical to account for usage in gateways than in the nodes themselves (as what might have to be done in cooperative forwarding).

- **Cooperative forwarding and gatewaying** [Fig. 1d]. This is the most flexible form of cooperation, which allows both providers to cooperatively forward packets through each other's nodes and infrastructure. However, it can also be the

most challenging approach in terms of usage accounting and ensuring QoS.

Any of the cooperative scenarios above may be further enhanced if nodes may concurrently exploit multiple paths through the additional resources of cooperating providers. As a simple example, suppose non-interfering gateways from Provider A and Provider B are both within range of a node from A. In a cooperative gatewaying scheme, node A can potentially benefit only if it is able to concurrently transmit to both gateways at an effective aggregate rate greater than the rate available via either gateway alone.

1.2. Multipath Bandwidth Scavenging

In both cooperative and non-cooperative scenarios, although current routing techniques allow packets from a single flow to be forwarded across different paths and gateways, naively striping packets onto multiple paths may cause problems for transport layer protocols with congestion control functionality and reliable in-order delivery. Heterogeneous path delays and loss characteristics may trigger timeouts and retransmissions, as well as head-of-line blocking at receiver buffers, forcing larger and longer buffering to be done [2–4]. A better alternative would be to partition application flows into subflows and enforce per-subflow congestion control and reliability mechanisms that adapt and respond to per-subflow path congestion and loss events [5]. Maintaining TCP-like flow semantics within individual subflows also yields better compatibility with stateful middleboxes [6]. These have been the general strategies taken by the Internet community with Multipath TCP (MPTCP) which is envisioned to provide TCP the capability to utilize multiple paths between source and destination for redundancy and better resource usage [5, 6].

While MPTCP can provide the multipath capability we require, TCP's (and MPTCP's) fairness characteristics however might not exactly sit well with competing providers who are primarily concerned with the SLAs of their own customers. Providers may not wish to fairly share bandwidth with competitors, especially when the latter merely want to opportunistically exploit bandwidth resources on top of what they already have within their own networks. In contrast, an IoT service provider might only allow a competitor to scavenge whatever remaining available bandwidth, if any, is available.

Although it may be relatively straightforward to impose differentiated QoS treatment at gateways, within the wireless network itself, a per-hop QoS approach that involves traffic classification, the management of multiple queues, and the enforcement of differentiated QoS policies might be too resource- and energy-intensive. Thus, alternative mechanisms to protect primary subflows, or flows that originate from a provider's own nodes (analogous to traffic from primary users in the context of spectrum whitespace usage by cognitive radios [7]), possibly through endpoint rather than per-hop behavior, need to be devised.

1.3. Less–Than–Best Effort Congestion Control as a Scavenging Mechanism

While we wish to have an ability to exploit and use multiple paths, opportunistic scavenging secondary subflows that are too aggressive may negatively impact the ability of other nodes to use the network. In scavenging scenarios, a paramount concern is to minimize negative impact on entities volunteering the use of idle resources [8]. This makes TCP's fairness incompatible with opportunistic endpoint-based scavenging behavior, for the following reasons:

- A secondary scavenging subflow will fairly compete for bandwidth with a primary subflow, and

- In a shared wireless medium, secondary scavenging TCP subflows traversing multiple paths may compete with primary TCP subflows over relatively wide areas of the network

These may be mitigated through the use of congestion control mechanisms that detect the onset of congestion more quickly than conventional packet loss-based ones, and yield network usage to primary subflows. The less-than-best effort (LBE) class of congestion control algorithms may offer this ability through rapid congestion detection, such as through delay measurement [9]. When mixed with TCP flows in a bottleneck link, LBE flows yield bandwidth.

Furthermore, an LBE flow will also attempt to maximize the use of the available bandwidth if there are no competing flows. These characteristics make LBE congestion control a good candidate mechanism for opportunistic bandwidth scavenging.

Building on current work by others on LBE congestion control and MPTCP, we developed a hybrid transport-layer approach to concurrent multipath bandwidth scavenging that combines MPTCP's multipath mechanisms with LBE congestion control. Section 2 of this paper starts with a discussion on its basic design, while Section 3 presents initial results from our effort to validate functionality and behavior. Section 4 briefly reviews related work, while Section 5 concludes and outlines future work.

2. MP-LBE Design

In MP-LBE, two communicating endpoints start by establishing a single primary subflow that uses standard TCP-like congestion control. In cooperative scenarios, we assume that underlying routing mechanisms ensure that the primary subflow's path consists of nodes and gateways of the same provider. Secondary subflows that use LBE congestion control mechanisms are then launched on any other discovered paths. These secondary subflows essentially perform bandwidth scavenging, opportunistically using its own and competing providers' resources.

In order to balance congestion among its subflows, MPTCP uses a coupled congestion control algorithm that influences the per-subflow congestion control. This way, MPTCP moves more of its traffic away from the more congested subflows, and it maintains TCP-friendliness when sharing bottleneck links with standard TCP-like traffic. Unlike MPTCP however, MP-LBE does not employ coupled congestion control because LBEs already avoid congested links by design.

2.1. Congestion Control in Secondary Subflows

We aim to achieve low-impact multipath bandwidth scavenging by exploring the use of the LBE class of congestion control methods in secondary subflows. Although there are several methods in this class, we started with a comparative evaluation of TCP-LP and LEDBAT, with a view of expanding these evaluations to other algorithms in the future.

TCP-LP. TCP-LP is a congestion control algorithm that manages the congestion window of the sender based on the one-way forward delay experienced by the traffic on a bottleneck [10]. These one-way delay (owd) measurements approximate queuing delay, and variations in these delays allow TCP-LP to infer congestion earlier than standard TCP through a simple threshold-based algorithm.

One-way Delay Calculation: Upon receiving an ACK, TCP-LP calculates owd from the difference between the receiver's timestamp in the ACK and the sender's timestamp from the original sent packet, which the receiver copies into the ACK and echoes back to the sender. A delay smoothing parameter γ prevents false early congestion indications due to large but short-term variations in network delay coming from bursty cross traffic. TCP-LP computes the exponentially weighted moving average of owd as

$$sd_i = (1 - \gamma)sd_i + \gamma d_i \qquad (1)$$

where d_i is the owd of the packet i and sd_i is the smoothed owd.

Delay Threshold: TCP-LP tracks the maximum owd (d_{max}) and minimum owd (d_{min}) measurements throughout the connection. Whenever owd is calculated, it is compared to last measured d_{max} and d_{min} values, which are then replaced with the new owd if needed, before calculating the smoothed owd. The d_{min} estimates propagation delay, and $d_{max} - d_{min}$ thus estimates the maximum queueing delay. When the smoothed owd exceeds the sum of the propagation delay plus a fraction of the maximum observed queueing delay on that path, congestion is inferred.

$$sd_i > d_{min} + (d_{max} - d_{min})\delta \qquad (2)$$

Congestion Avoidance Policy: When congestion inferred from the threshold formula above, TCP-LP cuts the congestion window by half and enters an inference phase wherein it awaits further congestion indication until the inference phase timeout expires. If congestion is detected during the inference phase, $cwnd$ is reduced to the size of 1. Otherwise, TCP-LP proceeds with an additive increase of $cwnd$.

LEDBAT. LEDBAT is very similar to TCP-LP in that it also uses owd measurements to infer congestion. Like TCP-LP, it measures owd using the timestamps carried by the ACKs received at the sender side. In place of TCP-LP's threshold-based algorithm for inferring congestion, LEDBAT makes use of a target queuing delay value. When queuing delay becomes higher than a specified target value, LEDBAT takes this as an indication that there is a large amount of traffic piling in a buffer somewhere in the network. Congestion is then assumed and LEDBAT reduces its sending rate to alleviate the potential congestion in the network.

Queuing Delay Measurement: On a typical noiseless path, end-to-end delay is generally composed of transmission delay (d_{tr}), propagation delay (d_p), queuing delay (d_q), and processing delay (d_{pr}). All delays are assumed constant except for the queuing

delay. The constant delays are measured through base delay. LEDBAT assumes that minimum owd results from a path with zero queues on its buffers. Continuous measurement of owd over a selected observation window must be done to account for route changes that can result to a change in the constant delays. In principle, the smaller the duration of the observation window, the more responsive the LEDBAT is. On the contrary, if the observation window is too large, the LEDBAT cannot account for frequent route changes. For every sampling, base delay is updated by getting the minimum between the current base delay and the owd:

$$d_{base} = min(owd, d_{base}) \qquad (3)$$

Queuing delay is computed by subtracting the base delay from the owd as shown in the equations below. Since LEDBAT can approximate base delays, all other delays aside from the base delay on an owd is assumed to be queuing delay. Delay measurements must be low-pass filtered to avoid unstable values that in turn may cause erratic sending rates [11].

$$owd \approx d + d_p + d_q + d_{tr}$$
$$d_{base} \approx d + d_p + d_{tr} \qquad (4)$$
$$d_q \approx owd - d_{base}$$

Avoiding Congestion: LEDBAT avoids congestion by regulating the $cwnd$ size using a proportional-integral-derivative (PID) controller. The controller varies the $cwnd$ proportional to the difference of the queueing delay and the target value. For every ACK received at the sender side, the $cwnd$ adjustment is calculated as

$$off_{TARGET} = (TARGET - d_q)/TARGET$$
$$(5)$$
$$cwnd+ = GAIN * off_{TARGET} * bytes_{newlyacked} * \frac{MSS}{cwnd}$$
$$(6)$$

When queuing delay is greater than the target delay, off_{TARGET} becomes a negative value, and $cwnd$ is reduced. When queueing delay is less than the target, $cwnd$ is increased.

LEDBAT normally infers congestion before loss-based TCP, thus backing off before packet loss events occur. However, if packet loss still occurs, LEDBAT must treat this loss as a strong sign of congestion. LEDBAT would then react like standard TCP where, instead of performing PID control, it does a multiplicative decrease of its $cwnd$.

2.2. Congestion Control in Primary Subflows

Our primary subflows use standard TCP SACK congestion control, with slow start, congestion avoidance, fast retransmit, and fast recovery congestion control algorithms. The congestion control mechanism is loss-based and does not detect congestion as early as the delay-based algorithms of TCP-LP and LEDBAT.

When TCP connections share a bottleneck, they react to congestion in a way that tends to divide the bottleneck link capacity evenly among all the connections. MP-LBE's primary subflows are expected to behave similarly, and in cases where there are no available links that secondary subflows can scavenge, the minimum throughput that MP-LBE should attain should be at least as much as a TCP-share of its primary subflow's link.

3. Evaluation

We modified Nishida's implementation of MPTCP for NS-2 [12] and disabled congestion control coupling between subflows. We configured the first subflow to use standard TCP congestion control, while succeeding subflows added to the connection used an LBE congestion control algorithm. We tested two versions of MP-LBE: one that used LEDBAT on secondary flows and another that used TCP-LP. Our LEDBAT implementation for MP-LBE used a target queueing delay value of 12ms, while our MP-LBE TCP-LP implementation used $\gamma=1/8$ and $\delta=0.25$ for the MP-LBE TCP-LP implementation.

The topology used in all the simulations is shown in Figure 2. An MP-LBE connection is configured with two subflows, one primary and one secondary subflow, and each of these share a bottleneck link with a TCP connection. The bottleneck links each have a capacity of 5Mbps and 5ms delay. The link used by the primary subflow will be referred to as the top link, while the link used by the secondary subflow will be referred to as the bottom link.

3.1. Bandwidth Scavenging Behavior

We first explore MP-LBE's ability to scavenge for additional bandwidth from idle links. In this simulation, both bottleneck links have competing standard TCP traffic. MP-LBE's primary subflow should share its link with the competing traffic (TCP-fashion), while the secondary link should give way to the competing traffic. Halfway into the simulation, the TCP traffic on the bottom link ends. This frees up the bottom link, and MP-LBE's secondary flow should react by maximizing the available bandwidth once the link becomes idle. Both MP-LBE (LEDBAT) and MP-LBE (TCP-LP) are able to achieve this behavior, as seen in Figures 3 and 4. When the secondary flow is using LEDBAT,

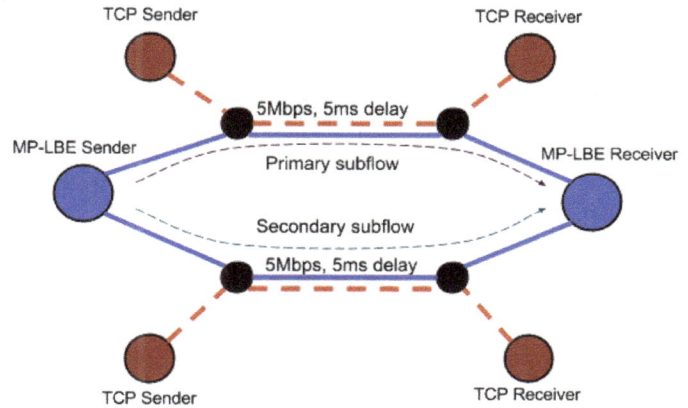

Figure 2. Simulation topology. Each access link directly connected to sender and receiver have 100Mbps capacity, with 5ms delay. Both bottleneck links have 50Mbps capacity with 5ms delay.

simulations show that it is able to maximize the available bandwidth better than TCP-LP. LEDBAT achieves a steadier throughput because its *cwnd* size does not change as drastically as that of TCP-LP.

3.2. LBE Behavior

To demonstrate the LBE behavior of the secondary flow, we used the same topology, but this time only the primary flow was made to compete with regular TCP at the start of the simulation. The bottom link had no competing traffic, which allowed the secondary flow to maximize 5Mbps capacity of the link. At 45 seconds, a regular TCP connection was started on the bottom link.

Figures 5 and 6 show the simulation results. MP-LBE using TCP-LP on its secondary link was able to back off more rapidly than in the case of LEDBAT, but both demonstrated correct LBE behavior when the competing TCP traffic on the bottom link was started.

3.3. Goodput

Even though bandwidth aggregation from multiple paths improves throughput, the goodput achieved may be less than the theoretical maximum due to out-of-order arrival of packets. To evaluate MP-LBE's goodput performance, we recorded the data-level sequence numbers (DSNs) received by the destination node as a function of time.

We used the same topology as in Figure 2, but eliminated all competing TCP traffic. We ran the simulation using regular MPTCP, in addition to the simulation runs for MP-LBE (LEDBAT) and MP-LBE (TCP-LP). Figure 7 shows the data obtained from our simulation. The y-axis is scaled to 1:536, as 536 is the data-level length and sequence numbers are in increments of 536.

(a) Throughput

(b) *cwnd* behavior

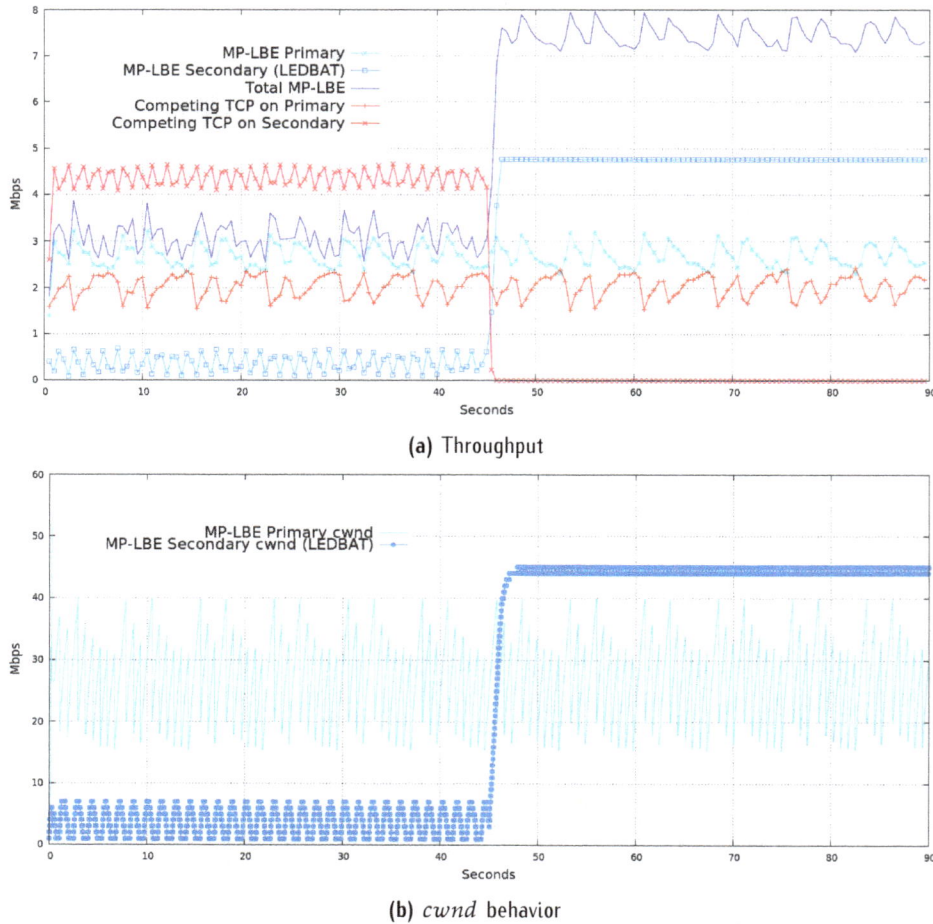

Figure 3. Bandwidth scavenging behavior of MP–LBE using LEDBAT.

Figure 7 shows almost-identical rates of DSN increase in both MPTCP and MP-LBE (LEDBAT). MP-LBE (TCP-LP) on the other hand registers significantly lower data sequence numbers received within the same timeframe.

Figure 8 shows that during the first 3 seconds of the simulation MP-LBE's primary subflow (for both LEDBAT and TCP-LP) received sequence numbers at a slower rate than the secondary subflow. This caused goodput to suffer because only the DSNs on the secondary subflow are arriving, while all the sequence numbers sent through the primary subflow are delayed. In the case of MP-LBE (LEDBAT), the delayed packets finally arrive a little after 3 seconds and the primary subflow picks up its pace. After this point, the slope of MP-LBE (LEDBAT)'s DSN curve matches that of MPTCP.

4. Related Work

Resource scavenging is not a new concept, having been previously used to harness idle computing resources to perform useful calculations for users other than the resource owner [8]. In more recent literature, bandwidth scavenging commonly refers to dynamic, opportunistic access to unused spectrum by cognitive radios [7]. Our approach is quite different in that it focuses on a solution at the transport layer through the use of LBE congestion control in a multipath fashion. While there has been some recent similar work on the development of a multipath version of LEDBAT called LEDBAT-MP [13], we are interested in the more general class of LBEs and intend to comparatively evaluate several of the representative algorithms for our intended application. Furthermore, our approach makes a crucial distinction between primary and secondary flows, ensuring that nodes can rely on at least one flow, the primary one, to compete fairly within the network.

In order to achieve cooperative gatewaying among providers, we need mechanisms to enable concurrent access to their respective fixed wireless infrastructure. BeWifi [14], a service rolled out by service provider Telefonica, allows users to use idle capacity through neighbors' access points within range. While BeWifi

(a) Throughput

(b) *cwnd* behavior

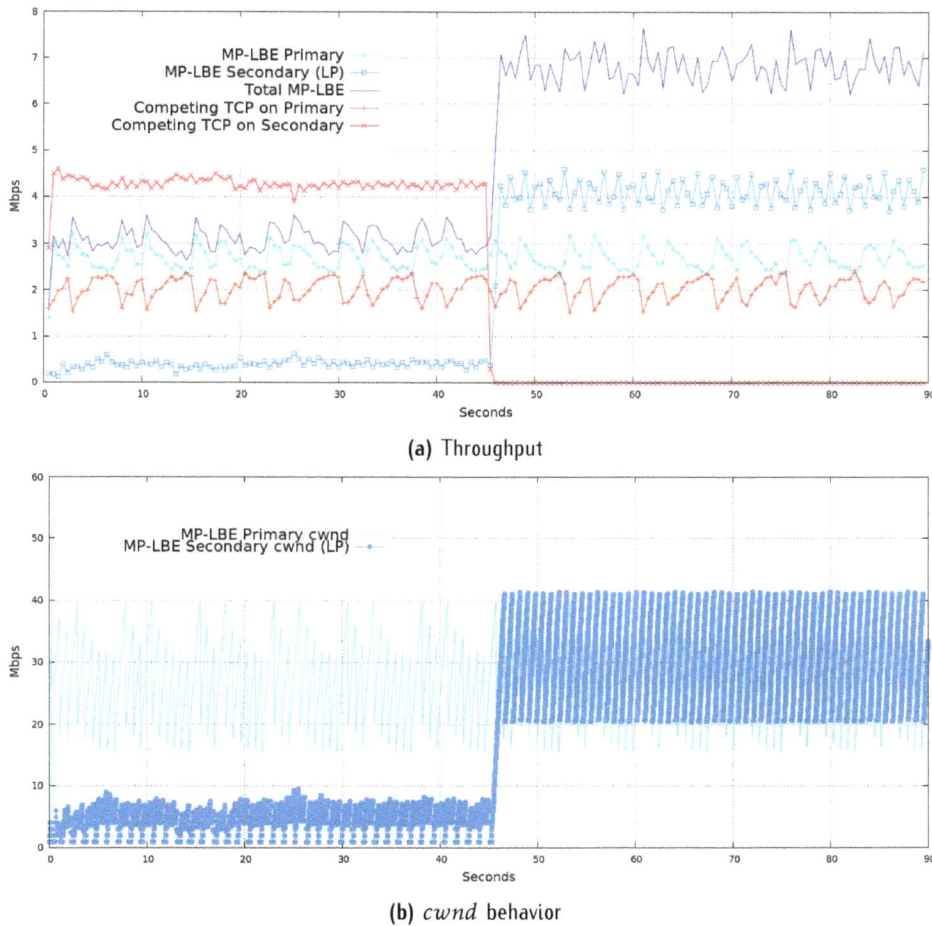

Figure 4. Bandwidth scavenging behavior of MP–LBE using TCP–LP.

applies to a single-provider model, it offers insight into the usefulness of the ability to scavenge idle bandwidth from cooperating peers. On the other hand, CableWiFi [15] employs a multi-provider model, allowing customers from five ISPs access to the consortium's infrastructure. From a technical point of view, one mechanism that can enable cooperative gatewaying is offered by BaPu (Bunching of Access Point Uplinks) [16] is a software-based approach that employs packet overhearing, using it to pool together WiFi uplinks that are in close proximity to one another. A BaPu-Gateway AP schedules which contributing BaPu-APs will send the packets through to the receiver. BaPu-APs are also configured to prioritize the home user's traffic over any background traffic that is generated when APs act as BaPu contributors. BaPu was designed primarily for uploading user-generated content over the Internet and cannot be used for downloads.

The ability to concurrently exploit multiple paths for bandwidth scavenging may also be viewed as a problem of bandwidth aggregation. Application layer solutions such as DBAS [17] typically do not require changes in the underlying protocols and instead rely on endpoint middleware to intercept traffic and manage scheduling, reordering, and transmission over multiple interfaces. DBAS' ability to deal with stateful middleboxes, as well as the ensuing fairness of its subflows is however not known. Alternatively, instead of placing the functionality within the endpoint itself, dedicated proxy middleboxes may be deployed within the network in order to do aggregation, delay equalization and packet scheduling [2, 3]. This seems to be more feasible to do within the fixed infrastructure, and represents additional cost and management overhead for providers. We preferred to take an endpoint-based approach since it offers an end-to-end solution, covering both the fixed and wireless portions of the network.

5. Conclusion and Future Work

To provide a low-impact mechanism that will encourage future IoT service providers to explore various models of cooperation, including, but not limited to,

(a) Throughput

(b) *cwnd* behavior

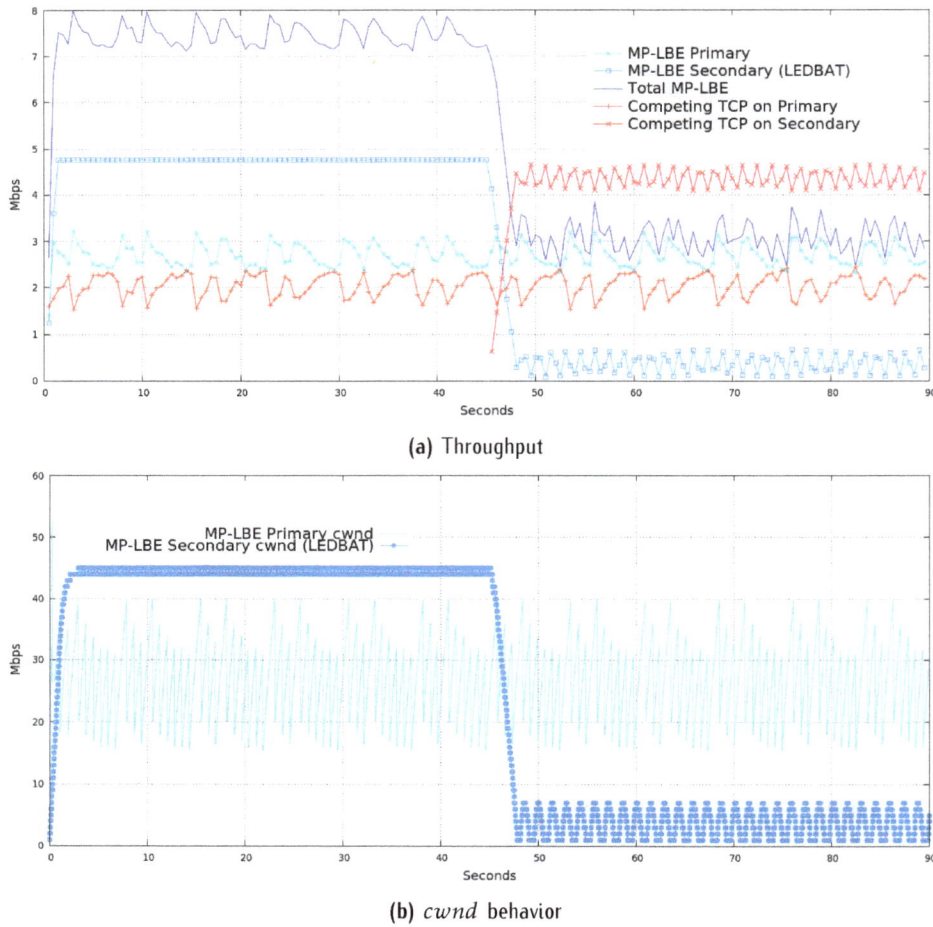

Figure 5. LBE Behavior of MP–LBE using LEDBAT.

cooperative forwarding and gatewaying, we propose a transport-layer approach for multipath bandwidth scavenging that uses TCP-like congestion control for primary subflows and less-than-best effort (LBE) congestion control for secondary subflows. The use of LBE for secondary subflows ensures that these back off and yield bandwidth in the face of other traffic, including primary subflows from other IoT devices.

Our MP-LBE design effectively improves throughput when one or more idle links become available for secondary subflows. When no additional links are available, the primary subflow achieves the throughput of a single TCP flow, and secondary flows are able to rapidly use capacities along paths that become idle.

MP-LBE for both LEDBAT and TCP-LP yield lower goodput than MPTCP, with MP-LBE (LEDBAT) having worse out-of-order packet arrivals at the beginning of its connection lifetime. Noting that MPTCP employs scheduling on subflows to mitigate the impact of non-uniform path delays on packet arrivals, and consequently buffer requirements and goodput [18],

we intend to work on improving goodput for MP-LBE by considering and possibly extending the various approaches that have been proposed for reducing the number of out-of-order packet arrivals in multipath connections, such as delay equalization [2], packet scheduling [3], congestion window adaptation [19].

Recognizing "less than best effort" for what it is, smart objects and devices should principally rely on primary flows to carry critical traffic. However, the ability to scavenge additional bandwidth and paths will enable IoT sensors and devices to opportunistically explore shortcut fast paths and accelerate local aggregation and processing of data, or temporarily transmit information at higher-than-fair levels of spatial and temporal resolution. Resource pooling by cooperative IoT service providers should expand these opportunities even further.

Smart objects and devices in the Internet of Things will undoubtedly dedicate most of their resources to sensing and aggregating data, and any local processing and cognitive functionality required. With any new functionality being introduced, such as multipath

(a) Throughput

(b) *cwnd* behavior

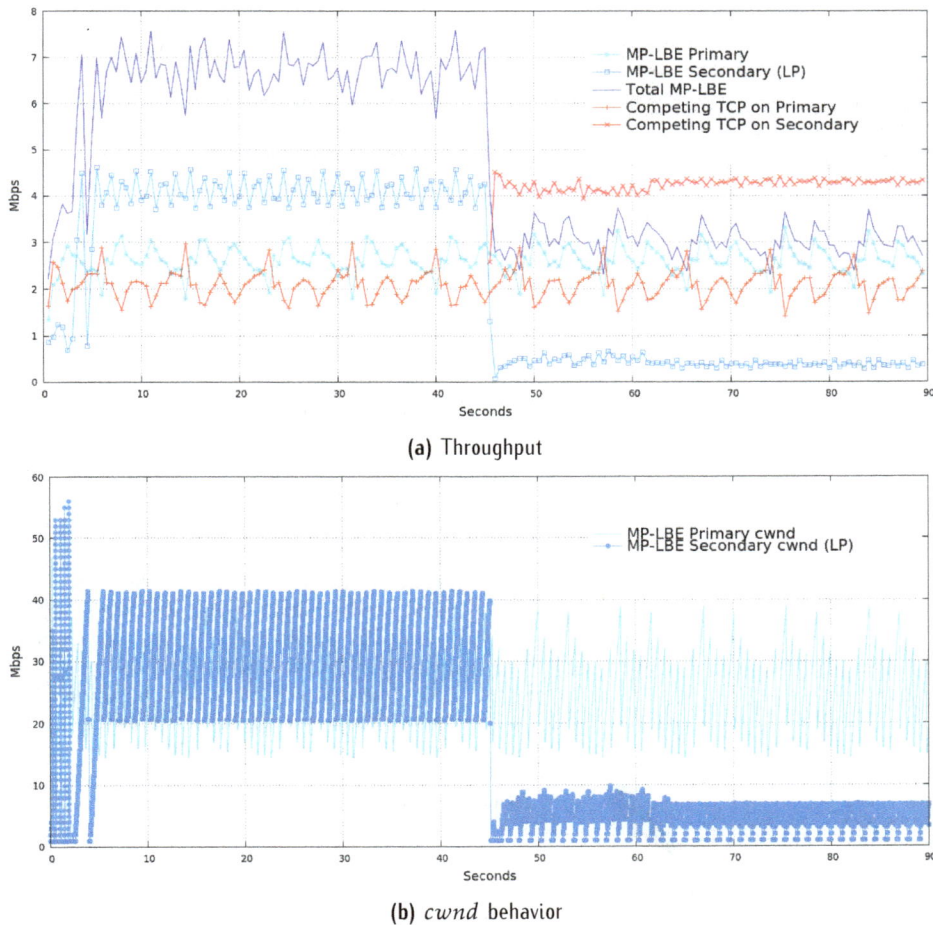

Figure 6. LBE Behavior of MP–LBE using TCP-LP.

bandwidth scavenging, prudent design dictates that there should be minimal impact on resource footprint. We intend to keep this as a guiding principle as our work moves forward.

Acknowledgement. This work has been supported by the Engineering Research and Development for Technology (ERDT) Consortium, Department of Science and Technology – Science Education Institute (DOST-SEI), Republic of the Philippines.

References

[1] Isabel Montes, Romel Parmis, Roel Ocampo, and Cedric Festin. Multipath Bandwidth Scavenging in the Internet of Things. In *Proceedings of the International Conference on Internet of Things as a Service, 2014*, 2014.

[2] Kristian Evensen, Dominik Kaspar, Paal Engelstad, Audun Fosselie Hansen, Carsten Griwodz, and Pål Halvorsen. A Network-layer Proxy for Bandwidth Aggregation and Reduction of IP Packet Reordering. In *Local Computer Networks, 2009. LCN 2009. IEEE 34th Conference on*, pages 585–592. IEEE, 2009.

[3] Kameswari Chebrolu, Bhaskaran Raman, and Ramesh R. Rao. A Network Layer Approach to Enable TCP over Multiple Interfaces. *Wirel. Netw.*, 11(5):637–650, September 2005.

[4] T. Zinner, K. Tutschku, A. Nakao, and P. Tran-Gia. Using Concurrent Multipath Transmission for Transport Virtualization: Analyzing Path Selection. In *Teletraffic Congress (ITC), 2010 22nd International*, pages 1–7, Sept 2010.

[5] Damon Wischik, Costin Raiciu, Adam Greenhalgh, and Mark Handley. Design, Implementation and Evaluation of Congestion Control for Multipath TCP. In *NSDI*, volume 11, pages 8–8, 2011.

[6] Costin Raiciu, Christoph Paasch, Sebastien Barre, Alan Ford, Michio Honda, Fabien Duchene, Olivier Bonaventure, Mark Handley, et al. How Hard Can It Be? Designing and Implementing a Deployable Multipath TCP. In *USENIX Symposium of Networked Systems Design and Implementation (NSDI'12)*, 2012.

[7] Anthony Plummer Jr., Mahmoud Taghizadeh, and Subir Biswas. Measurement-Based Bandwidth Scavenging in Wireless Networks. *IEEE Transactions on Mobile Computing*, 11(1):19–32, 2012.

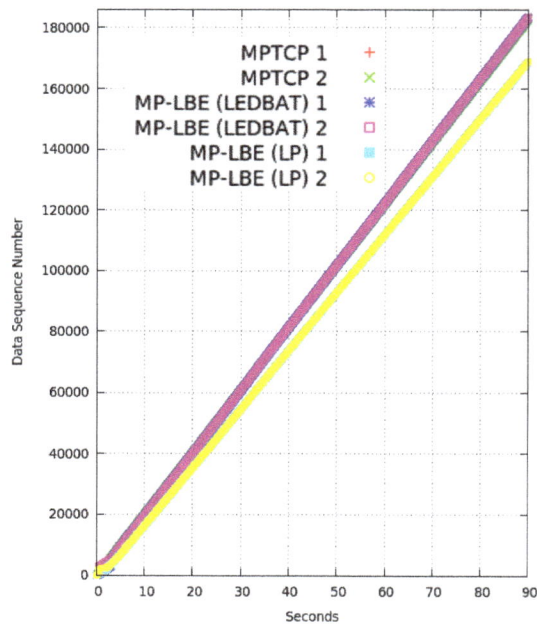

Figure 7. Data Sequence Numbers received at the destination node plotted against time.

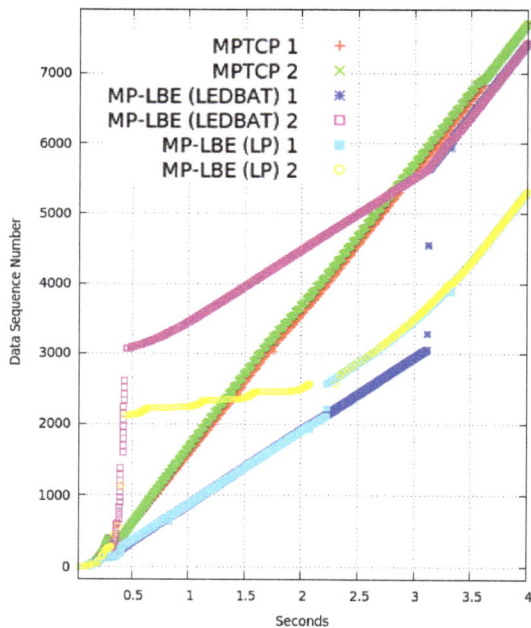

Figure 8. Data Sequence Numbers received at the destination node plotted against time, during the first 4 seconds of the simulation.

[8] Jonathan W Strickland, Vincent W Freeh, Xiaosong Ma, and Sudharshan S Vazhkudai. Governor: Autonomic Throttling for Aggressive Idle Resource Scavenging. In *Proceedings of the Second International Conference on Autonomic Computing, 2005.*, pages 64–75. IEEE, 2005.

[9] Michael Welzl and David Ros. A Survey of Lower-than-Best-Effort Transport Protocol. RFC 6297, RFC Editor, June 2011.

[10] Aleksandar Kuzmanovic and Edward W. Knightly. TCP-LP: Low-priority Service via End-point Congestion Control. *IEEE/ACM Trans. Netw.*, 14(4):739–752, August 2006.

[11] Stanislav Shalunov, Greg Hazel, Janardhan Iyengar, and Mirja Kuehlewind. Low Extra Delay Background Transport (LEDBAT). RFC 6817, RFC Editor, December 2012.

[12] Google Code Project. Multipath-TCP: Implement Multipath TCP on NS-2. http://code.google.com/p/multipath-tcp. Accessed June 30, 2014.

[13] Hakim Adhari, Sebastian Werner, Thomas Dreibholz, and Erwin Paul Rathgeb. LEDBAT-MP –On the Application of Lower-than-Best-Effort for Concurrent Multipath Transfer. In *Proceedings of the 4th International Workshop on Protocols and Applications with Multi-Homing Support (PAMS)*, Victoria, British Columbia/Canada, May 2014. ISBN 978-1-4799-2652-7.

[14] BeWifi. http://www.bewifi.es. Accessed June 30, 2014.

[15] Cable WiFi : Internet access brought to consumers through a collaboration among U.S. Internet Service Providers. http://www.cablewifi.com. Accessed June 30, 2014.

[16] Tao Jin, Triet Vo Huu, Erik-Oliver Blass, and Guevara Noubir. BaPu: Efficient and Practical Bunching of Access Point Uplinks. *CoRR*, abs/1301.5928, 2013.

[17] K. Habak, M. Youssef, and K.A. Harras. DBAS: A Deployable Bandwidth Aggregation System. In *New Technologies, Mobility and Security (NTMS), 2012 5th International Conference on*, pages 1–6, May 2012.

[18] Christoph Paasch, Simone Ferlin, Ozgu Alay, and Olivier Bonaventure. Experimental evaluation of multipath tcp schedulers. In *Proceedings of the 2014 ACM SIGCOMM workshop on Capacity sharing workshop*, pages 27–32. ACM, 2014.

[19] Dizhi Zhou, Wei Song, and Minghui Shi. Goodput Improvement for Multipath TCP by Congestion Window Adaptation in Multi-Radio Devices. In *Consumer Communications and Networking Conference (CCNC), 2013 IEEE*, pages 508–514. IEEE, 2013.

Cooperative MIMO Relaying with Orthogonal Space-Time Block Codes in Wireless Channels with and without Keyholes[*]

Tian Zhang[1,*], Wei Chen[2], Wei Zhang[3] and Zhigang Cao[2]

[1]School of Information Science and Engineering, Shandong University, Jinan 250100, China.
[2]State Key Laboratory on Microwave and Digital Communications, Tsinghua National Laboratory for Information Science and Technology (TNList), Department of Electronic Engineering, Tsinghua University, Beijing 100084, China.
[3]School of Electrical Engineering and Telecommunications, The University of New South Wales, Sydney, NSW 2052, Australia

Abstract

Cooperative multiple-input multiple-output (MIMO) relaying is investigated in the paper. We introduce DF-AF selection MIMO relaying, where the relay equipped with multiple antennas can adaptively switch between decode-and-forward (DF) and amplify-and-forward (AF) according to its decoding state of the source message. We consider two wireless environment scenarios: 1)The scenario with traditional channels are considered firstly. We analyze the outage performance of DF-AF selection MIMO relaying, and a closed-form expression is derived. In addition, the diversity order is obtained based on the expression. For comparison purpose, we also obtain the closed-form outage probability and the diversity order for the AF MIMO relaying and the DF MIMO relaying. 2)We investigate the cooperative MIMO relaying in the presnece of keyholes secondly. We present performance analysis of orthogonal space-time block coded transmission for a cooperative MIMO relaying system with keyholes. For DF MIMO relaying, exact outage probability and symbol error probability (SEP) are obtained. Regarding AF MIMO relaying and DF-AF selection MIMO relaying, the lower and upper bounds are derived. In both traditional and keyhole scenarios, theoretical analysis which has been further verified through Monte-Carlo simulations demonstrate that the DF-AF selection MIMO relaying has better performance than the AF MIMO relaying and the DF MIMO relaying.

Keywords: MIMO relaying, DF-AF selection, outage probability, SEP, keyhole

1. Introduction

Multiple-input multiple-output (MIMO) techniques have gained huge attention in the past decade because of their high spectral efficiency in both single-user and multi-user communications [2, 3]. Deploying multiple antennas at each node is a promising approach to solve the increasing demand for data-rate-intensive applications in wireless networks. Additionally, As a core idea in MIMO systems, space-time coding is an effective means for increasing the reliability of data transmission [4].

As important modi operandi of combating fading induced by multi-path propagation in wireless networks, cooperative diversity techniques have received much interest [5, 6]. In cooperative communications, in addition to the direct transmission from the source to the destination, some neighboring nodes can be used to relay the source signal to the destination, hence forming

[*]Invited paper.
The material in this paper was presented in part at the *IEEE ICCC'13*, Xi'an, China, Aug. 2013.[1]
*Corresponding author. Email: tianzhang.ee@gmail.com

a virtual antenna array to achieve spatial diversity. Several cooperative diversity protocols including amplify-and-forward (AF), decode-and-forward (DF), selection relaying and incremental relaying, were discussed in [6]. DF-AF selection relaying protocol, where each relay can adaptively switch between DF and AF according to its local SNR, has been developed and investigated in [7]-[10].

More recently, MIMO relaying technologies that exploit the cooperative diversity as well as the advantages of MIMO systems by accommodating multiple antennas at the relay nodes have been well developed [11, 12]. In [13], the authors obtained the bounds for the capacity of MIMO relay channels. A DF cooperative MIMO relay channel with orthogonal space-time block codes (OSTBC) was analyzed in [14]. MIMO cooperative diversity with scalar-gain AF relaying was studied in [15]. MIMO with non-coherent AF relaying was considered in [16] and [17]. In [18], the authors presented performance analysis of a AF cooperative MIMO relaying system based on Alamouti scheme [19]. The MIMO relay channels with the channel-state information (CSI)-assisted AF relaying technique was considered in [20]. DF MIMO relay channels employing Maximum Likelihood (ML) detection in Rayleigh fading was investigated in [21]. MIMO relaying with AF for UWB ad hoc networks was analyzed in [22]. In [23], DF relaying for MIMO ad hoc networks was studied, the authors demonstrated that the use of cooperative relay in a MIMO framework could bring in a significant throughput improvement.

It has been shown, both theoretically and experimentally, that degenerate channel phenomena termed "keyholes" may exist under realistic assumptions [24] - [27]. A spatial MIMO keyhole is a propagation scenario where the channel gain matrix has only unit rank, even when multiple uncorrelated antennas are employed. Thus, keyhole will degrade the MIMO channel capacity to that of a single-input single-output (SISO) channel.

To date, the effect of keyholes on performance of STBC over MIMO channels has been well investigated. The average symbol error rate (SER) of orthogonal space-time code (OSTBC) [28] with M-ary phase shift keying (M-PSK) and M-ray quadrature amplitude modulation (M-QAM) constellations over keyhole MIMO channels was analyzed in [29]. In [30], the authors derived exact analytical closed-form expressions for the ergodic capacity and information outage probability of keyhole MIMO channels in Nakagami-m fading environments. Exact expressions for the SER of OSTBC over a spatially correlated MIMO channel, in which the signal propagation suffers from a keyhole effect was derived in [31]. The performance of OSTBC in MIMO fading channels under keyhole condition was analyzed in [32]. The SER and BER of OSTBC with antenna selection over keyhole fading

Table 1. Summary of related papers

	Typical Papers
MIMO relaying	[11] - [23]
MIMO with keyholes	[29] - [40]
MIMO Relaying with keyholes	[41] - [44]

channels were examined in [33]. Exact expressions of SER of OSTBC in Nakagami-m keyhole channels with arbitrary fading parameters and the closed-form asymptotic expressions were derived in [34]. SER/BER and outage probability of OSTBC with M-PSK and M-QAM in keyhole MIMO fading channels were studied in [35]. Pairwise error probability (PEP) analysis of general space-time codes (STCs) in keyhole conditions was presented in [36]. In [37], the asymptotic PEP of STCs in generalized keyhole fading was obtained. In addition, the keyhole can be viewed as a special case of double-scattering [38]. In [38], analytical performance of Rayleigh-product MIMO channels (a special case of double scattering MIMO channels) was studied. Furthermore, the keyhole channel, which is regarded as a special case, was investigated. With respect to Rayleigh-product MIMO channels, the diversity-multiplexing tradeoff (DMT) analysis and the ergodic sum rate analysis can be found in [39] and [40], respectively.

There are a few works on keyhole MIMO relay channels. In [41], the authors studied MIMO relay channels in the presence of keyhole effect. The ergodic capacity is investigated when the source-relay channel is keyhole-free. Moreover, they demonstrated that cooperative diversity can mitigate keyhole effects. Hence it is important to study the cooperative MIMO relay channels with keyholes. In [42], the authors derived the exact ergodic capacity for MIMO AF relaying systems with a multi-keyhole effect on the relay-destination channel. In [43], the authors investigated the performance of MIMO AF relay networks with keyhole and spatial correlation. The SEP and outage probability were analyzed in Rayleigh fading environments when the source-relay link is keyhole-free.

Previous works related are presented in Table 1.

In this paper, we consider the cooperative MIMO relaying in the absence and presence of keyholes, respectively. First, DF-AF selection MIMO relaying is introduced in traditional (keyhole free) wireless channels. We investigate the outage probability of a cooperative DF-AF selection MIMO relaying system with OSTBC and selection diversity. A closed-form solution at arbitrary SNR is obtained and the diversity order is obtained based on the expression. Next, we investigate the MIMO relaying in the presence of keyholes. The outage probability and SEP of OSTBC over cooperative MIMO relay channels with keyholes in Nakagami-m fading environments are analyzed. DF MIMO relaying, AF MIMO relaying and DF-AF selection MIMO relaying are considered, respectively. Specifically, exact outage probability and symbol error probability of DF MIMO relaying over keyhole channels are obtained. The lower and upper bounds are derived for the AF MIMO relaying and the DF-AF selection MIMO relaying. Moreover, we prove by theoretical analysis and simulations that the DF-AF selection MIMO relaying has better performance than the DF MIMO relaying and the AF MIMO relaying over traditional and keyhole channels. To summarize, the contributions of this paper are as follows:

(i) DF-AF selection MIMO relaying is introduced in the cooperative MIMO channels. For the scenario without keyhole effect, we analyze the outage performance of DF-AF selection MIMO relaying. The closed-from outage probability and diversity order are derived.

(ii) We investigate MIMO relaying in the scenario that the keyholes exist.

- We consider the MIMO relay channels when the source-destination, the source-relay and the relay-destination channels all incur keyhole effect in this paper. As it makes assumptions on the mobility pattern or location of neither the relay nor the destination, this scenario is more practical and challenging.

- The outage probability and SEP of STBC over keyhole channels for AF MIMO relaying in Nakagami-m fading environments are investigated. Furthermore, the outage probability and SEP of keyhole DF MIMO relaying system and keyhole DF-AF selection MIMO relaying channels are considered.

(iii) We compare the performance of the three MIMO relaying schemes (i.e., the DF MIMO relaying, the AF MIMO relaying and the DF-AF selection MIMO relaying) in the scenarios with and without keyholes, respectively. We find that DF-AF selection MIMO relaying has the best performance in both scenarios.

Throughout this paper, the notations in Table 2 will be used.

2. System model

We consider a cooperative MIMO communication system as depicted in Figure 1, where the source, relay, and destination terminals have n_s, n_r and n_d antennas,

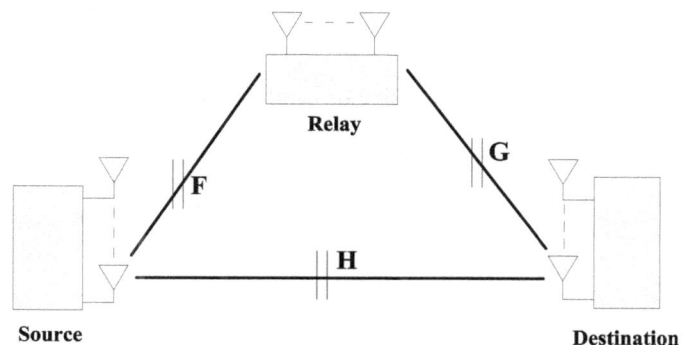

Figure 1. MIMO relay channel

respectively. The source-relay, relay-destination and source-destination channels are denoted by **H**, **F**, **G**. It is assumed that the instantaneous CSI is available at the receiver, i.e., **F** is available at the relay and the destination knows **F**, **H**, and **G**. We assume that the source and the relay can employ OSTBC encoding. A half-duplex relaying protocol where each transmission period is divided into two time slots is assumed. In the first time slot, the source transmits OSTBC coded signal to the destination as well as the relay. In the second time slot, the relay processes the received signal and forwards the processed signal to the destination according to some specific relaying protocol. In DF MIMO relaying, the relay transmits the decoded signal using OSTBC when the source message can be correctly decoded. Otherwise, the relay remains idle. In AF MIMO relaying, the relay simply amplifies and forwards the received signal. In this paper, we introduce an efficient MIMO relaying scheme referred to as DF-AF selection MIMO relaying. In the DF-AF selection MIMO relaying scheme, the relay equipped with multiple antennas could adaptively switch between DF protocol and AF protocol. Specifically, if the relay could fully decode the source message, it decodes the source message, and re-encodes the received signal by using OSTBC before forwarding the signal to the destination. Otherwise it amplifies and forwards the received signal to the destination. Maximum Likelihood (ML) detection is used at all receivers in the two time slots. The destination uses a certain combining technique to combine the signals of two time slots coming from the source and the relay to decode the information.

3. MIMO relaying without keyhole

In this section, we analyze the introduced DF-AF selection MIMO relaying in the cases that no keyhole exits. We assume that all channel matrices are assumed to undergo independent Rayleigh fading with elements obeying $\mathcal{CN}(0, 2)$. As outage probability is an important performance measure that is commonly used to characterize a wireless communication system, we focus

Table 2. Notations

$\Pr\{\cdot\}$	The probability of random event
$F_X(x)$	The cumulative density function (c.d.f.) of a random variable X
$f_X(x)$	The probability density function (p.d.f.) of a random variable X
$\mathbb{E}_X(\cdot)$	The expectation operator associated with X. Specially, $\overline{X} = \mathbb{E}_X(X)$.
$\Psi_X(s)$	The moment generating function (m.g.f.) associated with a random variable X, which is defined by $\Psi_X(s) = \mathbb{E}_X\left(e^{sX}\right)$
$\Gamma(\cdot)$	The gamma function
$\Gamma(\cdot, \cdot)$	The incomplete gamma function
$K_v(\cdot)$	The vth order modified Bessel function of the second kind
$W_{\eta,\xi}(\cdot)$	The Whittaker function
$X \sim \mathcal{G}(m)$	An random variable X has the p.d.f. given by $f_X(y) = \frac{1}{\Gamma(m)}\left(\frac{m}{\overline{X}}\right)^m y^{m-1} e^{-\frac{my}{\overline{X}}}$
$\|\mathbf{M}\|_F$	The Frobenius norm of a matrix $\mathbf{M}$
$\mathcal{CN}(\mu, \sigma^2)$	Circularly symmetric complex Gaussian distribution with mean μ and covariance σ^2
$\mathbb{C}^n$	The set of $n \times 1$ complex vectors

on the outage performance analysis. The closed-form outage probability is obtained, and the diversity order is derived thereafter. Since combining techniques do not affect the diversity order, we utilize selection combining (SC) for simplicity and conciseness in this scenario. In addition, we perform comparisons with the DF MIMO relaying and the AF MIMO relaying.

3.1. Outage probability of DF–AF selection MIMO relaying

A closed-form expression of outage probability valid at arbitrary SNR is obtained in the following theorem.

Theorem 1. The outage probability of the DF-AF selection MIMO relaying scheme P_{out} can be expressed as

$$P_{out} = \left[1 - \frac{\Gamma(n_d n_s, \alpha_0 \gamma_{th})}{\Gamma(n_d n_s)}\right]\left[\frac{\Gamma(n_r n_s, \alpha_1 \Delta)}{\Gamma(n_r n_s)} \times \right.$$
$$\left.\left(1 - \frac{\Gamma(n_d n_r, \alpha_2 \gamma_{th})}{\Gamma(n_d n_r)}\right) + \left(1 - \frac{\Gamma(n_r n_s, \alpha_1 \Delta)}{\Gamma(n_r n_s)}\right)\right],$$

(1)

where $\alpha_0 = \frac{Rn_s N_0}{2c_0 P_0}$, $\alpha_1 = \frac{Rn_s N_0}{2c_1 P_0}$, and $\alpha_2 = \frac{Rn_r N_0}{2c_2 P_1}$. R is the rate of the OSTBC.[1] $\Delta = \gamma_{th} = 2^{2R} - 1$. c_0, c_1 and c_2 represent the distance dependent power transfer factors for the source-destination, source-relay and relay-destination channels respectively. P_0, P_1 denote the transmit power of the source and the relay respectively. N_0 is the variance of the Gaussian noise at each receive antenna.

Proof. The equivalent instantaneous SNR per symbol of source-destination, source-relay and relay-destination channels are $\gamma_0 = \frac{c_0 P_0}{Rn_s N_0}\|\mathbf{H}\|_F^2$, $\gamma_1 = \frac{c_1 P_0}{Rn_s N_0}\|\mathbf{F}\|_F^2$ and $\gamma_2 = \frac{c_2 P_1}{Rn_r N_0}\|\mathbf{G}\|_F^2$ respectively [45]. From the assumption of the channel matrix, it can be derived that $\gamma_0 \sim \mathcal{G}(n_d n_s)$ with $\overline{\gamma_0} = \frac{2c_0 P_0}{Rn_s N_0}n_d n_s$, as well as $\gamma_1 \sim \mathcal{G}(n_r n_s)$ with $\overline{\gamma_1} = \frac{2c_1 P_0}{Rn_s N_0}n_r n_s$ and $\gamma_2 \sim \mathcal{G}(n_d n_r)$ with $\overline{\gamma_2} = \frac{2c_2 P_1}{Rn_r N_0}n_d n_r$. First, it can be derived that c.d.f. of $Y \sim \mathcal{G}(m)$ can be given by

$$F_Y(y) = 1 - \frac{\Gamma\left(m, \frac{my}{\overline{Y}}\right)}{\Gamma(m)}.$$

(2)

Consequently, the instantaneous equivalent end-to-end SNR per symbol at the destination is

$$\gamma = \max\left(\gamma_0, \xi\gamma_2 + (1-\xi)\frac{\gamma_1\gamma_2}{\gamma_1 + \gamma_2 + 1}\right),$$

(3)

where ξ denotes the decoding state at the relay,[2] and

$$\Pr\{\xi = 0\} = \Pr\{\gamma_1 < \Delta\} = F_{\gamma_1}(\Delta)$$

(4)

and

$$\Pr\{\xi = 1\} = 1 - \Pr\{\xi = 0\}.$$

(5)

The outage probability can be given by

$$P_{out} = \Pr\{\gamma < \gamma_{th}\}$$
$$= \Pr\{\max\left(\gamma_0, \xi\gamma_2 + (1-\xi)\frac{\gamma_1\gamma_2}{\gamma_1 + \gamma_2 + 1}\right) < \gamma_{th}\}$$
$$= \Pr\{\gamma_0 < \gamma_{th}\}\Pr\{\xi\gamma_2 + (1-\xi)\frac{\gamma_1\gamma_2}{\gamma_1 + \gamma_2 + 1} < \gamma_{th}\}.$$

(6)

---

[1] We consider the scenarios where the rates of the OSTBCs in the two hops are the same.

[2] Approximately, if the source-relay link is able to support a given transmission rate R, i.e., $\frac{1}{2}\log_2(1 + \gamma_1) \geq R$, or equivalently, if $\gamma_1 \geq 2^{2R} - 1$, the relay could fully decode the source message

By conditional probability and the Theorem of Total Probability, (6) can be rewritten as

$$P_{out} = \Pr\{\gamma_0 < \gamma_{th}\}\Big(\Pr\{\xi = 1\}\Pr\{\gamma_2 < \gamma_{th}|\xi = 1\}$$
$$+ \quad \Pr\{\xi = 0\}\Pr\Big\{\frac{\gamma_1\gamma_2}{\gamma_1 + \gamma_2 + 1} < \gamma_{th}|\xi = 0\Big\}\Big)$$
$$\overset{(a)}{=} \quad F_{\gamma_0}(\gamma_{th})\Big[\big(1 - F_{\gamma_1}(\Delta)\big)F_{\gamma_2}(\gamma_{th}) + F_{\gamma_1}(\Delta)\Big]. \quad (7)$$

(a) holds since when $\xi = 0$, i.e., $\gamma_1 < \Delta$, we have $\frac{\gamma_1\gamma_2}{\gamma_1+\gamma_2+1} < \gamma_1 < \Delta = \gamma_{th}$, i.e., $\Pr\Big\{\frac{\gamma_1\gamma_2}{\gamma_1+\gamma_2+1} < \gamma_{th}|\xi = 0\Big\} = 1$. Applying (2) and (12) along with some simple manipulations, we arrive at (1), which completes the proof. $\quad\square$

3.2. Diversity analysis of DF-AF selection MIMO relaying

Let $P = P_0 + P_1$, $P_0 = \theta P$. Define $\text{SNR} = \frac{P}{N_0}$. The diversity order

$$d = -\lim_{\text{SNR}\to\infty} \frac{\log P_{out}}{\log \text{SNR}}$$

can be give by the following theorem.

Theorem 2. The diversity order of the DF-AF selection MIMO relaying scheme is given by

$$d = n_d n_s + n_r \min\{n_s, n_d\}. \quad (8)$$

Proof. The lower gamma function

$$\gamma(a, b) \simeq (1/a)b^a \quad (9)$$

as $b \to 0$ [46], where $\simeq$ denotes asymptotic equality. Let $f(\text{SNR}) \sim \text{SNR}^d$ denote $0 < |\lim_{\text{SNR}\to\infty} \frac{f(\text{SNR})}{\text{SNR}^d}| < \infty$. It can be shown that $1 - \frac{\Gamma(n_d n_s, \alpha_0\gamma_{th})}{\Gamma(n_d n_s)} = \frac{\gamma(n_d n_s, \alpha_0\gamma_{th})}{\Gamma(n_d n_s)} \sim \text{SNR}^{-n_d n_s}$, $1 - \frac{\Gamma(n_d n_r, \alpha_2\gamma_{th})}{\Gamma(n_d n_r)} \sim \text{SNR}^{-n_d n_r}$, and $1 - \frac{\Gamma(n_r n_s, \alpha_1\Delta)}{\Gamma(n_r n_s)} \sim \text{SNR}^{-n_r n_s}$. Consequently, we obtain

$$P_{out} \sim \text{SNR}^{-(n_d n_s + n_r \min\{n_s, n_d\})},$$

i.e., the diversity order is $n_d n_s + n_r \min\{n_s, n_d\}$. $\quad\square$

Remark: If $n_s < n_d$, the diversity order is $n_s(n_d + n_r)$. Otherwise, the diversity order is $n_d(n_s + n_r)$.

3.3. Comparison of DF–AF selection MIMO relaying with AF MIMO relaying and DF MIMO relaying

First, we give the closed-form expressions of the outage probability for AF MIMO relaying and DF MIMO relaying.

Theorem 3. The outage probability of AF MIMO relaying is given by (10).

Proof. The instantaneous equivalent end-to-end SNR of AF MIMO relaying is used can be give by setting $\xi = 0$ in (3), i.e.,

$$\gamma_{af} = \max\Big(\gamma_0, \frac{\gamma_1\gamma_2}{\gamma_1 + \gamma_2 + 1}\Big).$$

Thus, the outage probability can be given by

$$P_{af} = \Pr\{\gamma_{af} < \gamma_{th}\}$$
$$= \Pr\{\gamma_0 < \gamma_{th}\}\Pr\Big\{\frac{\gamma_1\gamma_2}{\gamma_1 + \gamma_2 + 1} < \gamma_{th}\Big\}. \quad (11)$$

Meanwhile, c.d.f. of $\frac{\gamma_1\gamma_2}{\gamma_1+\gamma_2+1}$ can be expressed as [47]

$$F_{\frac{\gamma_1\gamma_2}{\gamma_1+\gamma_2+1}}(y) = 1 - \frac{2\alpha_2^{n_d n_r}(n_r n_s - 1)!e^{-(\alpha_1+\alpha_2)y}}{\Gamma(n_r n_s)\Gamma(n_d n_r)}$$
$$\times \sum_{i=0}^{n_r n_s - 1}\sum_{j=0}^{i}\sum_{k=0}^{n_d n_r - 1}\Big[\frac{1}{i!}\binom{i}{j}\binom{n_d n_r - 1}{k}\alpha_1^{\frac{2i-j+k+1}{2}}$$
$$\times \quad \alpha_2^{\frac{j-k-1}{2}}(1 + y)^{\frac{j+k+1}{2}}y^{\frac{2i+2n_d n_r-j-k-1}{2}}$$
$$\times \quad K_{j-k-1}\big(2\sqrt{\alpha_1\alpha_2 y(y + 1)}\big)\Big]. \quad (12)$$

With the help of (2) and (12), (10) can be obtained. $\quad\square$

Theorem 4. The diversity order of AF MIMO relaying is

$$d_{af} = n_d n_s + \min\{n_s, n_d\}n_r$$

Proof. First, we have

$$\frac{1}{2}\min\{\gamma_1, \gamma_2\} \leq \frac{\gamma_1\gamma_2}{\gamma_1 + \gamma_2 + 1} < \min\{\gamma_1, \gamma_2\}$$

when $\text{SNR} \to \infty$ [48]. Then, we obtain

$$\Pr\{\min\{\gamma_1, \gamma_2\} < \gamma_{th}\}$$
$$< \quad \Pr\Big\{\frac{\gamma_1\gamma_2}{\gamma_1 + \gamma_2 + 1} < \gamma_{th}\Big\}$$
$$\leq \quad \Pr\Big\{\frac{1}{2}\min\{\gamma_1, \gamma_2\} < \gamma_{th}\Big\}. \quad (13)$$

Using (2), it can be derived that

$$\Pr\{\min\{\gamma_1, \gamma_2\} < \gamma_{th}\}$$
$$= \quad 1 - \frac{\Gamma(n_r n_s, \alpha_1\gamma_{th})\Gamma(n_d n_r, \alpha_2\gamma_{th})}{\Gamma(n_r n_s)\Gamma(n_d n_r)}$$
$$= \quad \Big(1 - \frac{\Gamma(n_r n_s, \alpha_1\gamma_{th})}{\Gamma(n_r n_s)}\Big) + \Big(1 - \frac{\Gamma(n_d n_r, \alpha_2\gamma_{th})}{\Gamma(n_d n_r)}\Big)$$
$$- \quad \Big(1 - \frac{\Gamma(n_r n_s, \alpha_1\gamma_{th})}{\Gamma(n_r n_s)}\Big)\Big(1 - \frac{\Gamma(n_d n_r, \alpha_2\gamma_{th})}{\Gamma(n_d n_r)}\Big)$$
$$\sim \quad \text{SNR}^{-\min\{n_s, n_d\}n_r}. \quad (14)$$

$$P_{af} = \left[1 - \frac{\Gamma(n_d n_s, \alpha_0 \gamma_{th})}{\Gamma(n_d n_s)}\right]\left[1 - \frac{2\alpha_2^{n_d n_r}(n_r n_s - 1)! e^{-(\alpha_1 + \alpha_2)\gamma_{th}}}{\Gamma(n_r n_s)\Gamma(n_d n_r)} \sum_{i=0}^{n_r n_s - 1} \sum_{j=0}^{i} \sum_{k=0}^{n_d n_r - 1} \frac{1}{i!}\binom{i}{j}\right.$$
$$\left. \times \binom{n_d n_r - 1}{k}\alpha_2^{\frac{j-k-1}{2}}\gamma_{th}^{\frac{2i+2n_d n_r - j - k - 1}{2}}\alpha_1^{\frac{2i-j+k+1}{2}}(1+\gamma_{th})^{\frac{j+k+1}{2}}K_{j-k-1}\left(2\sqrt{\alpha_1 \alpha_2 \gamma_{th}(\gamma_{th}+1)}\right)\right] \quad (10)$$

Likewise, we can obtain that

$$\Pr\left\{\frac{1}{2}\min\{\gamma_1, \gamma_2\} < \gamma_{th}\right\} \sim \mathrm{SNR}^{-\min\{n_s, n_d\}n_r}. \quad (15)$$

Combining (13), (14), and (15), we get

$$1 - \frac{2\alpha_2^{n_d n_r}(n_r n_s - 1)! e^{-(\alpha_1 + \alpha_2)\gamma_{th}}}{\Gamma(n_r n_s)\Gamma(n_d n_r)} \sum_{i=0}^{n_r n_s - 1} \sum_{j=0}^{i} \sum_{k=0}^{n_d n_r - 1} \frac{1}{i!}$$
$$\times \binom{i}{j}\binom{n_d n_r - 1}{k}\alpha_2^{\frac{j-k-1}{2}}\alpha_1^{\frac{2i-j+k+1}{2}}\gamma_{th}^{\frac{2i+2n_d n_r - j - k - 1}{2}}$$
$$\times (1+\gamma_{th})^{\frac{j+k+1}{2}}K_{j-k-1}\left(2\sqrt{\alpha_1 \alpha_2 \gamma_{th}(\gamma_{th}+1)}\right)$$
$$= \Pr\left\{\frac{\gamma_1 \gamma_2}{\gamma_1 + \gamma_2 + 1} < \gamma_{th}\right\} \sim \mathrm{SNR}^{-\min\{n_s, n_d\}n_r}. (16)$$

Using (9), (10), and (16), we prove the lemma. □

Remark: The DF-AF selection MIMO relaying and the AF MIMO relaying have the same diversity order.

Theorem 5. The outage probability of DF MIMO relaying is given by

$$P_{df} = \left[1 - \frac{\Gamma(n_d n_s, \alpha_0 \gamma_{th})}{\Gamma(n_d n_s)}\right]\left[\frac{\Gamma(n_r n_s, \alpha_1 \Delta)}{\Gamma(n_r n_s)}\right.$$
$$\times \left.\left(1 - \frac{\Gamma(n_d n_r, \alpha_2 \gamma_{th})}{\Gamma(n_d n_r)}\right) + \left(1 - \frac{\Gamma(n_r n_s, \alpha_1 \Delta)}{\Gamma(n_r n_s)}\right)\right].$$
$$(17)$$

Proof. For DF MIMO relaying, the instantaneous equivalent end-to-end SNR can be expressed as $\gamma_{df} = \max(\gamma_0, \xi\gamma_2)$. Therefore, the outage probability can be obtained by

$$P_{df} = \Pr\{\gamma_{df} < \gamma_{th}\}$$
$$= \Pr\{\gamma_0 < \gamma_{th}\}\Pr\{\xi\gamma_2 < \gamma_{th}\} \quad (18)$$
$$= \Pr\{\gamma_0 < \gamma_{th}\}$$
$$\times \left(\Pr\{\xi = 1\}\Pr\{\gamma_2 < \gamma_{th}\} + \Pr\{\xi = 0\}\right). \quad (19)$$

Combining (4), (5), (2), and (19), (17) can be derived. □

Remark: The outage probability of the DF-AF selection MIMO relaying and that of the DF MIMO relaying are

the same when SC is utilized, and the diversity order is the same thereafter.

When MRC is used, $\gamma = \gamma_0 + \xi\gamma_2 + (1 - \xi)\frac{\gamma_1 \gamma_2}{\gamma_1 + \gamma_2 + 1} > \gamma_{df} = \gamma_0 + \xi\gamma_2$, and $\gamma > \gamma_{af} = \gamma_0 + \frac{\gamma_1 \gamma_2}{\gamma_1 + \gamma_2 + 1}$. Thus, the outage probability of DF-AF selection MIMO relaying is less than that of DF MIMO relaying and that of AF MIMO relaying, i.e., $P_{out} < P_{af}, P_{out} < P_{df}$.

4. MIMO relaying in the presence of keyholes

Keyhole effect (as illustrated in Fig. 2), under which a MIMO channel has uncorrelated spatial fading between antenna arrays but a rank-deficient transfer matrix, may exist in MIMO fading environments in realistic propagation environments. Keyhole effect will lead to significant performance degradation. Fortunately, recent researches demonstrate that cooperative diversity can mitigate keyhole effects [41]. Then we investigate the cooperative MIMO relaying in the keyhole scenario.

In this section, we consider the scenario that all MIMO channels incur keyholes. Due to the keyhole effects, $\mathbf{H} = \mathbf{h}_1 \mathbf{h}_2^H$, $\mathbf{h}_1 \in \mathbb{C}^{n_d}$, $\mathbf{h}_2 \in \mathbb{C}^{n_s}$. $\mathbf{F} = \mathbf{f}_1 \mathbf{f}_2^H$, $\mathbf{f}_1 \in \mathbb{C}^{n_r}$, $\mathbf{f}_2 \in \mathbb{C}^{n_s}$. $\mathbf{G} = \mathbf{g}_1 \mathbf{g}_2^H$, $\mathbf{g}_1 \in \mathbb{C}^{n_d}$, $\mathbf{g}_2 \in \mathbb{C}^{n_r}$. We assume independent Nakagami-m fading on both sides of the keyhole. Elements of $\mathbf{h}_1$, $\mathbf{h}_2$, $\mathbf{f}_1$, $\mathbf{f}_2$, $\mathbf{g}_1$, and $\mathbf{g}_2$ are statistically independent. The magnitudes of elements of $\mathbf{h}_1$, $\mathbf{h}_2$, $\mathbf{f}_1$, $\mathbf{f}_2$, $\mathbf{g}_1$, and $\mathbf{g}_2$ are modeled as Nakagami-m variants with general fading parameters $m_{h_{1,i}}, m_{h_{2,j}}, m_{f_{1,k}}, m_{f_{2,l}}, m_{g_{1,m}}$, and $m_{g_{2,n}}$ whereas the corresponding phases are uniformly distributed in $[0; 2\pi)$. The destination uses Maximal Ratio Combining (MRC) to combine the signals of two time slots coming from the source and the relay. The outage probability and SEP of OSTBC over MIMO relay channel with keyholes are analyzed in this section. First, some m.g.f.s are given as preliminary preparation. Next, m.g.f.-based method for computing the outage probability and SEP is introduced. Then we investigate the outage probability and SEP for the DF MIMO relaying, the AF MIMO relaying and the DF-AF selection MIMO relaying over keyhole channels. Furthermore, by comparison, we derive that DF-AF selection MIMO relaying has the best performance.

Denote $\sigma_0 = \frac{c_0 P_0}{R n_s N_0}$, $\sigma_1 = \frac{c_1 P_0}{R n_s N_0}$, and $\sigma_2 = \frac{c_2 P_1}{R n_r N_0}$, where R is the rate of the OSTBC, P_0, P_1 represent the transmit power of the source and the relay. c_0, c_1 and

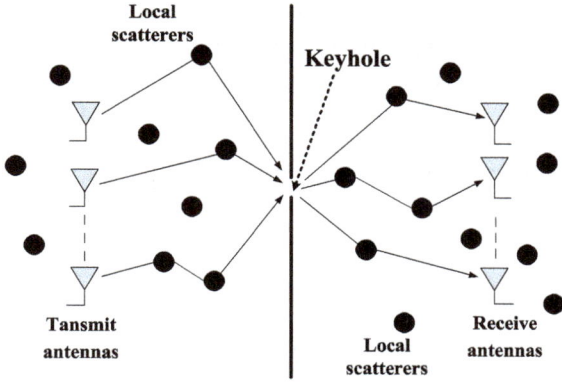

Figure 2. Keyhole effect in MIMO channel

c_2 are the distance dependent power transfer factors for the source-destination, source-relay and relay-destination channels respectively. N_0 is the variance of Gaussian noise at each receive antenna. The equivalent instantaneous SNRs per symbol of source-destination, source-relay and relay-destination channels can be given by

$$\gamma_0 = \frac{c_0 P_0}{R n_s N_0} \|\mathbf{H}\|_F^2 = \sigma_0 \|\mathbf{h}_1\|^2 \|\mathbf{h}_2\|^2, \qquad (20)$$

$$\gamma_1 = \frac{c_1 P_0}{R n_s N_0} \|\mathbf{F}\|_F^2 = \sigma_1 \|\mathbf{f}_1\|^2 \|\mathbf{f}_2\|^2 \qquad (21)$$

and

$$\gamma_2 = \frac{c_2 P_1}{R n_r N_0} \|\mathbf{G}\|_F^2 = \sigma_2 \|\mathbf{g}_1\|^2 \|\mathbf{g}_2\|^2 \qquad (22)$$

respectively [45, 49].

4.1. M.g.f. computation

Lemma 1. M.g.f. of γ_0, γ_1 and γ_2 are given by

$$\Psi_{\gamma_w}(s) = \sum_{p=1}^{\delta_w} \sum_{j=1}^{\kappa_{w,p}} \sum_{q=1}^{\tau_w} \sum_{l=1}^{\nu_{w,q}} \frac{\varrho_{w,p,j}\vartheta_{w,q,l} e^{-\frac{1}{2\sigma_w s \lambda_{w,p}\varepsilon_{w,q}}}}{(\lambda_{w,p}\varepsilon_{w,q})^{\frac{j+l-1}{2}}}$$

$$\times (-\sigma_w s)^{\frac{1-j-l}{2}} W_{\frac{1-j-l}{2},\frac{j-l}{2}}\left(-\frac{1}{\sigma_w s \lambda_{w,p}\varepsilon_{w,q}}\right), w = 0, 1, 2$$

$$(23)$$

and m.g.f. of $\theta := \min\{\gamma_1, \gamma_2\}$ is given by

$$\Psi_\theta(s) =$$

$$s \sum_{p=1}^{\delta_1} \sum_{j=1}^{\kappa_{1,p}} \sum_{q=1}^{\tau_1} \sum_{l=1}^{\nu_{1,q}} \sum_{k=0}^{l-1} \sum_{\hat{p}=1}^{\delta_2} \sum_{\hat{j}=1}^{\kappa_{2,\hat{p}}} \sum_{\hat{q}=1}^{\tau_2} \sum_{\hat{l}=1}^{\nu_{2,\hat{q}}} \sum_{\hat{k}=0}^{\hat{l}-1}$$

$$\frac{4\varrho_{1,p,j}\vartheta_{1,q,l}\varrho_{2,\hat{p},\hat{j}}\vartheta_{2,\hat{q},\hat{l}}\sigma_1^{-\frac{j+k}{2}}\sigma_2^{-\frac{\hat{j}+\hat{k}}{2}}}{\Gamma(j)\Gamma(k+1)\Gamma(\hat{j})\Gamma(\hat{k}+1)(\lambda_{1,p}\varepsilon_{1,q})^{\frac{j+k}{2}}(\lambda_{2,\hat{p}}\varepsilon_{2,\hat{q}})^{\frac{\hat{j}+\hat{k}}{2}}}$$

$$\times \int_0^\infty e^{sx} x^{\frac{j+k+\hat{j}+\hat{k}}{2}} K_{j-k}\left(2\sqrt{\frac{x}{\sigma_1\lambda_{1,p}\varepsilon_{1,q}}}\right)$$

$$\times K_{\hat{j}-\hat{k}}\left(2\sqrt{\frac{x}{\sigma_2\lambda_{2,\hat{p}}\varepsilon_{2,\hat{q}}}}\right) dx + 1, \qquad (24)$$

where δ_0, τ_0 denote the number of distinctive non-zero values of $\left\{\overline{|h_{1,i}|^2}m_{h_{1,i}}^{-1}\right\}_{i=1,\cdots,n_d}$ and $\left\{\overline{|h_{2,t}|^2}m_{h_{2,t}}^{-1}\right\}_{t=1,\cdots,n_s}$, respectively. The distinct values are denoted by $\{\lambda_{0,p}\}_{p=1,\cdots,\delta_0}$ and $\{\varepsilon_{0,q}\}_{q=1,\cdots,\tau_0}$. $\kappa_{0,p}$ and $\nu_{0,q}$ are defined as $\kappa_{0,p} = \sum_{m_{h_{1,i}}\in\Lambda_1} m_{h_{1,i}}$ with $\Lambda_1 = \left\{m_{h_{1,i}}|\overline{\|h_{1,i}\|^2} = \lambda_{0,p}m_{h_{1,i}}\right\}$, and $\nu_{0,q} = \sum_{m_{h_{2,t}}\in\Lambda_2} m_{h_{2,t}}$ with $\Lambda_2 = \left\{m_{h_{2,t}}|\overline{\|h_{2,t}\|^2} = \varepsilon_{0,q}m_{h_{2,t}}\right\}$. δ_1, τ_1 denote the number of distinctive non-zero values of $\left\{\overline{|f_{1,i}|^2}m_{f_{1,i}}^{-1}\right\}_{i=1,\cdots,n_r}$ and $\left\{\overline{|f_{2,t}|^2}m_{f_{2,t}}^{-1}\right\}_{t=1,\cdots,n_s}$ respectively. The distinct values are denoted by $\{\lambda_{1,p}\}_{p=1,\cdots,\delta_1}$ and $\{\varepsilon_{1,q}\}_{q=1,\cdots,\tau_1}$. $\kappa_{1,p}$ and $\nu_{1,q}$ are defined as $\kappa_{1,p} = \sum_{m_{f_{1,i}}\in\Lambda_3} m_{f_{1,i}}$ with $\Lambda_3 = \left\{m_{f_{1,i}}|\overline{\|f_{1,i}\|^2} = \lambda_{1,p}m_{f_{1,i}}\right\}$, and $\nu_{1,q} = \sum_{m_{f_{2,t}}\in\Lambda_4} m_{f_{2,t}}$ with $\Lambda_4 = \left\{m_{f_{2,t}}|\overline{\|f_{2,t}\|^2} = \varepsilon_{1,q}m_{f_{2,t}}\right\}$. δ_2, τ_2 denote the number of distinctive non-zero values of $\left\{\overline{|g_{1,i}|^2}m_{g_{1,i}}^{-1}\right\}_{i=1,\cdots,n_d}$ and $\left\{\overline{|g_{2,t}|^2}m_{g_{2,t}}^{-1}\right\}_{t=1,\cdots,n_r}$ respectively. The distinct values are denoted by $\{\lambda_{2,p}\}_{p=1,\cdots,\delta_2}$ and $\{\varepsilon_{2,q}\}_{q=1,\cdots,\tau_2}$. $\kappa_{2,p}$ and $\nu_{2,q}$ are defined as $\kappa_{2,p} = \sum_{m_{g_{1,i}}\in\Lambda_5} m_{g_{1,i}}$ with $\Lambda_5 = \left\{m_{g_{1,i}}|\overline{\|g_{1,i}\|^2} = \lambda_{2,p}m_{g_{1,i}}\right\}$, and $\nu_{2,q} = \sum_{m_{g_{2,t}}\in\Lambda_6} m_{g_{2,t}}$ with $\Lambda_6 = \left\{m_{g_{2,t}}|\overline{\|g_{2,t}\|^2} = \varepsilon_{2,q}m_{g_{2,t}}\right\}$. In addition, $\varrho_{w,p,j}$ and $\vartheta_{w,q,l}$ are given by

$$\varrho_{w,p,j} =$$

$$\frac{1}{(\kappa_{w,p}-j)!\lambda_{w,p}^{\kappa_{w,p}-j}} \frac{\partial^{\kappa_{w,p}-j}}{\partial y^{\kappa_{w,p}-j}}\left[\prod_{r=1,r\neq p}^{\delta_w} \frac{1}{(1+y\lambda_{w,r})^{\kappa_{w,r}}}\right]\Bigg|_{y=\frac{-1}{\lambda_{w,p}}}$$

and

$$\vartheta_{w,q,l} =$$

$$\frac{1}{(v_{w,q}-l)! \varepsilon_{w,q}^{v_{w,q}-l}} \frac{\partial^{v_{w,q}-l}}{\partial y^{v_{w,q}-l}} \left[\prod_{r=1,r\neq q}^{\tau_w} \frac{1}{(1+y\varepsilon_{w,r})^{v_{w,r}}} \right]\Bigg|_{y=\frac{-1}{\varepsilon_{w,q}}}$$

respectively.

Proof. Using (20), (21), and (22) in addition with Proposition 1 and Proposition 3 in Appendix, (23) can be obtained. C.d.f. of $\theta := \min\{\gamma_1, \gamma_2\}$ can be given by

$$
\begin{aligned}
F_\theta(y) &= 1 - \Pr\{\theta > y\} = 1 - \Pr\{\gamma_1 > y\}\Pr\{\gamma_2 > y\} \\
&= 1 - \left(1 - F_{\gamma_1}(y)\right)\left(1 - F_{\gamma_2}(y)\right). \quad (25)
\end{aligned}
$$

Consequently, m.g.f. of θ is derived as

$$
\begin{aligned}
\Psi_\theta(s) &= \int_0^\infty e^{sx} f_\theta(x)dx = \int_0^\infty e^{sx} dF_\theta(x) \\
&\overset{(a)}{=} \left[e^{sx}F_\theta(x)\right]_{x=0}^{x=\infty} - s\int_0^\infty e^{sx}F_\theta(x)dx, \\
&\overset{(b)}{=} -s\int_0^\infty e^{sx}F_\theta(x)dx, \; \mathcal{Re}\{s\} < 0, \quad (26)
\end{aligned}
$$

(a) is derived by using integration by parts, (b) holds since when $\mathcal{Re}\{s\} < 0$, we have $e^{sx}F_\theta(x) = 0$ for $x = 0$ and $x = \infty$. Using (A.5) and Proposition 2 in Appendix along with some rearrangement, (24) can be derived. $\qquad\square$

4.2. M.g.f.–based method

Let γ denote the total instantaneous received SNR. For M-PSK, the outage probability can be computed as [50]

$$P_{out} = \frac{1}{2\pi j} \int_{\sigma-j\infty}^{\sigma+j\infty} \frac{\Psi_\gamma(-s)}{s} e^{s\gamma_{th}} ds := f_{out}\left(\Psi_\gamma(\cdot)\right), \quad (27)$$

where $\gamma_{th} = 2^{(K+1)R} - 1$.

Meanwhile, SEP can be computed as

$$P_s(E) = \frac{1}{\pi} \int_0^{\frac{(M-1)\pi}{M}} \Psi_\gamma\left(-\frac{g_{psk}}{\sin^2\varphi}\right) d\varphi := f_{sep}\left(\Psi_\gamma(\cdot)\right), \quad (28)$$

where $g_{psk} = \sin^2\left(\frac{\pi}{M}\right)$.

Remark: $f_{out}()$ and $f_{sep}()$ are mappings from a function space to $[0,1]$.

4.3. Performance analysis

In this subsection, we first consider the DF MIMO relaying, the AF MIMO relaying, and the DF-AF selection MIMO relaying over keyhole channels respectively. Subsequently, we compare the three protocols.

The exact outage probability and SEP of DF MIMO relaying over keyhole channels is given by the following theorem.

Theorem 6. The outage probability and SEP of DF MIMO relaying over keyhole channels can be given by

$$P_{out} = f_{out}\Big(\quad\quad\quad\quad\quad\quad\quad\quad\quad\quad (29)$$

$$f_{sep}\left(\Psi_{\gamma_1}(s)\right)\Psi_{\gamma_0}(s) + \left(1 - f_{sep}\left(\Psi_{\gamma_1}(s)\right)\right)\Psi_{\gamma_0}(s)\Psi_{\gamma_2}(s)\Big)$$

and

$$P_s(E) = f_{sep}\Big(\quad\quad\quad\quad\quad\quad\quad\quad\quad\quad (30)$$

$$f_{sep}\left(\Psi_{\gamma_1}(s)\right)\Psi_{\gamma_0}(s) + \left(1 - f_{sep}\left(\Psi_{\gamma_1}(s)\right)\right)\Psi_{\gamma_0}(s)\Psi_{\gamma_2}(s)\Big),$$

where $P_e = f_{sep}\left(\Psi_{\gamma_1}(s)\right)$, $\Psi_{\gamma_0}(s)$, $\Psi_{\gamma_1}(s)$, and $\Psi_{\gamma_2}(s)$ can be given by Lemma 1.

Proof. By (28), the symbol error probability at the relay over source-relay channel is given by

$$P_e = f_{sep}\left(\Psi_{\gamma_1}(s)\right) \quad\quad\quad (31)$$

When MRC is used at the destination, the instantaneous total SNR at the destination is

$$\gamma_{df} = \gamma_0 + \mathcal{I}\gamma_2 \quad\quad\quad (32)$$

where $\mathcal{I}$ denotes the decoding state, $\mathcal{I} = 1$ if the relay could correctly decode, else $\mathcal{I} = 0$, i.e., $\Pr\{\mathcal{I} = 0\} = P_e$ and $\Pr\{\mathcal{I} = 1\} = 1 - P_e$.[3] So we can derive $f_{\mathcal{I}\gamma_2}(x) = (1 - P_e)f_{\gamma_2}(x) + P_e\delta(0)$. Consequently, we get

$$\Psi_{\mathcal{I}\gamma_2}(s) = (1 - P_e)\Psi_{\gamma_2}(s) + P_e. \quad\quad (33)$$

Considering the assumption of dependency, it yields that

$$\Psi_{\gamma_{df}}(s) = \Psi_{\gamma_0}(s)\Psi_{\mathcal{I}\gamma_2}(s). \quad\quad\quad (34)$$

Substituting (31) and (33) into (34) and making some manipulation, we have

$$\Psi_{\gamma_{df}}(s) = \quad\quad\quad\quad\quad\quad\quad\quad\quad\quad (35)$$

$$f_{sep}\left(\Psi_{\gamma_1}(s)\right)\Psi_{\gamma_0}(s) + \left(1 - f_{sep}\left(\Psi_{\gamma_1}(s)\right)\right)\Psi_{\gamma_0}(s)\Psi_{\gamma_2}(s).$$

Then, substituting (35) into (27) and (28), we arrive at (29) and (30). $\qquad\square$

Using (29), (30) and Lemma 1, we can obtain the exact expressions for outage and SEP. This will be useful in numerical evaluation for performance of DF keyhole MIMO relay channels.

The following theorem gives the lower and upper bounds on the outage probability and SEP of AF MIMO relaying over keyhole channels.

---

[3]Correctly decode or not is exactly accessed here and thereafter.

Theorem 7. The outage probability and SEP for the AF MIMO relaying over keyhole channels can be bounded by

$$f_{out}\left(\Psi_{\gamma_0}(s)\Psi_{\min\{\gamma_1,\gamma_2\}}(s)\right) < P_{out}$$
$$\leq f_{out}\left(\Psi_{\gamma_0}(s)\Psi_{\min\{\gamma_1,\gamma_2\}}\left(\tfrac{1}{2}s\right)\right) \quad (36)$$

and

$$f_{sep}\left(\Psi_{\gamma_0}(s)\Psi_{\min\{\gamma_1,\gamma_2\}}(s)\right) < P_s(E)$$
$$\leq f_{sep}\left(\Psi_{\gamma_0}(s)\Psi_{\min\{\gamma_1,\gamma_2\}}\left(\tfrac{1}{2}s\right)\right), \quad (37)$$

where $\Psi_{\gamma_0}(s)$ and $\Psi_{\min\{\gamma_1,\gamma_2\}}(s)$ are given by Lemma 1.

Proof. When MRC is used at the destination, the instantaneous total SNR at the destination is

$$\gamma_{af} = \gamma_0 + \frac{\gamma_1\gamma_2}{\gamma_1+\gamma_2+1} < \gamma_0 + \min\{\gamma_1,\gamma_2\} := \gamma_{1,up}. \quad (38)$$

On the other hand, when $\gamma_1 + \gamma_2 \gg 1$, i.e., high SNR, we have

$$\gamma_{af} \approx \gamma_0 + \frac{\gamma_1\gamma_2}{\gamma_1+\gamma_2} \geq \gamma_0 + \frac{1}{2}\min\{\gamma_1,\gamma_2\} := \gamma_{1,low}. \quad (39)$$

Using the dependency assumption and Proposition 3 in Appendix, we have

$$\Psi_{\gamma_{1,up}}(s) = \Psi_{\gamma_0}(s)\Psi_{\min\{\gamma_1,\gamma_2\}}(s) \quad (40)$$

and

$$\Psi_{\gamma_{1,low}}(s) = \Psi_{\gamma_0}(s)\Psi_{\min\{\gamma_1,\gamma_2\}}\left(\tfrac{1}{2}s\right). \quad (41)$$

Consequently, the outage probability and SEP for AF protocol can be bounded by

$$f_{out}\left(\Psi_{\gamma_{1,up}}(s)\right) < P_{out} \leq f_{out}\left(\Psi_{\gamma_{1,low}}(s)\right) \quad (42)$$

and

$$f_{sep}\left(\Psi_{\gamma_{1,up}}(s)\right) < P_s(E) \leq f_{sep}\left(\Psi_{\gamma_{1,low}}(s)\right). \quad (43)$$

Substituting (40) and (41) into (42) and (43), (36) and (37) can be obtained. □

It is not difficult to see that when $|\gamma_1 - \gamma_2|$ is sufficiently large, the lower bound will become tight, i.e., $P_{out} \approx f_{out}\left(\Psi_{\gamma_0}(s)\Psi_{\min\{\gamma_1,\gamma_2\}}(s)\right)$, $P_s(E) \approx f_{sep}\left(\Psi_{\gamma_0}(s)\Psi_{\min\{\gamma_1,\gamma_2\}}(s)\right)$. When $|\gamma_1 - \gamma_2|$ is sufficiently small and γ_1 is sufficiently large, the upper bound will become tight. Then we have $P_{out} \approx f_{out}\left(\Psi_{\gamma_0}(s)\Psi_{\min\{\gamma_1,\gamma_2\}}\left(\tfrac{1}{2}s\right)\right)$ and $P_s(E) \approx f_{sep}\left(\Psi_{\gamma_0}(s)\Psi_{\min\{\gamma_1,\gamma_2\}}\left(\tfrac{1}{2}s\right)\right)$.

The following theorem gives the lower and upper bounds for the outage probability and SEP of DF-AF selection MIMO relaying over keyhole channels.

Theorem 8. The outage probability and SEP for DF-AF selection MIMO relaying over keyhole channels can be bounded by

$$f_{out}\left(\Psi_{\gamma_{up}}(s)\right) < P_{out} \leq f_{out}\left(\Psi_{\gamma_{low}}(s)\right) \quad (44)$$

and

$$f_{sep}\left(\Psi_{\gamma_{up}}(s)\right) < P_s(E) \leq f_{sep}\left(\Psi_{\gamma_{low}}(s)\right) \quad (45)$$

respectively, where

$$\Psi_{\gamma_{up}} = \Psi_{\gamma_0}(s)\Big[\big(1- \quad (46)$$
$$f_{sep}\left(\Psi_{\gamma_1}(s)\right)\big)\Psi_{\gamma_2}(s) + f_{sep}\left(\Psi_{\gamma_1}(s)\right)\Psi_{\min\{\gamma_1,\gamma_2\}}(s)\Big],$$

and

$$\Psi_{\gamma_{low}} = \Psi_{\gamma_0}(s)\Big[\big(1- \quad (47)$$
$$f_{sep}\left(\Psi_{\gamma_1}(s)\right)\big)\Psi_{\gamma_2}(s) + f_{sep}\left(\Psi_{\gamma_1}(s)\right)\Psi_{\min\{\gamma_1,\gamma_2\}}\left(\tfrac{1}{2}s\right)\Big].$$

Proof. The instantaneous total SNR at the destination is

$$\gamma = \gamma_0 + \mathcal{I}\gamma_2 + (1-\mathcal{I})\frac{\gamma_1\gamma_2}{\gamma_1+\gamma_2+1}$$
$$< \gamma_0 + \mathcal{I}\gamma_2 + (1-\mathcal{I})\min\{\gamma_1,\gamma_2\} := \gamma_{up}. \quad (48)$$

Meanwhile, we have

$$\gamma \geq \gamma_0 + \mathcal{I}\gamma_2 + (1-\mathcal{I})\frac{\min\{\gamma_1,\gamma_2\}}{2} := \gamma_{low}. \quad (49)$$

Subsequently, applying independency as well as Proposition 3 in Appendix, we arrive at (46) and (47). Combining (48), (49) in addition with (27) and (28), (44) and (45) can be derived. □

It can be shown that $f_{sep}\left(\Psi_{\gamma_1}(s)\right) \to 0$ when γ_1 is sufficiently large. Then (46) and (47) become $\Psi_{\gamma_{up}} = \Psi_{\gamma_{low}} = \Psi_{\gamma_0}(s)\Psi_{\gamma_2}(s)$, we have $P_{out} = f_{out}\left(\Psi_{\gamma_0}(s)\Psi_{\gamma_2}(s)\right)$ and $P_s(E) = f_{sep}\left(\Psi_{\gamma_0}(s)\Psi_{\gamma_2}(s)\right)$.

Remark: We consider the MIMO relay channels when the source-destination, the source-relay, and the relay-destination channels all incur keyhole effect in this paper. This scenario is more complicated than the cases that only one or two channels incur keyhole effect. Moreover, we consider Nakagami-m fading environments. Since Nakagami-m fading is the generalization of Rayleigh fading, it is a more complex channel fading model than Rayleigh fading. Based on the above two reasons, we can explain why the derived results appear extremely complex. The complexity mainly exists in computing the MGFs (Lemma 1) and in the computation of the inverse Laplace transform (Eqns. (27), (28)). Regarding Lemma 1, the computations of the parameters $\varrho_{w,p,j}$ and $\vartheta_{w,q,l}$

are somewhat complicated. However, there are numerical methods to evaluate the partial derivatives [51], and an efficient method can be found in [52]. In Eq. (23), the Whittaker function $W_{\frac{1-j-l}{2}, \frac{j-l}{2}}\left(-\frac{1}{\sigma_w s \lambda_{w,p} \varepsilon_{w,q}}\right)$ appears complex. The computation complexity of such special function has been investigated in [53]. In Section 4, we use the "WhittakerW" function in the Matlab for the computation, and the execution time is acceptable. With respect to the inverse Laplace transform, there are a lot of numerical computation algorithms [54]. In [54], the comparisons of the accuracy and the computation time were performed. In the evaluations, we use the method proposed in [55]. This method can achieve better efficiency and accuracy by accelerating the convergence of the Fourier series obtained from the inversion integral using the trapezoidal rule.

Finally, we compare the three protocols as follows.

Lemma 2. In terms of the outage probability and symbol error probability over keyhole channels, the DF-AF selection MIMO relaying protocol is better than the AF MIMO relaying and the DF MIMO relaying.

Proof. Observe that $\gamma_2 > \frac{\gamma_1 \gamma_2}{\gamma_1 + \gamma_2 + 1}$. Comparing (32), (38) with (48), we have $\gamma > \gamma_{df}$ and $\gamma > \gamma_{af}$. Thus, DF-AF selection MIMO relaying has lower outage probability and symbol error probability. $\square$

5. Numerical results

In this section, computer simulations are conducted to verify the accuracy of our analytical results. We show Monte-Carlo simulation results and compare them with our analysis.

5.1. Keyhole free scenario

In the simulations, we set $c_0 = 0.9$, $c_1 = 0.95$, $c_2 = 0.85$, $N_0 = 1$, and the Alamouti code ($n_s = n_r = 2$, $R = 1$) [19] is used.

Figure 3 shows the outage probability of DF-AF selection MIMO relaying, AF MIMO relaying and DF MIMO relaying with different numbers of antennas at the destination, n_d. We assume equal power allocation, i.e., $P_0 = P_1 = 1/2P$. We can notice that DF-AF selection MIMO relaying has the same outage performance as DF MIMO relaying and has better outage performance than AF MIMO relaying. It can also be noted that the number of antennas at the destination has a strong impact of the performance enhancement, since the diversity order is $n_d n_s + n_r \min\{n_d, n_s\}$. Observe that simulation curves match in high accuracy with analytical ones.

Figure 4 plots the diversity order with respect to n_s and n_d. In the simulations, we set $n_r = 2$. We can observe the relations in the figure. For example, the diversity order is $2n_d + 4$ for $n_d \geq 2$ when $n_s = 2$.

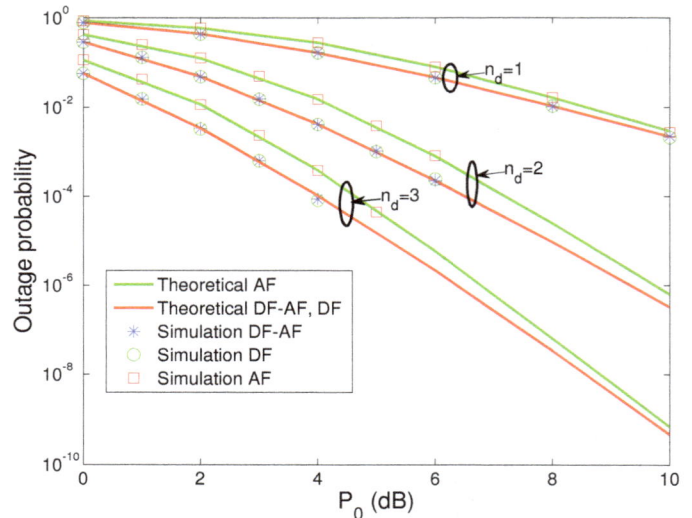

Figure 3. Outage performance of DF-AF selection MIMO relaying, DF MIMO relaying, and AF MIMO relaying with different values of n_d.

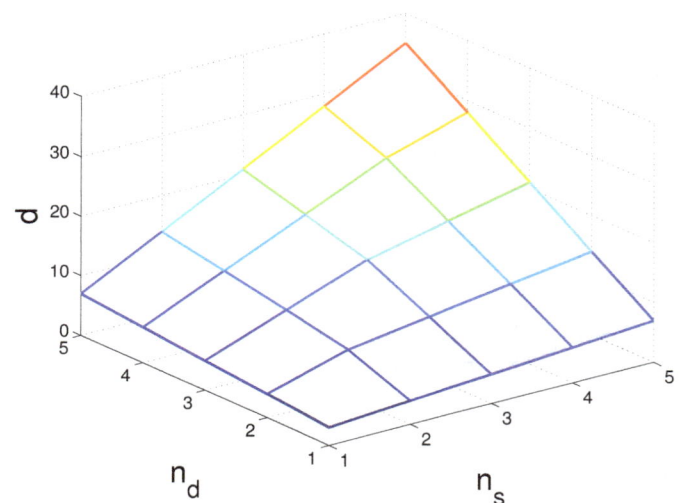

Figure 4. Diversity performance of DF-AF selection MIMO relaying

To further demonstrate the advantages of the DF-AF selection MIMO relay scheme, we show the outage performance of the three schemes when MRC is used in Figure 5. In the simulations, we utilize different values of power allocation, $\theta = \frac{P_0}{P}$. From the figure, we can see that the DF-AF selection MIMO relay scheme has better outage performance than DF MIMO relaying and AF MIMO relaying. It can also be observed that the outage probability first decreases and then increases with the increase of θ. It is because that when θ is small, the decoding at the relay fails with high probability,

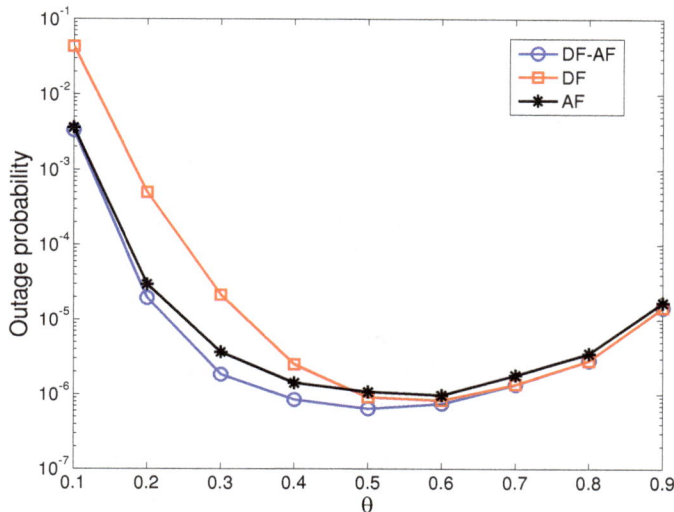

Figure 5. Outage performance of DF–AF selection MIMO relaying, DF MIMO relaying, and AF MIMO relaying with $n_d = 2$ and $P = 10dB$.

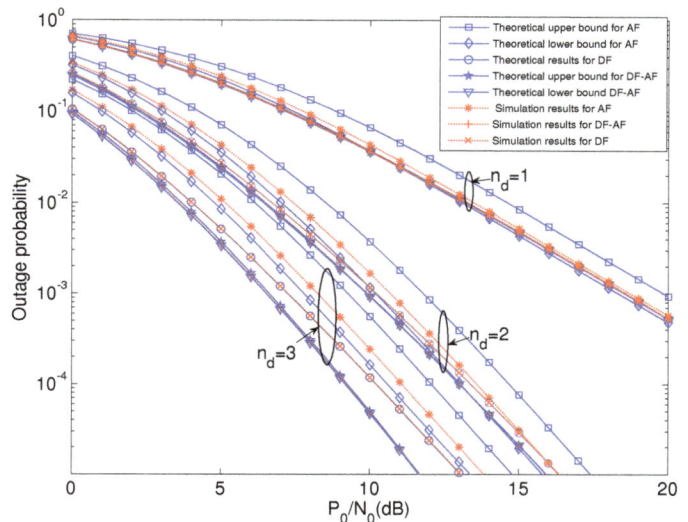

Figure 6. Outage probability performance of Alamouti code in Nakagami-m keyhole environments with fading parameters $m_{h_{21}} = 2, m_{h_{22}} = 3, m_{g_{22}} = 2$ and all other fading parameters equal to 1

i.e., AF protocol will be applied, and $\gamma = \gamma_0 + \frac{\gamma_1\gamma_2}{\gamma_1+\gamma_2+1}$. With the increase of θ, $|\gamma_1 - \gamma_2|$, i.e., the difference between γ_1 and γ_2 decreases and then $\frac{\gamma_1\gamma_2}{\gamma_1+\gamma_2+1}$ increases, so the outage probability decreases. Once θ is larger than a certain value, the relay could correctly decode the source message with high probability. Then DF protocol will be used and $\gamma = \gamma_0 + \gamma_2$. In this case, γ_2 with decreases with the increase of θ, then the outage probability will increase.

5.2. Keyhole scenario

The outage probability and SEP of different MIMO relaying schemes over keyhole channels are evaluated. We set $P_0 = P_1$, $c_0 = c_1 = c_2 = 1$ and employ BPSK modulation in the simulations. The source and the relay use the same OSTBCs. We consider two kinds of OSTBC schemes: Alamouti code and $\mathcal{G}_3$ [56].

Figure 6 and Figure 7 plot the outage probability performance and SEP performance of keyhole channels with DF-AF selection MIMO relaying, AF MIMO relaying and DF MIMO relaying when Alamouti code is used at the source and the relay ($n_s = n_r = 2$, $R = 1$), respectively. To compare the impact of the number of antennas at the destination (n_d), the outage probabilities and SEPs at different the numbers of antennas at the destination are presented.

Figure 8 and Figure 9 illustrate the outage probability performance and SEP performance of keyhole channels with DF-AF selection MIMO relaying, AF MIMO relaying and DF MIMO relaying when $\mathcal{G}_3$ is used at the source and the relay ($n_s = n_r = 3$, $R = 1/2$), respectively.

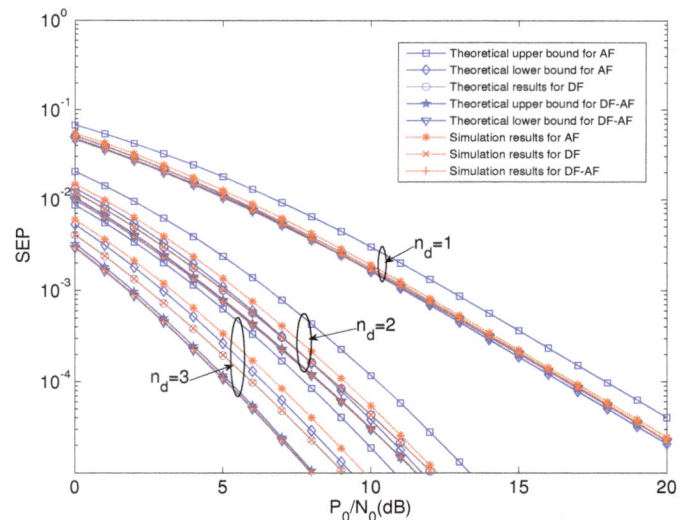

Figure 7. SEP performance of Alamouti code in Nakagami-m keyhole environments with fading parameters $m_{h_{21}} = 2, m_{h_{22}} = 3, m_{g_{22}} = 2$ and all other fading parameters equal to 1

Different numbers of antennas at the destination are also considered.

The observations from numerical results can be summarized as follows.

(i) The analytical results and the simulation results are in excellent agreement. For DF MIMO relaying, the analytical results and the simulation results match in high accuracy. For AF MIMO

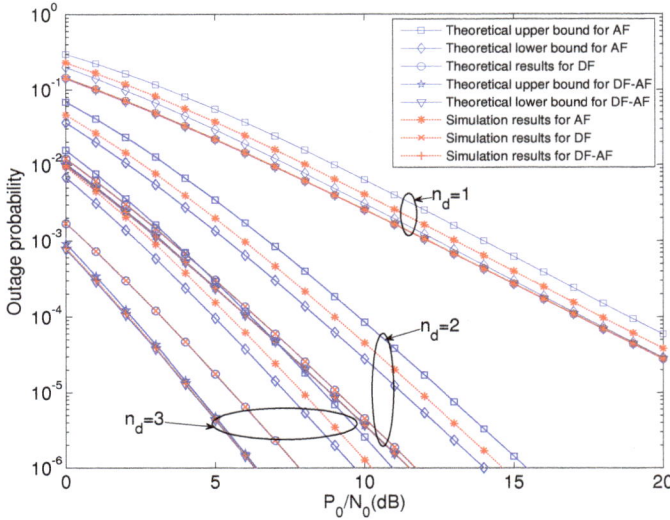

Figure 8. Outage probability performance of $\mathcal{G}_3$ code when all fading parameters equal to 1

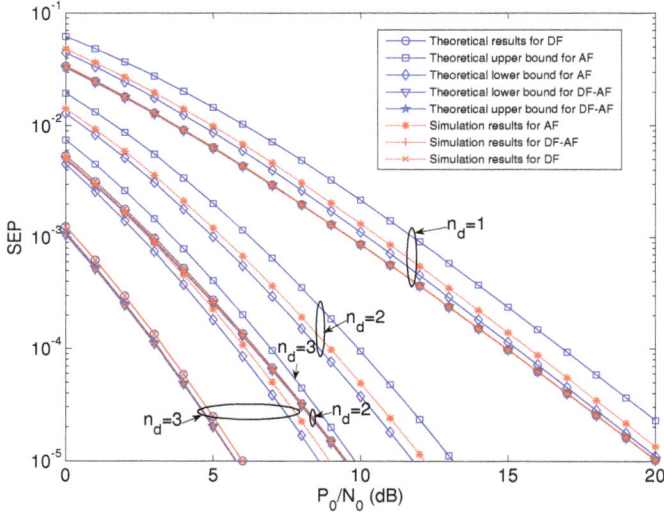

Figure 9. SEP performance of $\mathcal{G}_3$ code when all fading parameters equal to 1

relaying and DF-AF MIMO relaying, the analytical results give lower and upper bounds for the simulation results.

(ii) Regarding outage probability performance and SEP performance over keyhole channels, the DF-AF selection MIMO relaying is better than the DF MIMO relaying which is better than the AF MIMO relaying.

(iii) The OSTBC scheme used at the source and relay as well as the number of antennas at the destination plays an important role in the performance evaluations.

6. Conclusion

Cooperative MIMO relaying is investigated in the paper. We introduce the DF-AF selection MIMO relaying scheme. For the keyhole-free scenario, we investigate the outage probability of the DF-AF selection MIMO relaying. The closed-form outage probability and diversity order are derived. For comparison purpose, we also obtain the outage and diversity of the DF MIMO relaying and the AF MIMO relaying. For the scenario that the keyholes exist, the outage probability and symbol error probability of OSTBC over MIMO relay channels with keyholes are analyzed. Exact outage probability and symbol error probability are obtained for DF MIMO relaying. With respect to AF MIMO relaying and DF-AF selection MIMO relaying, lower and upper bounds are derived. In addition, we proved that DF-AF selection MIMO relaying protocol has the best performance in both scenarios. Numerical results verify our proposed analysis.

Appendix A.

Proposition 1. Let $\alpha := (\alpha_1, \cdots, \alpha_K)$, $\beta := (\beta_1, \cdots, \beta_L)$ with $|\alpha_i|^2 \sim \mathcal{G}(m_{\alpha_i})(i = 1, \cdots, K)$, $|\beta_t|^2 \sim \mathcal{G}(m_{\beta_t})(t = 1, \cdots, L)$ with all elements being independent. Then m.g.f. of $\zeta = \|\alpha\|^2 \|\beta\|^2$ is given by

$$\Psi_\zeta(s) = \sum_{p=1}^{\delta} \sum_{j=1}^{\kappa_p} \sum_{q=1}^{\tau} \sum_{l=1}^{\nu_q} \frac{\rho_{p,j} \vartheta_{q,l} e^{-\frac{1}{2s\lambda_p \varepsilon_q}}}{(\lambda_p \varepsilon_q)^{\frac{j+l-1}{2}}}$$

$$\times \quad (-s)^{\frac{1-j-l}{2}} W_{\frac{1-j-l}{2}, \frac{j-l}{2}} \left(-\frac{1}{s\lambda_p \varepsilon_q}\right). \quad (A.1)$$

Proof. P.d.f. of $\zeta = \|\alpha\|^2 \|\beta\|^2$ is given by [30]

$$f_\zeta(x) = \sum_{p=1}^{\delta} \sum_{j=1}^{\kappa_p} \sum_{q=1}^{\tau} \sum_{l=1}^{\nu_q} \frac{2\rho_{p,j} \vartheta_{q,l} x^{\frac{j+l}{2}-1}}{\Gamma(j)\Gamma(l)(\lambda_p \varepsilon_q)^{\frac{j+l}{2}}}$$

$$\times \quad K_{j-l}\left(2\sqrt{\frac{x}{\lambda_p \varepsilon_q}}\right), \quad (A.2)$$

where δ, τ denote the number of distinctive non-zero values of $\frac{|\alpha_i|^2}{m_{\alpha_i}}$ and $\frac{|\beta_t|^2}{m_{\beta_t}}$ respectively. The distinct values are denoted by λ_p and ε_q. κ_p and ν_q are defined as $\kappa_p = \sum_{m_{\alpha_i} \in \left\{m_{\alpha_i} \mid \frac{|\alpha_i|^2}{m_{\alpha_i}} = \lambda_p\right\}} m_{\alpha_i}$ and $\nu_q = \sum_{m_{\beta_t} \in \left\{m_{\beta_t} \mid \frac{|\beta_t|^2}{m_{\beta_t}} = \varepsilon_q\right\}} m_{\beta_t}$. In addition, $\rho_{p,j}$ and $\vartheta_{q,l}$ are given by $\rho_{p,j} = \frac{1}{(\kappa_p - j)! \lambda_p^{\kappa_p - j}} \frac{\partial^{\kappa_p - j}}{\partial y^{\kappa_p - j}} \left[\prod_{r=1, r \neq p}^{\delta} \frac{1}{(1+y\lambda_r)^{\kappa_r}}\right]\bigg|_{y=\frac{-1}{\lambda_p}}$ and $\vartheta_{q,l} = \frac{1}{(\nu_q - l)! \varepsilon_q^{\nu_q - l}} \frac{\partial^{\nu_q - l}}{\partial y^{\nu_q - l}} \left[\prod_{r=1, r \neq q}^{\tau} \frac{1}{(1+y\varepsilon_r)^{\nu_r}}\right]\bigg|_{y=\frac{-1}{\varepsilon_q}}$ respectively. By definition, m.g.f. of ζ is given by

$$\Psi_\zeta(s) = \int_{-\infty}^{\infty} e^{sx} f_\zeta(x) dx. \quad (A.3)$$

Substituting (A.2) into (A.3), after arranging terms, (A.3) can be rewritten as

$$\Psi_\zeta(s) = \sum_{p=1}^{\delta}\sum_{j=1}^{\kappa_p}\sum_{q=1}^{\tau}\sum_{l=1}^{\nu_q} \frac{2\rho_{p,j}\vartheta_{q,l}}{\Gamma(j)\Gamma(l)(\lambda_p\varepsilon_q)^{(j+l)/2}}$$

$$\times \int_0^\infty e^{sx}x^{(j+l)/2-1}K_{j-l}\left(2\sqrt{\frac{x}{\lambda_p\varepsilon_q}}\right)dx. \quad (A.4)$$

Using equation (6.643.3) in [57], (A.4) can be reexpressed as

$$\Psi_\zeta(s) = \sum_{p=1}^{\delta}\sum_{j=1}^{\kappa_p}\sum_{q=1}^{\tau}\sum_{l=1}^{\nu_q} \frac{2\rho_{p,j}\vartheta_{q,l}}{\Gamma(j)\Gamma(l)(\lambda_p\varepsilon_q)^{(j+l)/2}}\frac{\Gamma(j)\Gamma(l)}{2\sqrt{\frac{1}{\lambda_p\varepsilon_q}}}$$

$$\times e^{-\frac{1}{2s\lambda_p\varepsilon_q}}(-s)^{\frac{1-j-l}{2}}W_{\frac{1-j-l}{2},\frac{j-l}{2}}\left(-\frac{1}{s\lambda_p\varepsilon_q}\right) = \sum_{p=1}^{\delta}\sum_{j=1}^{\kappa_p}\sum_{q=1}^{\tau}\sum_{l=1}^{\nu_q}$$

$$\frac{\rho_{p,j}\vartheta_{q,l}e^{-\frac{1}{2s\lambda_p\varepsilon_q}}}{(\lambda_p\varepsilon_q)^{(j+l-1)/2}}(-s)^{\frac{1-j-l}{2}}W_{\frac{1-j-l}{2},\frac{j-l}{2}}\left(-\frac{1}{s\lambda_p\varepsilon_q}\right). \quad (A.5)$$

$\square$

Proposition 2. C.d.f. of ζ defined in Proposition 1 is given by

$$F_\zeta(x) = 1 - \sum_{p=1}^{\delta}\sum_{j=1}^{\kappa_p}\sum_{q=1}^{\tau}\sum_{l=1}^{\nu_q}\sum_{k=0}^{l-1} \frac{2\rho_{p,j}\vartheta_{q,l}x^{\frac{j+k}{2}}}{\Gamma(j)\Gamma(k+1)(\lambda_p\varepsilon_q)^{\frac{j+k}{2}}}$$

$$\times K_{j-k}\left(2\sqrt{\frac{x}{\lambda_p\varepsilon_q}}\right). \quad (A.6)$$

Proof. The proof is similar to the proof of Theorem 4 in [30]. By replacing $\Upsilon(R)$ with x, we arrive at (A.6). $\square$

Proposition 3. Let $Y = cX$, where X is random variable and $c \neq 0$ is constant. Then $\Psi_Y(s) = \Psi_X(cs)$.

Proof.

$$F_Y(x) = \Pr\{Y < x\} = \Pr\{cX < x\}$$

$$= \Pr\{X < \frac{1}{c}x\} = F_X\left(\frac{1}{c}x\right). \quad (A.7)$$

By differentiating (A.7) with respect to x, we have $f_Y(x) = \frac{1}{c}f_X\left(\frac{1}{c}x\right)$. The MGF of Y can be given by

$$\Psi_Y(s) = \int_{-\infty}^\infty f_Y(x)e^{sx}dx = \int_{-\infty}^\infty \frac{1}{c}f_X\left(\frac{1}{c}x\right)e^{sx}dx$$

$$= \int_{-\infty}^\infty f_X(t)e^{(cs)t}dt = \Psi_X(cs). \quad (A.8)$$

$\square$

Acknowledgement. This work is partially supported by the National Basic Research Program of China (973 Program) under Grants 2013CB336600 and 2012CB316001, the National Nature Science Foundation (NSF) of China under Grant 61322111, Beijing Nova Program, and New Century Talent Program of MoE.

References

[1] T. Zhang, W. Chen, W. Zhang, and Z. Cao, "DF-AF selection MIMO relaying with orthogonal space-time block codes in wireless networks," *Proc. IEEE ICCC'13*, Xi'an, China, Aug. 2013.

[2] G. J. Foschini, "Layered space-time architecture for wireless communication in fading environments when using multi-element antennas," *Bell Labs Tech. J.*, pp. 41-59, 1996.

[3] E. Telatar, "Capacity of multiantenna gaussian channels," *Eur. Trans. Telecommun.*, vol. 10, no. 6, pp. 585-596, 1999.

[4] D. Gesbert, M. Shafi, D. Shiu, P. J. Smith, and A. Naguib, "From theory to practice: an overview of MIMO space-time coded wireless systems," *IEEE J. Sel. Areas Commun.*, vol. 21, no. 3, pp. 281-302, Arp. 2003.

[5] A. Sendonaris, E. Erkip, and B. Aazhang, "User cooperation diversity-part I: system description," *IEEE Trans. Commun.*, vol. 51, no. 11, pp. 1927-1938, Nov. 2003.

[6] J. N. Laneman, D. N. C. Tse, and G. W. Wornell, "Cooperative diversity in wireless networks: efficient protocols and outage behavior," *IEEE Trans. Inf. Theory*, vol. 51, no. 12, pp. 3062-3080, Dec. 2004.

[7] B. Zhao and M. Valenti, "Some new adaptive protocols for the wireless relay channel," *Proc. Allerton Conf. Commun., Control, and Comp.*, Monticello, IL, Oct. 2003.

[8] M. R. Souryal and B. R. Vojcic, "Performance of amplify-and-forward and decode-and-forward relaying in Rayleigh fading with turbo codes," *Proc. IEEE ICASSP*, Toulouse, France, May 2006.

[9] W. Su, and X. Liu, "On optimum selection relaying protocols in cooperative wireless networks," *IEEE Trans. Commun.*, vol. 58, no. 1, pp. 52-57, Jan. 2010.

[10] T. Zhang, W. Chen, and Z. Cao, "Opportunistic DF-AF selection relaying with optimal relay selection in Nakagami-*m* fading environments," *Proc. the 1st IEEE International Conference on Communications in China (IEEE ICCC'12)*, Beijing, China, 2012, pp. 683-688.

[11] H. Muhaidat and M. Uysal, "Cooperative diversity with multiple-antenna nodes in fading relay channels," *IEEE Trans. Wireless Commun.*, vol. 7, no. 8, pp. 3036-3046, Aug. 2008.

[12] B. K. Chalise and L. Vandendorpe, "MIMO relay design for multipoint-to-multipoint communications with imperfect channel state information," *IEEE Trans. Signal Process.*, vol. 57, no. 7, pp. 2785-2796, Jul. 2009.

[13] B. Wang, J. Zhang, and A. Host-Madsen, "On the capacity of MIMO relay channels," *IEEE Trans. Inf. Theory*, vol. 51, no. 1, pp. 29-43, 2005.

[14] B. K. Chalise and L. Vandendorpe, "Outage probability of a MIMO relay channel with orthogonal space-time block codes," *IEEE Commun. Lett.*, vol. 12, no. 4, pp. 280-282, Apr. 2008.

[15] Y. Song, H. Shin, and E. Hong, "MIMO cooperative diversity with scalar-gain amplify-and-forward relaying," *IEEE Trans. Commun.*, vol. 57, no. 7, pp. 1932-1938, Jul. 2009.

[16] P. Dharmawansa, M. R. McKay, and R. K. Mallik, "Analytical performance of amplify-and-forward MIMO relaying with orthogonal space-time block codes," *IEEE Trans. Commun.*, vol. 58, no. 7, pp. 2147-2158, Jul. 2010.

[17] S. Muhaidat, J. K. Cavers, and P. Ho, "Transparent amplify-and-forward relaying in MIMO relay channels," *IEEE Trans. Wireless Commun.*, vol. 9, no. 10, pp. 3144-3154, Oct. 2010.

[18] A. Abdaoui, S. S. Ikki, M. H. Ahmed, and E. Chatelet, "On the performance analysis of MIMO relaying scheme with space time block codes," *IEEE Trans. Veh. Technol*, vol. 59, no. 7, pp. 3604-3609, Sep. 2010.

[19] S. M. Alamouti, "A simple transmit diversity technique for wireless communications," *IEEE J. Sel. Areas Commun.*, vol. 16, no. 8, pp. 1451-1458, Oct. 1998.

[20] L. Yang and Q. T. Zhang, "Performance analysis of MIMO relay wireless networks with orthogonal STBC," *IEEE Trans. Veh. Technol*, vol. 59, no. 7, pp. 3668-3674, Sep. 2010.

[21] G. V. V. Sharma, V. Ganwani, U. B. Desai, and S. N. Merchant, "Performance analysis of maximum likelihood detection for decode and forward MIMO relay channels in Rayleigh fading," *IEEE Trans. Wireless Commun.*, vol. 9, no. 9, pp. 2880-2889, Sep. 2010.

[22] S. Zhu, K. K. Leung, "Cooperative orthogonal MIMO-relaying for UWB ad-hoc networks," *Proc. IEEE GLOBE-COM'07*, 2007.

[23] S. Chu and X. Wang, "Adaptive exploitation of cooperative relay for high performance communications in MIMO ad hoc networks," *Proc. The 7th IEEE International Conference on Mobile Ad-hoc and Sensor Systems (IEEE MASS'10)*, San Francisco, CA, 2010.

[24] D. Chizhik, G. J. Foschini, and R. A. Valenzuela, "Capacities of multielement transmit and receive antennas: Correlations and keyholes," *Electron. Lett.*, vol. 36, no. 13, pp. 1099-1100, June 2000.

[25] D. Gesbert, H. Bölcskei, D. Gore, and A. Paulraj, "MIMO wireless channels: Capacity and performance prediction," in *Proc. IEEE GLOBECOM'00*, San Francisco, CA, 2000.

[26] D. Chizhik, G. J. Foschini, M. J. Gans, and R. A. Valenzuela, "Keyholes, correlations, and capacities of multi-element transmit and receive antennas," *IEEE Trans. Wireless Commun.*, vol. 1, no. 2, pp. 361-368, Apr. 2002.

[27] P. Almers, F. Tufvesson, and A. F. Molisch, "Keyhole effect in MIMO wireless channels: measurements and theory," *IEEE Trans. Wireless Commun.*, vol. 5, no. 12, pp. 3596-3604, Dec. 2006

[28] V. Tarokh, H. Jafarkhani, and A. R. Calderbank, "Space-time codes from orthogonal designs," *IEEE Trans. Inform. Theory*, vol. 45, no. 5, pp. 1456-1467, Jul. 1999.

[29] H. Shin and J. H. Lee, "Effect of keyholes on the symbol error rate of space-time block codes," *IEEE Commun. Lett.*, vol. 7, no. 1, pp. 27-29, Jan. 2003.

[30] A. Muller and J. Speidel, "Ergodic capacity and information outage probability of MIMO Nakagami-m keyhole channels with general branch parameters," in *Proc. IEEE WCNC'07*, Hong Kong, 2007, pp. 2184-2189.

[31] P. Yahampath and A. Hjϕungnes, "Symbol error rate analysis of spatially correlated keyhole MIMO channels with space-time block coding and linear precoding," in *Proc. IEEE GLOBECOM'07*, Washington, DC USA, 2007, pp. 5367-5371.

[32] Y. Gong and K. B. Letaief, "On the error probability of orthogonal space-time block codes over keyhole MIMO channels," *IEEE Trans. Wireless Commun.*, vol. 6, no. 9, pp. 3402-3409, Sep. 2007.

[33] N. H. Tran, H. H. Nguyen, and T. Le-Ngoc, "Symbol and bit rrror probabilities of orthogonal space-time block codes with antenna selection over keyhole fading channels," *IEEE Trans. Wireless Commun.*, vol. 7, no. 12, pp. 4818-4824, Dec. 2008.

[34] H. Zhao, Y. Gong, Y. L. Guan, and S. Li, "Performance analysis of space-time block codes in Nakagami-m keyhole channels with arbitrary fading parameters," in *Proc. IEEE ICC'08*, Beijing, China, 2008, pp. 4090-4094.

[35] H. Zhao, Y. Gong, Y. L. Guan, and Y. Tang, "Performance analysis of M-PSK/M-QAM modulated orthogonal space-time block codes in keyhole channels," *IEEE Trans. Veh. Technol*, vol. 58, no. 2, pp. 1036-1043, Feb. 2009

[36] S. Sanayei, A. Hedayat, and A. Nosratinia, "Space time codes in keyhole channels: Analysis and design," *IEEE Trans. Wireless Commun.*, vol. 6, no. 6, pp. 2006-2011, June 2007.

[37] A. Nezampour, A. Nasri, and R. Schober, "Asymptotic analysis of space-time codes in generalized keyhole fading ahannels," *IEEE Trans. Wireless Commun.*, vol. 10, no. 6, pp.1863- 1873, Jun. 2011.

[38] C. Zhong, S. Jin, and K.-K. Wong, "MIMO Rayleigh-Product Channels with Co-Channel Interference," *IEEE Trans. Commun.*, vol.57, no.6, pp. 1824-1835, Jun. 2009.

[39] S. Yang and J.-C. Belfiore, "On the diversity of Rayleigh product channels,"*Proc. ISIT2007*, Nice, France, Jun., 2007

[40] C. Zhong and T. Ratnarajah, "Ergodic Sum Rate Analysis of Rayleigh Product MIMO Channels with Linear MMSE Receiver," *Proc. ISIT2011*, Saint-Petersburg, Russia, 2011.

[41] O. Souihli and T. Ohtsuki, "The MIMO relay channel in the presence of keyhole effects," in *Proc. IEEE ICC'10*, Captown, South Africa, 2010, pp. 1-5.

[42] A. Firag, H. A. Suraweera, P. J. Smith, and C. Yuen, "Dual-hop MIMO amplify-and-forward relay channel capacity with keyhole effect," *IEEE Commun. Lett.*, vol. 15, no. 10, pp. 1050-1052, Oct. 2011

[43] T. Q. Duong, H. A. Suraweera, T. A. Tsiftsis, H. -A. Zepernick, and A. Nallanathan, "OSTBC transmission in MIMO AF relay systems with keyhole and spatial correlation effects," *Proc. IEEE ICC'11*, 2011.

[44] T. Zhang, "Performance analysis of cooperative multiple-input multiple-output relaying with orthogonal space-time block codes (STBCs) in the presence of keyhole effects," *IET Commun.*, vol. 6, no. 13, pp. 1943-1951, Sept. 2012.

[45] S. Sandhu and A. Paulraj, "Space-time block codes: a capacity perspective," *IEEE Commun. Lett.*, vol. 4, pp. 384-386, Dec. 2000.

[46] S. Savazzi and U. Spagnolini, "Cooperative space-time coded transmissions in Nakagami-*m* fading channels," *Proc. IEEE GLOBECOM'07*, 2007, pp. 4334-4338.

[47] T. A. Tsiftsis, G. K. Karagiannidis, P. T. Mathiopoulos, and S. A. Kotsopoulos, "Nonregenerative dual-hop cooperative links with selection diversity," *EURASIP J. Wireless Commun. Networking*, vol. 2006, article ID 17862, pp. 1-8, 2006.

[48] P. A. Anghel and M. Kaveh, "Exact symbol error probability of a cooperative network in a Rayleigh-fading environment," *IEEE Trans. Wireless Commun.*, vol. 3, no. 5, pp. 1416-1421, Sep. 2004.

[49] H. Shin and J. H. Lee, "Exact symbol error probability of orthogonal space-time block codes," in *Proc. IEEE GLOBECOM'02*, Taipei, Taiwan, 2002, pp. 1197-1121.

[50] M. K. Simon and M.-S. Alouini, *Digital Communication Over Fading Channels*, 2nd ed. New York: John Wiley and Sons, 2005.

[51] R. L. Burden and J. D. Faires, *Numerical Analysis*, 7th ed. Brooks/Cole. ISBN 0-534-38216-9, 2000.

[52] R. D. Neidinger, "An efficient method for the numerical evaluation of partial derivatives of arbitrary order," *ACM Transactions on Mathematical Software (TOMS)*, vol. 18, iss. 12, Jun. 1992

[53] J. M. Borwein and P. B. Borwein, *Pi and the AGM: A study in analytic number theory and computational complexity*, John Wiley, 1987.

[54] B. Davies and B. Martin, "Numerical inversion of Laplace transforms: A survey and comparison of methods," *J. Comput. Phys.*, 33 (1), pp.1-32, 1979.

[55] F. R. De hoogs, J. H. Knightt, and A. N. Stokes, "An improved method for numerical inversion of Laplace transforms, " *SIAM J. ScI. STAT. COMPUT.*, vol.3, no.3, Sep. 1982.

[56] V. Tarokh, H. Jafarkhani, and A. R. Calderbank, "Space-time block coding for wireless communications: Performance results," *IEEE J. Sel. Areas Commun.*, vol. 17, no. 3, pp. 451-460, Mar. 1999.

[57] I. Gradshteyn and I. Ryzhik, *Table of Integrals, Series, and Products*, 7th ed. New York: Academic Press, Inc, 2007.

Multiple Access Techniques for Next Generation Wireless: Recent Advances and Future Perspectives

Shree Krishna Sharma[1,*], Mohammad Patwary[2], Symeon Chatzinotas[1]

[1]SnT - securityandtrust.lu, University of Luxembourg, Luxembourg
[2]FCES, Staffordshire University, United Kingdom

Abstract

The advances in multiple access techniques has been one of the key drivers in moving from one cellular generation to another. Starting from the first generation, several multiple access techniques have been explored in different generations and various emerging multiplexing/multiple access techniues are being investigated for the next generation of cellular networks. In this context, this paper first provides a detailed review on the existing Space Division Multiple Access (SDMA) related works. Subsequently, it highlights the main features and the drawbacks of various existing and emerging multiplexing/multiple access techniques. Finally, we propose a novel concept of clustered orthogonal signature division multiple access for the next generation of cellular networks. The proposed concept envisions to employ joint antenna coding in order to enhance the orthogonality of SDMA beams with the objective of enhancing the spectral efficiency of future cellular networks.

Keywords: 5G, Multiple Access, Multiplexing, SDMA, Cellular Networks, Orthogonal Multiple Access

1. Introduction

The next generation of wireless networks is expected to provide better quality of service, lower latency, low energy consumption, and higher throughput [1]. In order to meet these requirements, several techniques such as ultra-high cell densification, bandwidth extension beyond 6 GHz, i.e., mm-wave and the increased spectral efficiency utilizing massive Multiple Input Multiple Output (MIMO) techniques have been considered as the promising solutions in the recent literature.

The advances in multiple access techniques has been one of the main drivers towards moving to a new generation of wireless starting from the first generation (1G) to the current fourth generation (4G). The 1G cellular system, i.e., Advanced Mobile Phone Service (AMPS), was based on Frequency Division Multiple Access (FDMA) technology while the second generation system (2G) was based on Time Division Multiple Access (TDMA). In parallel to the TDMA, Code Division Multiple Access (CDMA) based systems were developed and eventually they were adopted in third generation (3G) cellular systems. Subsequently, Orthogonal Frequency Division Multiplexing Access (OFDMA) came as a candidate technique for 4G due to its several advantages such as intrinsic orthogonality, receiver circuit simplicity, better spatial diversity and multiplexing capabilities. The current 4G technology, also called LTE (Long Term Evolution), and is evolving towards LTE-Advanced (LTE-A) with the inclusion of several advanced features such as coordinated multi-point transmission and carrier aggregation [2].

In order to meet the increasing capacity demands of future wireless networks, there has been a great interest of moving towards higher frequency range such as millimeter-wave (mm-wave) frequencies [3]. Although a significant capacity gain is expected to achieve with mm-wave communication systems over the current 4G wireless networks, several aspects of cellular systems need to be redesigned in order to fully achieve the desired capacity gain. Among these aspects, multiple access scheme is of significant importance and the investigation of suitable multiplexing/multiple access schemes is crucial in designing future high-frequency wireless networks [4].

The OFDMA, which is the multiuser version of the OFDM scheme, is a widely used technology in today's 4G wireless networks. Despite its significant benefits, the main drawback for making it less attractive for future 5G wireless is that each single subcarrier in an OFDM system is shaped using a rectangular window in the time domain, leading to sinc-shaped sub-carriers in the frequency domain [5]. Furthermore, its important characteristic that the spectrum can be divided into multiple parallel orthogonal sub-bands with the highest possible efficiency, is applicable only for the case of frequency synchronization and perfect time alignment within the duration of the cyclic prefix. The LTE technology is able to partially address this issue by employing a closed loop ranging mechanism and demanding oscillator requirements. However, from the energy efficiency point of view, this is not an efficient approach.

In addition to the commonly used time, and frequency multiplexing/multiple access schemes, the concept of Space Division Multiple Access (SDMA) has received important attention in the cellular literature [6, 8–10]. In this scheme, multiple users can be served simultaneously in the same channel by the superposition of the beam, thus leading to the better utilization of the available spectrum. In this paper, we first provide a review on the existing works in the area of the SDMA technique. We then discuss the emerging multiplexing/multiple access techniques such as Orbital Angular Momentum Multiplexing (OAMM), Polarization Division Multiple Access (PDMA), Interweave Division Multiple Access (IDMA) and Sparse Code Multiple Access (SCMA). Subsequently, we propose a novel clustered orthogonal signature multiple access scheme for the next generation of cellular networks.

The remainder of this paper is organized as follows: Section 2 provides an overview of the existing works on the applications of SDMA technique in wireless networks. Section 3 highlights the main features of the emerging multiplexing/multiple access techniques while Section 4 proposes a novel clustered orthogonal signature division multiple access. Finally, Section 5 concludes this paper.

2. Space Division Multiple Access

2.1. Introduction

As compared to the traditional mobile communication systems, the radio capacity in an SDMA-based system can be increased by employing antenna arrays at the Base Station (BS) side, and subsequently forming adaptive directed beams in both uplink and downlink directions [6]. This scheme provides the possibility to serve multiple users simultaneously in the same channel by the superposition of the beams and thus allows to enhance the capacity of the system. The importance of SDMA in wireless is not only because of its multiple access capability but also it enhances the spectral efficiency by enabling the frequency reuse within a particular cell [7].

The SDMA technology can be combined with the Orthogonal Frequency Division Multiple Access (OFDMA) technology in order to enhance the spectral spectral efficiency of future wireless systems, resulting in a joint SDMA-OFDMA system. The allocation of time, frequency and space resources to different user terminals in a joint SDMA-OFDMA system is a highly complex resource allocation problem [11]. However, in an SDMA system, the pre-processing of users' signals assuming the channel knowledge at the BS can enable the simultaneous transmission of several data streams to many users while minimizing the multiuser interference. Furthermore, SDMA based MIMO-OFDMA system can support multiple users in both frequency domain and the spatial domain, hence providing finer granularity of the resource allocation than the pure FDMA or pure SDMA based systems [12]. Moreover, the capacity gain which can be achieved due to multiuser diversity can be more significant in SDMA-based systems. However, in practice, SDMA techniques have to deal with the following two main problems [13].

i   In the scenarios where two or more users come close to each other or the spatial signatures of the users become almost identical due to the underlying scattering environment (insufficient scattering, keyhole channels [14]), the channel matrices of these users may become highly correlated. Furthermore, channel correlation may also arise due to mutual coupling between the transmit and/or receive antenna elements [15]. In these correlated scenarios, channel correlation may become a source of link failure or outage while employing multiuser detection techniques.

ii Since the mobile users are generally located at different distances from the BS, a near-far problem may arise in an uplink SDMA based system. This, in turn, causes the channel matrix observed by the BS to be heavily unbalanced. Such an unbalanced channel matrix may result in the degradation of the total system capacity due to a high eigenvalue spread or the condition number.

With regard to the aforementioned problems, the two main constraints that limit the performance of SDMA based wireless systems are [16]: (i) users sharing the same radio frequency channel, i.e., co-channel users, should be located in different angular locations, and (ii) the difference in their received power levels should not be very large. With respect the the first constraint, the angular separation between co-channel users should exceed the angular resolution of the employed directional antenna in order to assign orthogonal SDMA beams.

Considering the first constraint, the contribution in [17] analyzed the capacity of an adaptive SDMA system for a given angular user density distribution which can be either obtained from the measurements or from the scenarios dealing with the user mobility. It has been concluded that the capacity of a wireless system can be enhanced by creating multiple independent beams per traffic channel with the help of an adaptive antenna array at the BS. Furthermore, it has been illustrated that the consideration of the expected user density and an appropriate selection of the BS sites are of significant importance while planning an SDMA-based cellular network in order to enhance its overall spectrum efficiency.

On the other hand, the implication of the second constraint refers to the near-far problem as encountered in CDMA systems. In this regard, the contribution in [16] investigates the grouping of the mobile users into power classes. In CDMA systems, this objective is usually achieved by means of the power control mechanism, since the users' grouping may further deteriorate the system performance. On the other hand, in a joint TDMA-SDMA system, individual TDMA channels can be assigned to different power classes and hence additional power control mechanisms are unnecessary. A power class can be either static or dynamic and the users belonging to the same class can share the same set of channels [16].

Towards the direction of enabling an efficient use of SDMA technique in future wireless networks, several techniques are being investigated in the literature. The key enabling techniques for future SDMA-based wireless networks along with their corresponding references are listed in Table 1. In the following, we review the existing works in the area of SDMA-based wireless systems considering several aspects such as transmission and reception techniques, resource allocation and scheduling, and research challenges.

2.2. Transmission and Reception Schemes

Multiple-input multiple-output (MIMO) wireless systems, which employ multiple antennas at both the transmitter and receiver, have received significant attention due to the promising capacity improvement provided by these systems. In MIMO systems, a transmitter can split the information symbols into multiple streams and sends each data stream via a single antenna. Each receive antenna then receives a different linear combination of the signals from the different transmit antennas. This technology provides a significant capacity benefit and the achievable capacity asymptotically increases linearly with the increasing number of transmit and receive antennas [39].

Multiuser MIMO, which is the multiuser version of the MIMO system, relies on the principle that multiplexing streams for different users on different antennas can achieve large gains, even if each

user device contains only a few antennas [40]. Multiuser MIMO provides several advantages such as no requirement of the scattering environment, cheap single-antenna terminals, and simple resource allocation mechanism [33]. In order to further enhance the capacity gains, the concept of Massive MIMO has recently emerged [33, 34]. Massive MIMO system uses a very large number of service antennas at the BS which helps to eliminate the multiuser interference with the help of very sharp beams. This is known with several names in the literature such as Large MIMO, Hyper MIMO, and Full-Dimension MIMO. The main advantages of massive MIMO systems are [33]: (i) system throughput improvement, (ii) reduced latency, (iii) higher energy efficiency, (iv) robustness against jamming, and (v) simplification of the medium access layer. Despite its aforementioned advantages, this technology has to address the following main challenges [33, 34]: (i) pilot contamination becomes more problematic in massive MIMO systems than in the conventional MIMO, (ii) the effect of hardware impairments becomes more severe for massive MIMO systems due to the requirement of low-cost terminals, (iii) significant performance degradation occurs in the presence of channel correlation, and (iv) suitable calibration methods are required for the Time Division Duplex (TDD) operation mode.

In the context of LTE-Advanced and IEEE 802.16m networks, Coordinated Multiple Point (CoMP) transmission (also referred to as collaborative MIMO, network MIMO, etc.) has received important attention in the literature [22, 23]. This CoMP transmission scheme is capable of enhancing cell-edge user performance in an interference-limited environment. Besides enhancing the cell edge user performance, the CoMP approach can also improve the system capacity. In this context, the contribution in [22] evaluates the performance of various CoMP methods considering the ray-tracing based path loss calculation. Furthermore, the authors in [23] specify different scenarios of the CoMP based on the following information sharing levels: (a) no Channel State Information (CSI) sharing, (b) partial CSI sharing, (c) full CSI sharing, (d) no data sharing, (e) partial data sharing, and (f) full data sharing.

Based on the aforementioned scenarios, the CoMP techniques can be broadly categorized into the following two types.

i CoMP Joint Processing/Transmission (CoMP-JPT) [24, 25]: In this scheme, multiple BSs collaborate to convert the interfering signal into a desired signal in the downlink and subsequently transmit data to the edge users in a cooperative manner in order to enhance the cell-edge throughput.

ii CoMP Coordinated Scheduling/Beamforming (CoMP-CSB) [26]: In this scheme, multiple BSs

Table 1. Key Enabling Techniques for Future SDMA-based Wireless Networks

Techniques	References
Multiuser MIMO	[10, 18–21]
Coordinated Multiple Point (CoMP)/Collaborative MIMO/Network MIMO	[22, 23]
CoMP Joint Processing/Transmission (CoMP-JPT)	[24, 25]
CoMP Coordinated Scheduling/Beamforming (CoMP-CSB)	[26–28]
Opportunistic Scheduling (OS)	[29–32]
Massive MIMO	[33, 34]
Three dimensional (3D) Beamforming	[35–37]

collaborate in order to mitigate the Inter-Cell Interference (ICI). Although most of the existing research on BS cooperation schemes focus on CoMP-JPT, CoMP-CSB is more practical since only the exchange of partial information such as CSI over the backhaul is required. In this context, the contribution in [26] proposes a CoMP-CSB scheme with user selection in order to enhance the cell-edge users' throughput considering the scenario of partial CSI and no data sharing scheme.

One of the main enablers of the CoMP technology is Cloud Radio Access Network (C-RAN) architecture in which the baseband processing is centralized, i.e., Baseband Units (BBUs) from the multiple BSs are pooled in such a way that they can be shared among these BSs. [38]. As compared to the traditional RAN architecture, C-RAN needs fewer BBUs, thus decreasing the overall cost of the network operation. Furthermore, C-RAN can support non-uniform traffic and makes the efficient use of the network resources such as BSs. Despite its several advantages, it has to address several challenges from the deployment perspective such as the need of suitable BBU cooperation and visualization techniques, requirement of high bandwidth, low delay and low cost transport network [38].

The main problems in designing transmission schemes for SDMA-based systems are: (i) difficulty in acquiring CSI, and (ii) poor synchronization. In this regard, authors in [42] have studied the performance degradation in SDMA networks that may result due to poor synchronization and the imperfect channel knowledge. Subsequently, a signal model for OFDM-based SDMA networks has been presented whose transmissions are impaired by the carrier frequency offset and the sampling frequency offset. It has been concluded that the poor knowledge of the channel state severely limits the maximum number of users that can be served, and is more important than the fine synchronization in terms of serving maximum number of users. In order to address the CSI acquisition difficulty, the contribution in [43] has studied the the optimal statistical precoder design for a simple multiuser case in which the transmitter has two antennas serving only two single-antenna users.

Users' data at the receiver of a multiuser SDMA system can be separated on the basis of their unique spatial signatures in the form of Channel Impulse Responses (CIRs). In this direction, several SDMA Multi-User Detectors (MUDs) have been proposed in the literature [10, 20, 21]. For an MUD to achieve near-single-user performance, the CIRs need to be accurately estimated. In order to achieve a near optimal performance, joint channel estimation and signal detection schemes have recently received significant attention [8, 9]. Among the existing linear MUD techniques, Minimum Mean Square Error (MMSE)-MUD [20] and the Constrained Least Square (CLS)-MUD [21] have been widely investigated in the literature in the context of Mean Square Error (MSE) performance analysis. Since minimizing the MSE doses not necessarily guarantee that the minimum Bit Error Rate (BER) or Symbol Error Rate (SER) of a communication system, the trend is towards exploring techniques which employ the minimum BER constraint such as in [44]. In the above context, the contribution in [10] proposes a differential evolution algorithm-aided iterative channel estimation and turbo MUD scheme for MIMO-aided OFDM/SDMA systems.

Besides, the contribution in [45] studies and compares various MUD schemes such as Zero Forcing (ZF), MMSE, Maximum Likelihood (ML), QR Decomposition (QRD), and Minimum Bit Error Rate (MBER) considering correlated MIMO channel models based on IEEE 802.16n standard. The ML detection provides the optimal performance but its complexity increases exponentially with the constellation size of the employed modulation and the number of users. On the other hand, the QRD-based MUD scheme can be a substitute to the ML detection due its low complexity and near optimal performance. Although the MMSE MUD minimizes the MSE, this may not guarantee the minimum BER of the system. In [45], it has been concluded that the MBER MUD performs better than the classic MMSE MUD in term of the minimum probability of error by directly minimizing the BER cost function.

Recently, the concept of three dimensional (3D) beamforming has received important attention in order to enhance the capacity of future wireless networks [35–37]. In contrast to 2D beamforming, the 3D beamforming controls the radiation pattern in both elevation and azimuth planes, thus providing additional degrees of freedom while planning a cellular network.

2.3. Resource Allocation and Scheduling

In multiuser MIMO-OFDMA systems, adaptive resource allocation in different dimensions such as frequency, time, and space becomes challenging due to the inclusion of the space dimension and a large number of resources to be managed. In this context, the authors in [18] have investigated the performance, complexity, and fairness of suboptimal resource allocation strategies with the objective of maximizing the sum rate. Furthermore, the contribution in [19] analyzes the Symbol Error Rate (SER) performance of a capacity-aware adaptive MIMO beamforming scheme, which iteratively finds the beamforming weight vectors that enhance the capacity of OFDM-SDMA systems. Moreover, closed-form expressions for the SER performance of OFDM-SDMA systems have been derived with MIMO-Maximum Ratio Combining (MRC) and the proposed capacity-aware MIMO beamforming scheme.

The system fairness can be measured in terms of Jain's Index of Fairness (JIF), given by [46]

$$\text{JIF} = \frac{(\sum_{k=1}^{K} \bar{R}_k)^2}{K \sum_{k=1}^{K} \bar{R}_k^2}, \qquad (1)$$

where $\bar{R}_k$ is the mean rate of user k and the value of JIF ranges from 0 to 1. The higher the JIF, the more fair is the throughput distribution among users. Considering this fairness metric, the contribution in [46] studies the fair resource allocation problem of SDMA/ Multiple Input Single Output (MISO)/OFDMA systems. The Proportional Rate Greedy (PRG) algorithm proposed in [46] allocates powers among the selected users for each subcarrier considering user fairness into account.

Coordinated Beamforming (CBF) or coordinated scheduling is regarded as an effective way of mitigating ICI in OFDM-based systems. In order to take full advantage of multiuser diversity, an efficient scheduler should be able to schedule a set of users which experience favorable channel realizations in each time slot. In a cell containing multiple users, only the users having strong channel norms are usually selected as candidates for scheduling. In this context, the contribution in [27] proposes a CBF scheme based on leakage-controlled MMSE precoding. In addition, a regularized factor to maximize the Signal to Interference-plus-Noise Ratio (SINR) of the leakage controlled-MMSE precoding has been derived and the achievable throughput loss has been analyzed.

Furthermore, the contribution in [28] analyzes the sum-rate performance of joint opportunistic scheduling and the receiver design for multiuser MIMO-SDMA

downlink systems. In this approach, the BS exploits the limited feedback on the effective SINRs, and schedules simultaneous data transmission on multiple beams to the user terminals which have the largest effective SINRs. Moreover, considering smart antennas at the access points and single antennas at the user terminals, the authors in [47] investigate the use of joint optimal downlink beamforming, power control and access point allocation in a multicell SDMA system. Additionally, the contribution in [48] provides an overview of the scheduling algorithms proposed for multiuser MIMO based 4G wireless networks.

In an opportunistic transmission scheme, channel fluctuations need to be tracked in order to schedule the BS transmissions for the user with the best channel. The main concept behind opportunistic beamforming is to induce rapid channel fluctuations by employing multiple antennas at the BS, which subsequently helps to improve the multiuser diversity gain. In general, the use of spatial diversity is harmful to the multiuser diversity gain. In this regard, the multi-channel multiuser diversity scheduling scheme proposed in [41] simultaneously exploits both diversities. In addition, the contribution in [13] investigates the effect of distributed BS antennas on the reduction of intra-user correlations while analyzing the performance of a MIMO-OFDM system.

Opportunistic Scheduling (OS) can be considered as another promising technique in an SDMA-based system in order to enhance the system throughput by exploiting multiuser diversity with the limited channel feedback [32]. Existing OS schemes can be classified into two categories, namely, Time-Sharing (TS) and SDMA-based OS schemes. In a TS-OS scheme, only the user terminal with the best instantaneous channel conditions is scheduled in one slot being independent of the number of beams employed by the BS. On the other hand, an SDMA-based OS serves multiple terminals simultaneously with multiple orthonormal beams in each time slot. The sum-rate of SDMA-based OS grows linearly with M whereas for the TS-OS, it increases only linearly with $min(M, N)$, M and N being the number of transmit and receive antennas, respectively [29]. In this context, the contribution in [30] proposes the SDMA-based OS for systems with single-antenna mobile terminals while for the mobile terminals having multiple receive antennas, the contribution in [29] proposes to allow each antenna compete for its desired beam as if it was an individual terminal. In the latter case, each beam is assigned to a specific receive antenna of a chosen terminal but the signals captured by the undesired antennas of the mobile terminal are discarded, thus leading to inefficient utilization of multiple antennas.

To address the aforementioned issue, the contribution in [31] proposes various linear combining techniques exploiting signals received by all receive antennas and considers the improved effective SINR as a scheduling metric. In the similar context, the contribution in [32] provides a a systematic approach for deriving asymptotic throughput and scaling laws using SINR based on the extreme value theory. Consequently, with the help of a comparison between the Signal-to-Interference Ratio (SIR) and SINR-based analyses, it has been argued that the SIR-based analysis is more computationally efficient for SDMA-based systems, and subsequently more effective in order to capture the high-order behavior of the asymptotic system performance.

A comprehensive overview on SDMA/OFDMA scheduling challenges are highlighted in [11], which further proposes an SDMA-OFDMA Greedy Scheduling Algorithm (sGSA) for WiMAX systems. The proposed solution in [11] considers feasibility constraints in order to allocate resources for multiple mobile terminals on a per packet basis by employing the following two approaches: a) a cluster-based SDMA grouping algorithm, and b) a computationally efficient frame layout scheme. The later approach allocates multiple SDMA groups per frame based on their packet Quality of Service (QoS) utility. In addition, the contribution in [49] considers Opportunistic Beamforming (OB) with finite number of single-antenna users under the constraint that the feedback overhead from the mobile terminals to the BS is constant. The impact of the fading variances of the users and the spatial correlation on the sum rate of TDMA and SDMA based OB has been analyzed. It has been concluded that for a small number of spread out users and moderate to high Signal to Noise Ratio (SNR) values, SDMA-OB scheme performs worse than the TDMA-OB. In addition, authors in [50] investigate a multiuser two-way relay system using SDMA communications and proposes an optimal scheduling method that maximizes the sum rate while ensuring fairness among users. Subsequently, rate and angle-based sub-optimal scheduling methods have been studied in order to reduce the computational load at the relay.

2.4. Research Challenges

The performance of an SDMA based system may degrade due to the imperfect channel knowledge and poor synchronization. Furthermore, the difficulty of obtaining instantaneous CSI knowledge at the transmitters may prevent the practical implementation of many multiuser SDMA systems. In the following, we highlight the main research challenges in SDMA-based wireless networks.

i A major challenge, common to all SDMA systems, is the requirement of CSI knowledge at the transmitter to enable the transmission of multiple streams without any harmful interference.

ii Under the scenarios that the receivers have their perfect CSIs but the transmitter knows only the statistical CSI, the optimization for downlink multiuser MIMO is still not well understood in the literature. This problem has been recently studied in [43] but only for the case of two users having single antennas. Therefore, the generalization of these results for arbitrary number of antennas with arbitrary number of users is an important future research topic.

iii Massive MIMO has been considered as one of the key enablers for future 5G wireless networks. As the number of BS antennas in future SDMA networks increases, the system gets almost entirely limited from the reuse of pilots in neighboring cells. This leads to the pilot contamination and antenna correlation problems, which appear to be fundamental challenges for designing very large MIMO systems.

iv In order to take full advantage of the SDMA scheme, it is important to investigate suitable techniques which can simultaneously improve both the spatial and multiuser diversity gains.

v In order to make the best use of the available resources in different dimensions, suitable optimal scheduling algorithms need to be investigated by employing a cross-layer design with the cooperation between the Medium Access (MAC) and physical layers.

vi Inter-user interference is the main limiting factor in multiuser SDMA systems. In order to mitigate this, suitable resource allocation (joint carrier and power allocation) strategies need to be investigated.

vii Besides the effects of co-channel interference, channel fading and the noise, there may arise distortion in the system performance due to the presence of residual hardware impairments caused due to several reasons such as phase noise, analog to digital converter inaccuracies, oscillator mismatch, etc [51]. In this context, it's important to investigate the residual hardware aware adaptive beamforming schemes in order to improve the system performance in the presence of practical impairments.

viii Opportunistic or cognitive radio communication has been considered as one of the key techniques to enhance the spectral efficiency of future

5G networks [52, 53]. In this context, SDMA-based cognitive wireless networks should employ suitable opportunistic spectrum access/user selection/shceduling algorithms while providing sufficient protection to the already existing licensed systems.

ix The investigation of suitable transceiver design strategies such as precoding/beamforming for physical layer multicasting and multi-group multicasting [54] considering a large number of antennas at the BS side is another important research challenge.

3. Emerging Multiplexing/Multiple Access Schemes

In this section, we briefly discuss the main emerging multiplexing/multiple access schemes.

3.1. Orbital Angular Momentum Multiplexing

Recently, Orbital Angular Momentum Multiplexing (OAMM)) has been shown as an important candidate for high capacity millimeter wave communications [4]. In this multiplexing method, one important property of an electromagnetic wave that each beam has a unique helical phase front is utilized in order to obtain multiplex multiple beams. The orthogonality of the beams is defined by a different Orbital Angular Momentum (OAM) state number which is the amount of phase front "twisting". Authors in [4] demonstrated that this scheme can enhance the system capacity as well as the spectral efficiency of mm-wave wireless communication links by transmitting multiple data streams with a single aperture tranmit/receive pair.

The OAMM implementation is completely different from the implementation of the traditional radio frequency spatial multiplexing and therefore it requires a significant architectural change [4]. The traditional spatial multiplexing scheme requires multiple spatially separated transmitter and receiver aperture pairs for the transmission of multiple data streams whereas the multiplexed beams in the OAM scheme are completely coaxial throughout the transmission medium and it uses only single transmitter and receiver aperture.

3.2. Polarization Division Multiplexing/Multiple Access

There is an emerging concept of polarization modulation technique for carrying information bearing signals [55]. This approach uses circular polarization of the propagating electromagnetic carrier as a modulation characteristic in contrast to amplitude, frequency and/or phase modulation attributes used in the conventional schemes. The circular modulation techniques have the capability of providing inherent benefits of circular polarization as well as the diversity

gain in wireless fading channels. Moreover, the concepts of Polarization Division Multiplexing (PDM) and phase division multiplexing widely used in the optical communications can be regarded as other promising approaches in order to enhance the multiplexing gain of 5G wireless systems on the top of the currently used frequency/time/code multiplexing schemes.

In a PDM scheme, two data streams can be multiplexed with orthogonal polarizations at the transmitter side in order to enhance the channel capacity. The main problem that may arise in wireless fading channels with this approach is that channel depolarization may induce correlation between two corresponding data streams at the receiver which are expected to be received on orthogonal polarizations [56]. This effect can be partially mitigated by utilizing PDM in combination with multi-antenna techniques such as space-time block coding and beamforming [57].

Like PDM, the main concept behind Polarization Division Multiple Access (PDMA) is to transmit two independent data streams at the same time and at the same frequency to two different users by employing orthogonal polarizations [56]. Recently, the contribution in [58] analyzed the capacity of the PDMA scheme and expressed the relation between PDMA channel capacity and Cross Polar Discrimination (XPD) in a mathematical form. Furthermore, authors in [56] have investigated a PDMA scheme for the downlink of a cellular system by employing collaborative transmit-receive polarization and polarization filtering detection for non-line of sight wireless fading channels. It has been concluded that the proposed PDMA scheme has a great potential to be utilized as a new multiple access scheme in the next generation of cellular wireless communication systems.

3.3. Interweave Division Multiple Access (IDMA)

Interweave Division Multiple Access (IDMA) is an asynchronous multiple access scheme in which different interleavers are used to distinguish users in contrast to the use of different codes in a conventional CDMA system. In a conventional CDMA scheme, interleavers are placed before the spreaders and they are effective only when used in conjunction with channel coding. Interleavers, which are usually placed between Forward Error Correction (FEC) coding and spreading, are used to combat the fading effect in CDMA, whereas the arrangement of interleaving and spreading is reversed in IDMA, and different interleavers distinguish distinct data streams.

IDMA can be considered as a special case of random waveform CDMA, and the accompanying chip-by-chip estimation algorithm is essentially a low-cost iterative soft cancellation technique. Furthermore, the computational cost per user is independent of the number of users, which is significantly lower than that of the MMSE technique. Authors in [59] have shown that IDMA with equal power level can achieve near single user performance in multiuser environments.

IDMA inherits many benefits from CDMA such as path diversity, mitigation of intra-cell interference, and a common spreading sequence. In [60], authors investigated the performance of the MIMO assisted multicarrier IDMA scheme with multiuser detection and showed that this scheme can provide better performance with the aid of VBLAST/ZF/MAP detection technique. In addition, authors in [61] have investigated the following three design aspects for multicarrier IDMA technique: (i) multiplexing versus diversity tradeoff, (ii) coding versus spreading tradeoff, and (iii) complexity versus performance tradeoff. Moreover, authors in [62] have recently studied a quantize and forward strategy for the half-duplex IDMA relay channel considering multiple users, single relay, and single destination.

Grouped IDMA. Grouped-IDMA is a version of the IDMA scheme in which active users are arranged into several groups and each group is characterized by an orthogonal code. This scheme inherits the advantages of IDMA and orthogonal CDMA, utilizing the group specific orthogonal spreading code for group separation. It has been shown in [63] that the grouped-IDMA achieves better performance than the simple IDMA when the number of users is relatively large, especially in low/medium SNR region.

In the grouped IDMA scheme, each group of users works in the same way as the IDMA and each group is assigned to a group-specific orthogonal code. Therefore, inter-group interference can be eliminated by carrying out de-spreading at the receiver. Furthermore, interference among different users in the same group can be reduced by a chip-by-chip detection algorithm used in IDMA [63].

For a wireless system with K number of users divided into G groups, the grouped IDMA can be interpreted as

- IDMA when $G = 1$

- Orthogonal CDMA when $G = K$

- Grouped IDMA when $1 \leq G \leq K$

3.4. Universal Filtered Multi-Carrier (UFMC)

As stated earlier, OFDM technology is more demanding in terms of synchronization and energy consumption. To alleviate the drawbacks, Filter Bank based Multi-Carrier (FBMC) has been recently considered as an alternative to OFDM. In contrast to the OFDM, this technique applies a filtering functionality to each of the subcarriers. Therefore, side-lobes with FBMC become much lower and thus the intercarrier interference issue is far less harmful than in OFDM. Despite several benefits of FBMC technique, practical system configurations renders most of them [5]. In this context, there is an emerging concept of Universal Filtered Multi-Carrier (UFMC) which can benefit from the advantages of FBMC while addressing its drawbacks.

UFMC is a filtering operation applied to a group of consecutive subcarriers (e.g., a given allocation of a single user) in order to reduce out-of-band sidelobe levels. This subsequently minimizes the potential inter-channel interference between adjacent users in the case of asynchronous transmissions. In this technique, filtering operation is applied to a group of consecutive subcarriers instead of per subcarrier filtering employed in the FBMC technique [64]. The UFMC can significantly reduce the effect of sidelobe interference and is better suitable for fragmented spectrum operation. It has been shown in [64] that the UFMC outperforms the cyclic prefix-based OFDM for both perfect and imperfect frequency synchronization between user equipments and the BSs.

3.5. Sparse Code Multiple Access (SCMA)

Due to the closed loop nature of the multiuser MIMO system, it suffers from different practical limitations such as channel aging and high feedback overhead required to feedback CSI from the serving users to the BS. To address these limitations, open loop multiplexing schemes such as non-orthogonal code domain multiple access can be considered as promising solutions [65]. In the category of non-orthogonal multiple access schemes, Sparse Code Multiple Access (SCMA) has been recently considered as a promising candidate. In this scheme, input information bits are mapped to multi-dimensional complex codewords which are selected from the predefined sets of the codebook. Subsequently, code-domain layers are allocated to different users without requiring the CSI knowledge of the paired user terminals [65, 66].

In comparison to MU-MIMO, SCMA-based multiuser wireless system is more robust in the presence of time varying channel. Furthermore, the CSI feedback problem is removed due to its open loop nature. In addition, higher data rate and the robustness to mobility are two major advantages of multiuser SCMA. Moreover, compared to the spatial domain processing in MU-MIMO schemes, code-domain multiplexing has a significant advantage in terms of the transmit-side computational complexity [65]. On the other hand, the main drawback of SCMA approach is that it requires a non-linear receiver in order to detect the corresponding layer of each user, thus resulting in the decoding complexity. However, the sparsity feature of SCMA codewords allows to utilize low complexity detection algorithms such as message passing algorithm [67].

Table 2. Main Features and drawbacks of Several Existing and Emerging Multiple Access/Multiplexing Schemes

Techniques	Main Features	Drawbacks
Frequency Division Multiple Access (FDMA)	i. All users can transmit in parallel ii. No need of time synchronization	i. Each station only gets a fraction of total bandwidth ii. Need of tunable transmitters and receivers
Time Division Multiple Access (TDMA)	i. Each user can use the total bandwidth ii. No need of tunable receivers	i. Need of time synchronization ii. Transmission over a fraction of the total time
Code Division Multiple Access (CDMA)	i. Each user can transmit over total bandwidth all the time ii. More users per MHz of bandwidth	i. Higher receiver complexity ii. Near-far effect
Orthogonal Frequency Division Multiple Access (OFDMA))	i. Intrinsic orthogonality ii. Receiver circuit simplicity	i. High peak-to-average ratio ii. Sensitive to frequency offset iii. Poor performance in highly asynchronous access scenarios
Space Division Multiple Access (SDMA)	i. Higher spectral efficiency and multiple access capabilities ii. Useful in combination with TDMA, FDMA, CDMA or OFDMA	i. Near-far effect ii. Poor synchronization iii. Loss of orthogonality in the presence of practical imperfections
Orbital Angular Momentum Multiplexing (OAMM)	i. Orthogonality of the beams defined by a different OAM state number ii. Suitable for mmwave communications iii. Higher system capacity	i. Requires a significant architectural change to implement ii. Intermodal crosstalk
Polarization Division Multiple Access (PDMA)	i. Two independent data streams transmitted with orthogonal polarizations ii. Can be combined with any existing multiple access schemes	i. Depolarization effect in fading channel ii. Need of extra receiver circuitry for polarization filtering detection
Interweave Division Multiple Access (IDMA)	i. Chip-interleaving process is used for user separation ii. Reverse arrangement of interleaving and spreading as compared to CDMA iii. Increased diversity against fading iv. Low receiver cost v. Suitable for low rate transmissions	i. Higher receiver complexity for wideband systems ii. Iterative processing required iii. Entire interleaver matrix need to be transmitted to the receiver iv. Need of memory optimization in transmitter and receiver side
Universal Filtered Multi-Carrier (UFMC)	i. Blockwise filtering provides additional flexibility ii. Shorter filter lengths and reduced sidelobe interference iii. Highly suitable for CoMP transmission, and Internet of Things iv. Better suited for fragmented spectrum than OFDM	i. Need of complex receiver circuitry ii. Not significant gains over OFDM for low rate transmissions
Sparse Code Multiple Access (SCMA)	i. Non-orthogonal code domain multiple access ii. Open loop multiplexing and no need of feedback channel iii. More robust in presence of time varying channel and the CSI feedback problem	i. Requires non-linear receiver ii. Complex codebook design since multiple layers are multiplexed with different codebooks

4. Proposed Clustered Orthogonal Signature Division Multiple Access

In order to fully realize the potentials of recently emerging wireless technologies such as massive MIMO and mm-wave communications, existing multiple access techniques may not be sufficient and we need to investigate new multiple access schemes for future wireless networks. In this context, several multiplexing/multiple access techniques discussed in the previous section are under investigation. Since the beams become very narrower in mm-wave communications and sufficient antenna spacing becomes an issue in massive MIMO systems, there is a high probability of the orthogonality loss between two cochannel users separated in the spatial domain. In this regard, the conventional concept of the SDMA technique should be adapted in future massive MIMO systems which will be possibly implemented in the mm-wave frequency range.

Herein, our proposition is to employ clustered orthogonal signature/power/code division multiple access on the top of the existing multiple access schemes such as SDMA. Unlike the grouped IDMA scheme, the idea here is to group users located within a cluster or a beam and to provide orthogonal codes to these groups. By employing the orthogonality over the conventional SDMA system, we can significantly enhance the number of users which can be supported by a given set of frequency resources in a particular cluster/beam. The proposed clustered orthogonal signature division multiple access scheme envisions to address the aforementioned drawbacks of several approaches highlighted in Table 2. This will further address the problem of loss of orthogonality which may arise in many existing SDMA-based approaches due to the time varying nature of the wireless channel.

One of the enablers for the proposed approach is joint antenna coding approach. By employing the orthogonal coding in combination with the dynamic 3D beamforming approach, we can separate users in different groups depending on the available radio resources. For future dense networks, the conventional SDMA may not be able to guarantee orthogonality between the beams where one beam implies one specific signature. In this context, the proposed idea is to enhance the orthogonality of the SDMA beams with the help of suitable orthogonal coding scheme by employing a joint antenna coding scheme. In order to design the orthogonal codes, there exist several possibilities in the literature such as Hadamard code [68], Gold code [69], and Polyphase orthogonal codes [70].

The proposed approach can be implemented in a two tier manner, meaning that one multiple access scheme in the first tier and another in the second tier. For example, in a heterogeneous network comprising of macro cells and small cells, the backhaul part, i.e., the link from the small cell BSs to the macro cell BS, may employ one multiple access scheme and the access part, i.e., from the end users to the small cell BS, may use another multiple access scheme. In this way, we suggest the following two approaches for implementing the proposed clustered orthogonal signature division multiple access.

i **Code/signature division multiple access after employing SDMA**: In this approach, the first tier uses signature (code) division multiple access and the second tier uses SDMA.

ii **SDMA after employing code/signature division multiple access**: In this approach, the first tier uses SDMA and the second tier uses code/signature division multiple access.

While devising a good multiple access scheme in a time varying wireless environment, the main objective should be to maximize the overall orthogonality. Depending on the deployed environment, if the beam sharpness and the surrounding environment cannot provide sufficient orthogonality, a suitable orthogonal coding can be implemented to enhance the overall orthogonality. The following advantages are foreseen in future wireless networks by employing the proposed multiple access scheme: (i) better flexibility to wireless design engineers, (ii) better performance in the presence of time varying wireless channels, (iii) easier to implement from the practical perspectives, and (iv) allocation of the available resources in an optimized way.

However, from the practical perspectives, the following factors need to be further investigated while realizing the proposed multiple access concept.

i Availability of the required digital signal processing hardware

ii Energy efficiency of the system and its implementation complexity

iii Flexibility of accommodating users/services into the system: In most of the current wireless systems, we need to tune system parameters according to the requirements.

iv Orthogonality of the code: Higher the orthogonality of the code, better becomes the system performance at the cost of the restriction in the number of users/services that can be supported.

5. Conclusions

In order to address the issue of spectrum shortage in future wireless networks, investigation of suitable multiple access/multiplexing scheme is of significant importance. In this regard, this paper has reviewed various features of the widely discussed SDMA scheme. In addition, it has highlighted the main features and the drawbacks of other several emerging multiple access/multiplexing schemes. More importantly, it has proposed a novel concept of clustered orthogonal multiple access scheme as an important candidate for future dense cellular networks.

In our future work, we plan to validate the proposed concept with the help of system level simulations. Furthermore, the comparison of the proposed two-tier approaches in terms of the overall orthogonality improvement is the part of our ongoing works.

References

[1] ANDREWS, J., BUZZI, S., CHOI, W., HANLY, S., LOZANO, A., SOONG, A. and ZHANG, J. (2014) What will 5G be? *IEEE J. Sel. Areas in Commun.* **32**(6): 1065–1082.

[2] VISWANATHAN, H. and WELDON, M. (2014) The past, present, and future of mobile communications. *Bell Labs Technical J.* **19**(8-21).

[3] RANGAN, S., RAPPAPORT, T. and ERKIP, E. (2014) Millimeter-wave cellular wireless networks: Potentials and challenges. *Proc. IEEE* **102**(3): 366–385.

[4] YAN, Y., *et al* (2014) High-capacity millimeter-wave communications with orbital angular momentum multiplexing. *Nature Communications* **5**(4876).

[5] SCHAICH, F. and WILD, T. (2014) Waveform contenders for 5G-OFDM vs. FBMC vs. UFMC. In *6th Int. Symp. on Communications, Control and Signal Process.*: 457–460.

[6] RAPAJIC, P. (1998) Information capacity of the space division multiple access mobile communication system. In *IEEE 5th Int. Symp. on Spread Spectrum Techniques and Applications*, **3**: 946–950 vol.3.

[7] ZETTERBERG, P. and OTTERSTEN, B. (1995) The spectrum efficiency of a base station antenna array system for spatially selective transmission. In *IEEE Trans. Vehicular Technol.*, **44**(3): 651–660.

[8] JIANG, M., AKHTMAN, J. and HANZO, L. (2007) Iterative joint channel estimation and multi-user detection for multiple-antenna aided OFDM systems. *IEEE Trans. Wireless Commun.* **6**(8): 2904–2914.

[9] YLIOINAS, J. and JUNTTI, M. (2009) Iterative joint detection, decoding, and channel estimation in turbo-coded MIMO-OFDM. *IEEE Trans. on Vehicular Technol.* **58**(4): 1784–1796.

[10] ZHANG, J., CHEN, S., MU, X. and HANZO, L. (2012) Turbo multi-user detection for OFDM/SDMA systems relying on differential evolution aided iterative channel estimation. *IEEE Trans. Commun.* **60**(6): 1621–1633.

[11] ZUBOW, A., MAROTZKE, J., CAMPS-MUR, D. and PEREZ-COSTA, X. (2012) Sgsa: An SDMA-OFDMA scheduling solution. In *18th European Wireless Conf.*: 1–8.

[12] CHAN, P. and CHENG, R. (2007) Capacity maximization for zero-forcing MIMO-OFDMA downlink systems with multiuser diversity. *IEEE Trans. on Wireless Commun.* **6**(5): 1880–1889.

[13] DAWOD, N., HAFEZ, R. and MARSLAND, I. (2006) A multiuser zeroforcing system with reduced near-far problem and MIMO channel correlations. In *Canadian Conf. on Electrical and Computer Engg.*: 936–939.

[14] LEVIN, G., and LOYKA, S. (2008) On the Outage Capacity Distribution of Correlated Keyhole MIMO Channels. In *IEEE Trans. Info. Theory* **54**(7): 3232–3245.

[15] SHARMA, S.K., CHATZINOTAS, S., and OTTERSTEN, B. (2013) SNR Estimation for Multi-dimensional Cognitive Receiver under Correlated Channel/Noise. In *IEEE Trans. Wireless Commun.* **12**(12): 6392–6405.

[16] TANGEMANN, M. (1995) Near-far effects in adaptive SDMA systems. In *IEEE Int. Symp. PIMRC*, **3**: 1293–1297.

[17] TANGEMANN, M. (1994) Influence of the user mobility on the spatial multiplex gain of an adaptive SDMA system. In *IEEE Int. Symp. PIMRC*: 745–749 vol.2.

[18] MACIEL, T. and KLEIN, A. (2010) On the performance, complexity, and fairness of suboptimal resource allocation for multiuser MIMO-OFDMA systems. *IEEE Trans. on Veh. Technol.* **59**(1): 406–419.

[19] SULYMAN, A. and HEFNAWI, M. (2010) Performance evaluation of capacity-aware MIMO beamforming schemes in OFDM-SDMA systems. *IEEE Trans. on Commun.* **58**(1): 79–83.

[20] VANDENAMEELE, P., VAN DER PERRE, L., ENGELS, M., GYSELINCKX, B. and DE MAN, H. (2000) A combined OFDM/SDMA approach. *IEEE J. Sel. Areas in Commun.* **18**(11): 2312–2321.

[21] THOEN, S., DENEIRE, L., VAN DER PERRE, L., ENGELS, M. and DE MAN, H. (2003) Constrained least squares detector for OFDM/SDMA-based wireless networks. *IEEE Trans. on Wireless Commun.* **2**(1): 129–140.

[22] BERGER, S., LU, Z., IRMER, R. and FETTWEIS, G. (2013) Modelling the impact of downlink CoMP in a realistic scenario. In *IEEE Wireless Communications and Networking Conference*: 3932–3936.

[23] BELL, A.S. and LUCENT, A. (2008) *Collaborative MIMO for LTE-A downlink*. Tech. Rep. R1-082501, 3GPP TSG RAN WG1 Meeting 53b.

[24] ZHANG, R. (2010) Cooperative multi-cell block diagonalization with per-base-station power constraints. *IEEE J. on Sel. Areas in Commun.* **28**(9): 1435–1445.

[25] CHATZINOTAS, S., IMRAN, M. A., and HOSHYAR, R. (2009) On the multicell processing capacity of the cellular MIMO uplink channel in correlated rayleigh fading environment. *IEEE Trans. Wireless Commun.* **8** (7): 3704–3715.

[26] JANG, U., SON, H., PARK, J. and LEE, S. (2011) CoMP-CSB for ICI nulling with user selection. *IEEE Trans. Wireless Commun.* **10**(9): 2982–2993.

[27] ZHOU, T., PENG, M., WANG, W. and CHEN, H.H. (2013) Low-complexity coordinated beamforming for downlink multicell SDMA/OFDM systems. *IEEE Trans. on Veh. Technol.* **62**(1): 247–255.

[28] PUN, M.O., KOIVUNEN, V. and POOR, H. (2011) Performance analysis of joint opportunistic scheduling and receiver design for MIMO-SDMA downlink systems. *IEEE Trans. on Commun.* **59**(1): 268–280.

[29] SHARIF, M. and HASSIBI, B. (2007) A comparison of time-sharing, DPC, and beamforming for MIMO broadcast channels with many users. *IEEE Trans. on Commun.* **55**(1): 11–15.

[30] SHARIF, M. and HASSIBI, B. (2005) On the capacity of MIMO broadcast channels with partial side information. *IEEE Trans. on Info. Theory* **51**(2): 506–522.

[31] PUN, M.O., KOIVUNEN, V. and POOR, H. (2007) Opportunistic scheduling and beamforming for MIMO-SDMA downlink systems with linear combining. In *IEEE Int. Symp. PIMRC*: 1–6.

[32] PUN, M.O., KOIVUNEN, V. and POOR, H. (2008) SINR analysis of opportunistic MIMO-SDMA downlink systems with linear combining. In *IEEE Int. Conf. on Commun.*: 3720–3724.

[33] LARSSON, E., EDFORS, O., TUFVESSON, F. and MARZETTA, T. (2014) Massive MIMO for next generation wireless systems. *IEEE Communications Mag.* **52**(2): 186–195.

[34] LU, L., LI, G., SWINDLEHURST, A., ASHIKHMIN, A. and ZHANG, R. (2014) An overview of massive MIMO: Benefits and challenges. *IEEE J. Sel. Topics in Signal Process.* **8**(5): 742–758.

[35] MOHAMMAD RAZAVIZADEH, S., AHN, M. and LEE, I. (2014) Three-dimensional beamforming: A new enabling technology for 5G wireless networks. *IEEE Signal Process. Mag.* **31**(6): 94–101.

[36] HALBAUER, H., SAUR, S., KOPPENBORG, J. and HOEK, C. (2013) 3D beamforming: Performance improvement for cellular networks. *Bell Labs Technical J.* **18**(2): 37–56.

[37] SHARMA, S.K., CHATZINOTAS, S. and OTTERSTEN, B. (2015) 3D beamforming for spectral coexistence of satellite and terrestrial networks. In *IEEE Vehicular Technol. Conf.*.

[38] CHECKO, A. and ET AL (2015) Cloud RAN for Mobile Networks-A Technology Overview. *IEEE Communications Surveys & Tutorials*, **17**(1): 405-426.

[39] LOZANO, A. and TULINO, A. (2002) Capacity of multiple-transmit multiple-receive antenna architectures. *IEEE Transactions on Information Theory* **48**(12): 3117–3128.

[40] CHOI, R.U., IVRLAC, M., MURCH, R. and UTSCHICK, W. (2004) On strategies of multiuser MIMO transmit signal processing. *IEEE Transactions on Wireless Communications* **3**(6): 1936–1941.

[41] AKTAS, D. and EL GAMAL, H. (2003) Multiuser scheduling for MIMO wireless systems. In *Vehicular Technology Conference, 2003. VTC 2003-Fall. 2003 IEEE 58th*, **3**: 1743–1747 Vol.3.

[42] OBERLI, C. and RIOS, M. (2007) OFDM-based SDMA networks: Signal model under imperfect synchronization and channel state information. In *2th International OFDM Workshop*.

[43] RAGHAVAN, V., HANLY, S. and VEERAVALLI, V. (2013) Statistical beamforming on the grassmann manifold for the two-user broadcast channel. *IEEE Trans. on Info. Theory* **59**(10): 6464–6489.

[44] ALIAS, M., CHEN, S. and HANZO, L. (2005) Multiple-antenna-aided OFDM employing genetic-algorithm-assisted minimum bit error rate multiuser detection. *IEEE Trans. Veh. Technol.* **54**(5): 1713–1721.

[45] DAS, S. and BAGADI, K.P. (2011) Comparative analysis of various multiuser detection techniques in SDMA-OFDM system over the correlated MIMO channel model for IEEE 802.16n. *World Academy of Science, Engg. and Technol.* **53**(1).

[46] LU, W., JI, F. and YU, H. (2011) A general resource allocation algorithm with fairness for SDMA/MISO/OFDMA systems. *IEEE Commun. Letters* **15**(10): 1072–1074.

[47] STRIDH, R., BENGTSSON, M. and OTTERSTEN, B. (2001) System evaluation of optimal downlink beamforming in wireless communication. In *IEEE Veh. Technol. Conf.*, **1**: 343–347.

[48] AJIB, W. and HACCOUN, D. (2005) An overview of scheduling algorithms in MIMO-based fourth-generation wireless systems. *IEEE Network* **19**(5): 43–48.

[49] JORSWIECK, E., SVEDMAN, P. and OTTERSTEN, B. (2008) Performance of TDMA and SDMA based opportunistic beamforming. *IEEE Trans. Wireless Commun.* **7**(11): 4058–4063.

[50] JOUNG, J. and SAYED, A. (2010) User selection methods for multiuser two-way relay communications using space division multiple access. *IEEE Trans. Wireless Commun.* **9**(7): 2130–2136.

[51] PAPAZAFEIROPOULOS, A.K., SHARMA, S.K. and CHATZINOTAS, S. (2015) Impact of transceiver impairments on the capacity of dual-hop relay massive MIMO systems. In *IEEE GLOBECOM Workshop*.

[52] SHARMA, S., BOGALE, T., CHATZINOTAS, S., OTTERSTEN, B., LE, L.B. and WANG, X. (2015) Cognitive radio techniques under practical imperfections: A survey. *IEEE Commun. Surveys Tutorials* **17**(4): 1858–1884.

[53] SHARMA, S., PATWARY, M., CHATZINOTAS, S., OTTERSTEN, B. and ABDEL-MAGUID, M. (2015) Repeater for 5G wireless: A complementary contender for spectrum sensing intelligence. In *IEEE Int. Conf. on Commun.*: 1416–1421.

[54] CHRISTOPOULOS, D., CHATZINOTAS, S. and OTTERSTEN, B. (2014) Weighted Fair Multicast Multigroup Beamforming Under Per-antenna Power Constraints. *IEEE Trans. Signal Process.* **62**(19): 5132–5142.

[55] UL ABIDIN, Z., XIAO, P., AMIN, M. and FUSCO, V. (2012) Circular polarization modulation for digital communication systems. In *Int. Symp. on Commun. Systems, Networks Digital Signal Process.*: 1–6.

[56] KWON, S.C. and STUBER, G. (2014) Polarization division multiple access on NLOS wide-band wireless fading channels. *IEEE Trans. Wireless Commun.* **13**(7): 3726–3737.

[57] DENG, Y., BURR, A. and WHITE, G. (2005) Performance of MIMO systems with combined polarization multiplexing and transmit diversity. In *IEEE Veh. Technol. Conf.*, **2**: 869–873 Vol. 2.

[58] KWON, S.C. (2014) Optimal power and polarization for the capacity of polarization division multiple access channels. In *IEEE Globecom*: 4221–4225.

[59] PING, L., LIU, L., WU, K. and LEUNG, W. (2006) Interleave division multiple-access. *IEEE Trans. Wireless Commun.* **5**(4): 938–947.

[60] NAGARADJANE, P., CHANDRASEKARAN, S., VISHVAKSENAN, K. and RAMAKRISHNAN, M. (2010) MIMO multi carrier interleave division multiple access system with multiuser detection. In *Int. Conf. on Wireless Commun. and Sensor Comput.*: 1–4.

[61] ZHANG, R. and HANZO, L. (2008) Three design aspects of multicarrier interleave division multiple access. *IEEE Trans. on Veh. Technol.* **57**(6): 3607–3617.

[62] LIU, L., LI, Y., SU, Y. and SUN, Y. (2015) Quantize-and-forward strategy for interleave division multiple-access relay channel. *IEEE Trans. on Veh. Technol.* **PP**(99): 1–1.

[63] TU, Y., FAN, P. and ZHOU, G. (2006) Grouped interleave-division multiple access. In *First Int. Conf. on Commun. and Networking in China*: 1–5.

[64] VAKILIAN, V., WILD, T., SCHAICH, F., TEN BRINK, S. and FRIGON, J.F. (2013) Universal-filtered multi-carrier technique for wireless systems beyond LTE. In *IEEE Globecom Workshops*: 223–228.

[65] NIKOPOUR, H., YI, E., BAYESTEH, A., AU, K., HAWRYLUCK, M., BALIGH, H. and MA, J. (2014) SCMA for downlink multiple access of 5G wireless networks. In *IEEE Globecom*: 3940–3945.

[66] TAHERZADEH, M., NIKOPOUR, H., BAYESTEH, A. and BALIGH, H. (2014) SCMA codebook design. In *IEEE Veh. Technol. Conf.*: 1–5.

[67] HOSHYAR, R., RAZAVI, R. and AL-IMARI, M. (2010) LDS-OFDM an efficient multiple access technique. In *IEEE Veh. Technol. Conf.*: 1–5.

[68] DEL RIO, A. and RIFA, J. (2013) Families of Hadamard $BBZ_2BBZ_4Q_8$-codes. *IEEE Trans. on Info. Theory* **59**(8): 5140–5151.

[69] DAS, B., SARMA, M., SARMA, K. and MASTORAKIS, N. (2015) Design of a few interleaver techniques used with Gold codes in faded wireless channels. In *Int. Conf. on Signal Process. and Integrated Networks*: 237–241.

[70] LIU, Y.C., CHEN, C.W. and SU, Y. (2013) New constructions of zero-correlation zone sequences. *IEEE Trans. on Info. Theory* **59**(8): 4994–5007.

4

Spectrum Sensing: To Cooperate or Not to Cooperate?

Dongliang Duan[1], Liuqing Yang[2*], and Shuguang Cui[3]

1. Department of Electrical and Computer Engineering, University of Wyoming, Laramie, WY
2. Department of Electrical and Computer Engineering, Colorado State University, Fort Collins, CO
3. Department of Electrical and Computer Engineering, Texas A&M University, College Station, TX

Abstract

While it is well accepted that cooperative spectrum sensing will significantly improve the sensing performance, the necessity of cooperation is not sufficiently appreciated. In this paper, by analyzing the spectrum sensing problem from the system perspective, we show that without cooperation, the performance will suffer from a fundamental tradeoff between reliability and efficiency. However, if cooperation is incorporated in the spectrum sensing process and the threshold is selected appropriately, the efficiency-reliability tradeoff in the non-cooperative case can be largely overcome by exploiting the cooperative diversity. These results show that cooperation in spectrum sensing is not just a luxury but a necessity.

Keywords: Cognitive Radio, Cooperative Spectrum Sensing, Diversity, Efficiency and Reliability

1. Introduction

In recent years, wireless services grow explosively both in variety and in quantity. While the current spectrum resources have already been assigned to various licensed services, the wireless industry faces the bottleneck of spectrum exhaustion. In the meantime, it is also reported that the current spectrum usage is highly inefficient [1] and many spectrum holes can be utilized at certain time or location. Under this condition, cognitive radio systems [2] provide an opportunity for unlicensed users (a.k.a. secondary users) to detect and initiate communications on these spectrum holes unused by licensed users (a.k.a. primary users) [3].

To achieve this objective, cognitive radio systems need to both maximally utilize opportunities of the spectrum holes and minimally interfere with the primary users. These are facilitated by the secondary user spectrum sensing capability to detect the spectrum holes, which corresponds to the "cognitive" part of the secondary system. Extensive research has already been conducted to improve the performance for both single-user sensing and multi-user cooperative sensing (see e.g., [4–21]). However, most of these only treat spectrum sensing as a detection problem without considering the context of the overall communications system.

In several recent papers, the interference from the secondary user system to the primary user system is evaluated (see e. g., [22–24]). In particular, [22] and [24] both report the observation that improved detection does not necessarily lead to reduced interference. Such inconsistency between detection and interference performance has motivated our work in this paper. In essence, the spectrum sensing schemes need to be designed in the context of the overall communications system, including both the primary and secondary ones.

Specifically, there are two error events in spectrum sensing, namely false alarm and missed detection. When false alarms occur, the secondary users will lose the opportunity of utilizing the spectrum, thus leading to reduced *efficiency* of spectrum utilization. When missed detections occur, the secondary users will initiate inappropriate communications over the spectrum in use and incur interference to the primary users, thus jeopardizing the *reliability* of the cognitive system. In this regard, we need to analyze spectrum sensing in terms of both the system efficiency and reliability.

It is usually believed that cooperation in spectrum sensing is just some luxury performance enhancement scheme over non-cooperative ones. However, our

---

*Corresponding author. lqyang@engr.colostate.edu

analyses and comparisons of non-cooperative sensing (NCoS) vs. cooperative sensing (CoS) with soft and hard information fusion will show that NCoS leads to a fundamental efficiency-reliability tradeoff, which can be largely overcome by CoS for any reasonable schemes. In essence, we claim that cooperation is actually a necessity rather than an option, in order to overcome the fundamental tradeoff.

The rest of the paper is organized as follows: the basic signal model and performance metrics under study are introduced in Section 2. The analysis of the fundamental efficiency-reliability tradeoff for non-cooperative sensing is given in Section 3. Then, the necessity of cooperative spectrum sensing to overcome this fundamental tradeoff is demonstrated with numerical results for cooperative sensing with soft information fusion in Section 4 and for cooperative sensing with hard information fusion in Section 5, followed by concluding remarks in Section 6.

Notation: Subscripts 'R' and 'E' refer to reliability-oriented and efficiency-oriented schemes, respectively; subscripts 'S' and 'H' refers to cooperative sensing with soft information fusion and hard information, respectively. $x \sim \mathcal{CN}(\mu, \sigma^2)$ denotes a circular symmetric complex Gaussian random variable x with mean μ and variance σ^2; $x \sim \text{Ben}(p_1)$ denotes a random variable x Bernoulli distributed with $\Pr(x = 1) = p_1$. $g(\gamma) = o(f(\gamma))$ means $\lim_{\gamma \to +\infty} \frac{g(\gamma)}{f(\gamma)} = 0$. $g(\gamma) \sim f(\gamma)$ means $\lim_{\gamma \to +\infty} \frac{g(\gamma)}{f(\gamma)} = c$ where c is a non-zero constant with respect to γ. $Q_N(x) = \int_x^{+\infty} \lambda^{N-1} \frac{e^{-\lambda}}{(N-1)!} d\lambda$ is the tailed cumulative distribution function of chi-square distribution with $2N$ degrees of freedom.

2. Problem Formulation

2.1. Signal Model

In the spectrum sensing process, the sensing users will encounter the signals under the following two hypotheses:

$$H_0 : \text{absence of primary user}$$
$$H_1 : \text{presence of primary user.}$$

As illustrated in [5, 14], after normalization by the noise power, the received signal is:

$$r|H_0 = n \sim \mathcal{CN}(0, 1),$$
$$r|H_1 = h_{p,s}x + n \sim \mathcal{CN}(0, \gamma + 1),$$
(1)

where x is the transmitted symbol of the primary user, n is the normalized additive white Gaussian noise (AWGN), $h_{p,s}$ is the channel coefficient from the primary user to the secondary user and $\gamma = \frac{\sigma_{p,s}^2 \mathcal{E}_p}{N_0}$ is the signal-to-noise ratio (SNR) at the secondary user, where N_0 is

the power of AWGN, $\mathcal{E}_p$ is the transmitted power of the primary user, and $\sigma_{p,s}^2$ is the variance or equivalently the average strength of channel $h_{p,s}$. Due to path loss, we know that $\gamma \sim \gamma_1 d_{p,s}^{-K}$, where $d_{p,s}$ is the distance between the primary user and the sensing secondary user, γ_1 is the SNR when $d_{p,s} = 1$, and K is the path-loss exponent. In the remaining of this paper, a large SNR implies a small distance from the sensing secondary user to the primary user, and vice versa.

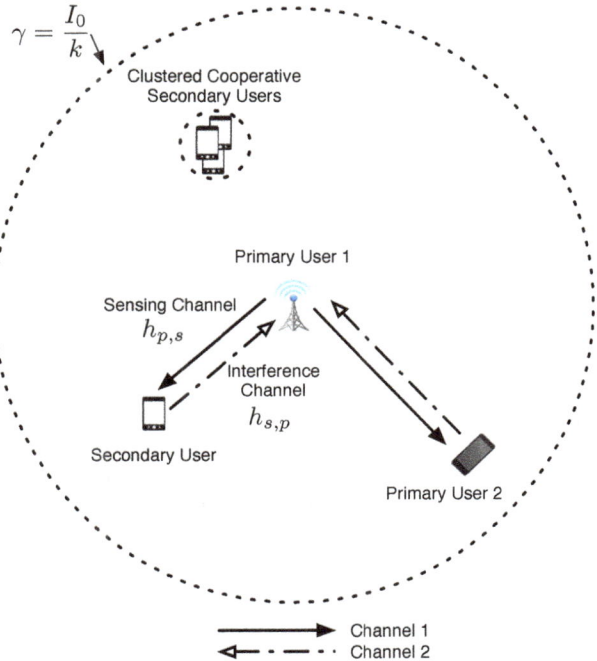

Figure 1. The system model.

In general cognitive radio systems, a secondary user senses the signal from a transmitter of the primary system and then incurs interference to the intended primary receiver, whose location can be quite unpredictable. This renders the interference analysis intractable and downgrades the usefulness of spectrum sensing, since the secondary user is not sensing the signal strength at the passive receiver side of the primary user system. To overcome this disagreement between sensing and interfering, and to simplify the analysis, [24] considered the setup where the primary network operating in the frequency division duplex mode as shown in Fig. 1[1]. In this case, the secondary user senses the downlink signals (channel 1 in the figure) and then transmits through the uplink channel (channel 2 in the figure), or vice versa. Accordingly, the location of the primary transmitter being sensed will also be the location of the receiver subject to the

---

[1]For simplicity, the communications among secondary users are not shown in the figure.

interference from the secondary user (both at primary user 1 in Fig. 1). Hence the interference channel $h_{s,p}$ has the same average strength as the sensing channel $h_{p,s}$ since they undergo the same path-loss effect with $d_{p,s} = d_{s,p}$. That is, $\sigma_{s,p}^2 = \sigma_{p,s}^2$, where $\sigma_{s,p}^2$ and $\sigma_{p,s}^2$ are the variances of $h_{s,p}$ and $h_{p,s}$, respectively.

2.2. Performance Metrics

In the context of the overall cognitive radio system, efficiency and reliability could be characterized by the capability of the secondary users in exploiting the unused spectrum and the average interference generated to disturb the active primary users. Hence, we will use the spectrum loss factor η and the average interference I as the performance metrics.

The spectrum loss factor is defined as the rate at which a false alarm occurs; thus

$$\eta = P_f . \tag{2}$$

For the average interference, one needs to consider both the occurrence rate and the average strength of the interference. The occurrence rate of interference is defined as the missed detection probability P_{md}. The average interference strength depends on both the secondary user transmit power and the channel from the secondary user to the primary one. As a result, the average interference strength is $\mathcal{E}_s \sigma_{s,p}^2$, where $\mathcal{E}_s$ is the signal strength of the secondary users. With the signal model given in Section 2.1, this can be rewritten as $\mathcal{E}_s \sigma_{p,s}^2$. By the definition of γ, this is equal to $k\gamma$, where $k = \frac{\mathcal{E}_s}{\mathcal{E}_p} N_0$. It is reasonable to assume that the transmit power levels of the primary and secondary users are both kept constant for them to maintain a certain coverage range. Therefore, k is a constant and the resultant average interference is given by

$$I = k\gamma P_{md} . \tag{3}$$

From this expression, we notice that when a secondary user is far away from the primary user (low SNR), the detection performance must be bad; however, the average interference is not necessarily high due to the low interference strength. On the other hand, when a secondary user is close to the primary user (high SNR), the average interfering signal strength can be really high due to the high interference strength even though the detection performance may be better. In our performance analyses, we will focus on the high SNR range, and the performance at low SNR will be shown numerically.

With these performance metrics, one can have two spectrum sensing strategies emphasizing either the system reliability or the system efficiency. For reliability-oriented systems, the average interference I is minimized under the constraint of a preset spectrum

loss factor η; while for efficiency-oriented systems, the spectrum loss factor η is minimized under the constraint of a tolerable average interference I.

3. Analysis of Non-Cooperative Sensing

For NCoS, the optimal detector is the energy detector under our setup [25]:

$$\lambda = |r|^2 \underset{H_0}{\overset{H_1}{\gtrless}} \theta , \tag{4}$$

where θ is the decision threshold. Accordingly, the false alarm and missed detection probabilities are given as

$$P_f = e^{-\theta} , \tag{5}$$

and

$$P_{md} = 1 - e^{-\frac{\theta}{\gamma+1}} , \tag{6}$$

respectively.

3.1. Reliability-Oriented Scheme with $\eta \leq \eta_0$

Under this scheme, the threshold should be chosen to minimize the average interference I while ensuring that the efficiency is maintained at a preset level $P_f = \eta_0$. From Eq. (5) we obtain

$$\theta_R = -\ln \eta_0 . \tag{7}$$

Thus, the average interference to the primary user due to inappropriate secondary communications is

$$\begin{aligned} I_R &= k\gamma P_{md} \\ &= k\gamma (1 - (\eta_0)^{\frac{1}{\gamma+1}}) . \end{aligned} \tag{8}$$

As $\gamma \to +\infty$, $\frac{1}{\gamma+1} \to 0$. Taking Taylor series expansion of $\eta_0^{\frac{1}{\gamma+1}}$ with respect to $\frac{1}{\gamma+1}$ around 0, we have

$$\begin{aligned} I_R &= k\gamma \left(1 - \left(1 + \frac{1}{\gamma+1} \ln \eta_0 + o\left(\frac{1}{1+\gamma} \right) \right) \right) \\ &= -k \ln \eta_0 \frac{\gamma}{\gamma+1} + k\gamma o\left(\frac{1}{1+\gamma} \right) \\ &\approx k \ln \frac{1}{\eta_0} . \end{aligned} \tag{9}$$

That is, with the reliability-oriented scheme, the average interference I_R approaches constant $k \ln \eta_0$ as γ increases. The numerical plot is shown in Fig. 2.

3.2. Efficiency-Oriented Scheme with $I \leq I_0$

Similarly, in order to minimize the spectrum loss factor η under the interference constraint, the threshold should be selected such that $P_{md} = \min \left(\frac{I_0}{k\gamma}, 1 \right)$. Notice

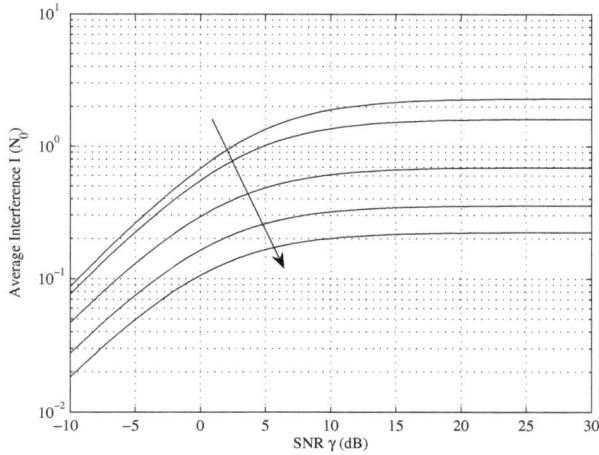

Figure 2. The reliability-oriented scheme for NCoS. $k = N_0$. Along the direction of the arrow, $\eta_0 = 0.1, 0.2, 0.5, 0.7, 0.8$.

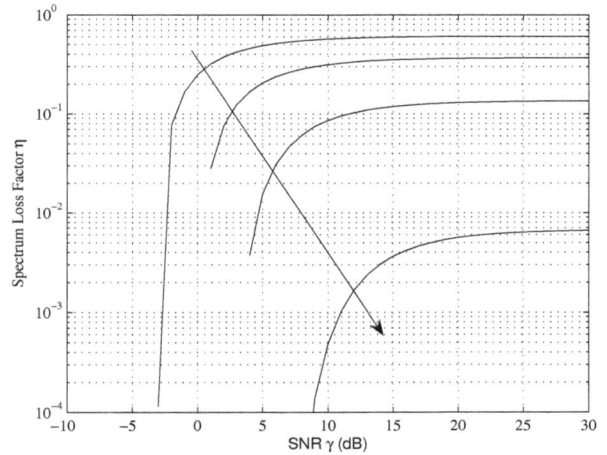

Figure 3. The efficiency-oriented scheme for NCoS. $k = N_0$. Along the direction of the arrow, $I_0 = 0.5N_0, N_0, 2N_0, 5N_0$.

that when $\gamma < \frac{I_0}{k}$ (out of the circle in Fig. 1), the spectrum loss factor can approach zero since the average interference constraint is satisfied even when $P_{md} = 1$. In other words, it does not matter whether or not the presence of the primary user is detected, simply because the interference induced by the secondary users is negligible.

For $\gamma \geq \frac{I_0}{k}$, setting $P_{md} = \frac{I_0}{k\gamma}$ together with Eq. (6), we obtain

$$\theta_E = -(\gamma + 1) \ln\left(1 - \frac{I_0}{k\gamma}\right). \tag{10}$$

With this threshold, the spectrum loss factor is

$$\eta_E = e^{(\gamma+1)\ln\left(1 - \frac{I_0}{k\gamma}\right)}. \tag{11}$$

As $\gamma \to +\infty$, by Taylor series expansion, we have

$$\ln\left(1 - \frac{I_0}{k\gamma}\right) = -\frac{I_0}{k\gamma} + o\left(\frac{1}{\gamma}\right). \tag{12}$$

Hence we obtain

$$\eta_E = e^{-\frac{I_0}{k}\frac{\gamma+1}{\gamma} + (\gamma+1)o\left(\frac{1}{\gamma}\right)} \\ \approx e^{-\frac{I_0}{k}}. \tag{13}$$

This means that with the efficiency-oriented scheme, the system performance also saturates at high SNR. The numerical plot is shown in Fig. 3[2].

From the analyses above, we see that for NCoS, the reliability and efficiency exhibit a persistent tradeoff: when one is lower, the other is inevitably higher. The saturating effects in Figs. 2 and 3 confirm that such tradeoff cannot be improved by the SNR increase.

---

[2]Without loss of generality, we set $k = N_0$ in all our numerical studies.

4. Analysis of Cooperative Sensing with Soft Information Fusion (SCoS)

With cooperative sensing, a fusion center is collecting information from geographically distributed local secondary sensing users to make a global decision. If the channels from the local secondary users to the fusion center have sufficiently high bandwidth, we can assume that the fusion center can receive roughly exact sensed signal values $r_1, r_2, \ldots, r_N$ from different sensing users, where N is the total number of sensing users. For practical considerations, we study the case where the N secondary users are geographically clustered to cooperate. This is based on the consideration that the cooperative detection of the primary user presence is only meaningful when the cooperating secondary users are subject to the same primary user activity and are thus posing similar interferences to the primary receiver. As a result, the cooperating users share the same level of large-scale path loss as shown in Fig. 1. In this case, the SNRs in their received signals are the same and r_i's are independent identically distributed (i.i.d). In this case, the optimal detector is again an energy detector [25]:

$$\lambda_S = \sum_{i=1}^{N} |r_i|^2 \underset{H_0}{\overset{H_1}{\gtrless}} \theta_S, \tag{14}$$

where θ_S is the decision threshold at the fusion center.

Since r_i's are independent circular symmetric complex Gaussian variables, the false alarm and missed detection probabilities are respectively

$$P_{f,S}(N) = Q_N(\theta_S), \tag{15}$$

and

$$P_{md,S}(N) = 1 - Q_N\left(\frac{\theta_S}{\gamma + 1}\right). \tag{16}$$

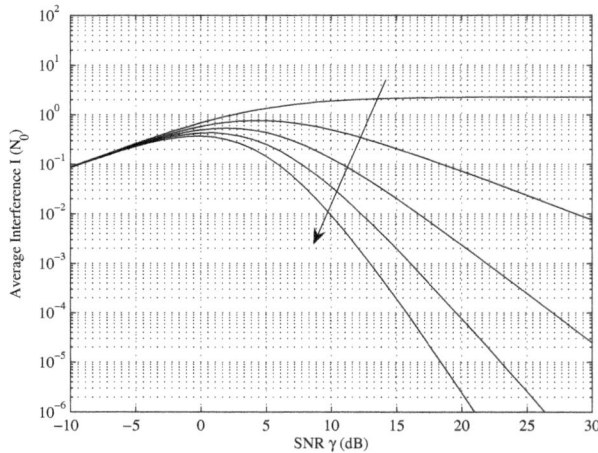

Figure 4. The reliability-oriented scheme for SCoS with $\eta \leq 0.01$. $k = N_0$. Along the direction of the arrow, $N = 1, 2, 3, 4, 5$.

4.1. Reliability-Oriented Scheme with $\eta \leq \eta_0$

Similar to the case of NCoS, in the reliability-oriented scheme, we set $P_{f,S} = \eta_0$ and obtain the threshold according to Eq. (15) as:

$$\theta_{R,S}(N) = Q_N^{-1}(\eta_0) . \qquad (17)$$

With this threshold, we have

$$I_{R,S}(N) = k\gamma \int_0^{\frac{Q_N^{-1}(\eta_0)}{\gamma+1}} \lambda^{N-1} \frac{e^{-\lambda}}{(N-1)!} d\lambda .$$

As $\gamma \to +\infty$, $\frac{Q_N^{-1}(\eta_0)}{\gamma+1} \to 0$. Thus, within the integration interval $\lambda \in \left[0, \frac{Q_N^{-1}(\eta_0)}{\gamma+1}\right]$, we have $e^{-\lambda} \approx 1$ for large γ. Thus,

$$I_{R,S}(N) \approx k\gamma \int_0^{\frac{Q_N^{-1}(\eta_0)}{\gamma+1}} \frac{\lambda^{N-1}}{(N-1)!} d\lambda$$

$$= k\gamma \cdot \frac{\lambda^N}{N!} \Big|_0^{\frac{Q_N^{-1}(\eta_0)}{\gamma+1}}$$

$$= k\gamma \frac{\left(Q_N^{-1}(\eta_0)\right)^N}{N!(\gamma+1)^N}$$

$$\sim \gamma^{-(N-1)} .$$

This result indicates that system reliability has a diversity order of $(N-1)$. The numerical plot is given in Fig. 4.

4.2. Efficiency-Oriented Scheme with $I \leq I_0$

Similar to the NCoS case, in the efficiency-oriented scheme, we can set $P_{md,S}(N) = \min\left(\frac{I_0}{k\gamma}, 1\right)$, which means that $P_{md,S}(N) = 1$ for $\gamma < \frac{I_0}{k}$.

When $\gamma \geq \frac{I_0}{k}$, setting $P_{md,S}(N) = \frac{I_0}{k\gamma}$ together with Eq. (16), we have the decision threshold as

$$\theta_{E,S}(N) = (\gamma+1)Q_N^{-1}\left(1 - \frac{I_0}{k\gamma}\right) . \qquad (18)$$

When $\gamma \to +\infty$, $P_{md,S}(N) = \frac{I}{k\gamma} \to 0$; then the integral upper bound $\frac{\theta_{E,S}(N)}{\gamma+1} \to 0$. Hence, within the integration interval $\left[0, \frac{\theta_{E,N}}{\gamma+1}\right]$, similar approximation as in Section 5.1 can be utilized. As a result:

$$\frac{I}{k\gamma} = P_{md,N} \approx \frac{(\theta_{E,S}(N))^N}{N!(\gamma+1)^N} .$$

That is,

$$\theta_{E,S}(N) \approx (\gamma+1)\left(\frac{N!I}{k\gamma}\right)^{\frac{1}{N}} \sim \gamma^{1-\frac{1}{N}} . \qquad (19)$$

With threshold $\theta_{E,S}(N)$, the spectrum loss factor is

$$\eta_{E,S}(N) = P_{f,S}(N) = \int_{\theta_{E,S}(N)}^{+\infty} \lambda^{N-1} \frac{e^{-\lambda}}{(N-1)!} d\lambda$$

$$= \left(\sum_{i=0}^{N-1} \frac{(\theta_{E,S}(N))^i}{i!}\right) e^{-\theta_{E,S}(N)} .$$

Recall that we have shown that $\theta_{E,S}(N)$ increases as γ increases. Hence as $\gamma \to +\infty$, the summation in $\eta_{E,S}(N)$ will be dominated by the term with the highest order. Together with the expression of $\theta_{E,S}(N)$ in Eq. (19), we obtain

$$\eta_{E,S}(N) \approx \frac{(\theta_{E,S}(N))^{N-1}}{(N-1)!} e^{-\theta_{E,S}(N)}$$

$$\sim \gamma^{\left((N-1)(1-\frac{1}{N})\right)} e^{-\gamma^{\left(1-\frac{1}{N}\right)}}$$

$$= o\left(\gamma^{-p}\right) ,$$

where p is an arbitrary positive real number. This implies that the system efficiency has a diversity order of infinity. The numerical plot is given in Fig. 5.

In this section, we see that thanks to cooperative sensing with soft information fusion, for both efficiency- and reliability-oriented schemes, these figures of merit can be consistently improved as SNR increases. This is in sharp contrast to the persistent tradeoff that we observed in the non-cooperative case.

5. Analysis of Cooperative Sensing with Hard Information Fusion (HCoS)

In order for the soft-information based sensing scheme described in Section 4 to work, the bandwidth of the channel between the sensing secondary users and the fusion center has to be very large. This usually requires a backbone wired communication system to

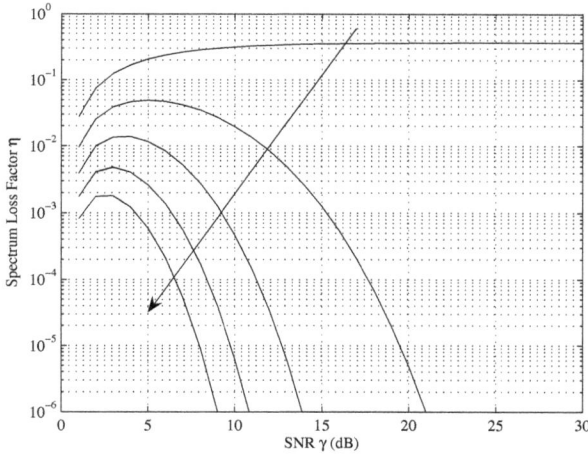

Figure 5. The efficiency-oriented scheme for SCoS with $I \leq N_0$. $k = N_0$. Along the direction of the arrow, $N = 1, 2, 3, 4, 5$.

connect the fusion center and the secondary users, which is impractical for mobile secondary devices. Hence, usually quantizations are applied to the received signal at the local secondary user and only a limited number of bits will be sent to the fusion center for a global decision. Although multi-bit quantization can be made (see e.g. [13, 18]), in this paper, we only study the extreme case where the local quantizations are simply binary.

In this case, each sensing secondary user makes a local one-bit decision d_i according to the optimal rule in the NCoS case,

$$|r|^2 \overset{d_i=1}{\underset{d_i=0}{\gtrless}} \theta_l \,, \qquad (20)$$

where θ_l is the local decision threshold adopted by all secondary sensing users. Hence, at the fusion center, the received local decisions follow an i.i.d. Bernoulli distribution with parameters $P_{f,l}$ and $1 - P_{md,l}$ under H_0 and H_1, respectively:

$$d_i|H_0 \sim \text{Ber}(P_{f,l}) \,, \qquad (21)$$
$$d_i|H_1 \sim \text{Ber}(1 - P_{md,l}) \,,$$

where

$$P_{f,l} = e^{-\theta_l} \qquad (22)$$

is the local false alarm probability, and

$$P_{md,l} = 1 - e^{-\frac{\theta_l}{\gamma+1}} \qquad (23)$$

is the local missed detection probability. Then, the decision rule at the fusion center is cast as

$$\lambda_H = \sum_{i=1}^{N} d_i \overset{H_1}{\underset{H_0}{\gtrless}} \theta_H \,, \qquad (24)$$

where λ_H and θ_H are the decision statistics and the decision threshold at the fusion center, respectively. Naturally, the HCoS involves two thresholds, namely the local threshold θ_l and the fusion threshold θ_H in the performance optimization process. However, in our previous work [20], we have shown that as $\gamma \rightarrow +\infty$, the optimal fusion rule is the 'OR' rule, i.e., the fusion center will declare the absence of the primary user only when all local sensing secondary users make universal decisions $d_i = 0$ and will declare the presence of primary user as long as at least one of the local sensing secondary users make decision $d_i = 1$. Hence, the optimal fusion threshold is set as $\theta_H^o = 1$.

Under the 'OR' rule, the global false alarm and missed detection probabilities depend on the local threshold θ_l as

$$P_{f,H}(N) = 1 - \left(1 - P_{f,l}\right)^N \qquad (25)$$
$$= 1 - \left(1 - e^{-\theta_l}\right)^N \,,$$

and

$$P_{md,H}(N) = (P_{md,l})^N \qquad (26)$$
$$= \left(1 - e^{-\frac{\theta_l}{\gamma+1}}\right)^N \,.$$

5.1. Reliability-Oriented Scheme with $\eta \leq \eta_0$

Similar to the cases of NCoS and SCoS, in the reliability-oriented scheme, we set $P_{f,H} = \eta_0$ and obtain the local threshold according to Eqs. (22) and (25):

$$\theta_{R,l}(N) = -\ln\left(1 - (1 - \eta_0)^{\frac{1}{N}}\right) \,. \qquad (27)$$

With this local threshold and by Eq. (26), we have

$$I_{R,H}(N) = k\gamma P_{md,H}(N)$$
$$= k\gamma \left(1 - e^{-\frac{\theta_{R,l}(N)}{\gamma+1}}\right)^N$$
$$= k\gamma \left(-\frac{1}{\gamma+1} \ln\left(\theta_{R,l}(N)\right) + o\left(\frac{1}{\gamma+1}\right)\right)^N$$
$$\approx k\gamma \left(\frac{1}{\gamma+1}\right)^N \left(-\ln\left(-\ln\left(1 - (1 - \eta_0)^{\frac{1}{N}}\right)\right)\right)^N$$
$$\sim \gamma^{-(N-1)} \,,$$

where Taylor series expansion is applied as in Eq. (9) for $\gamma \rightarrow +\infty$.

This result indicates that the system reliability for HCoS exhibits the same diversity order of $(N - 1)$ as for SCoS. The numerical plot is given in Fig. 6.

5.2. Efficiency-Oriented Scheme with $I \leq I_0$

Similar to NCoS and SCoS cases, in the efficiency-oriented scheme, we set $P_{md,H}(N) = \min\left(\frac{I_0}{k\gamma}, 1\right)$, which similarly means that $P_{md,H}(N) = 1$ for $\gamma < \frac{I_0}{k}$.

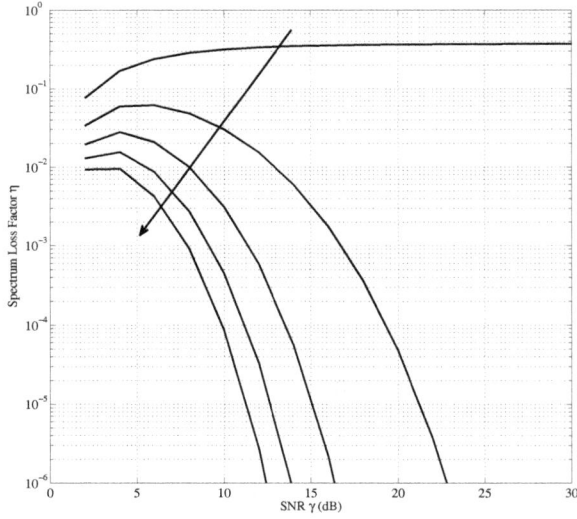

Figure 6. The reliability–oriented scheme for HCoS with $\eta \leq 0.01$. $k = N_0$. Along the direction of the arrow, $N = 1, 2, 3, 4, 5$.

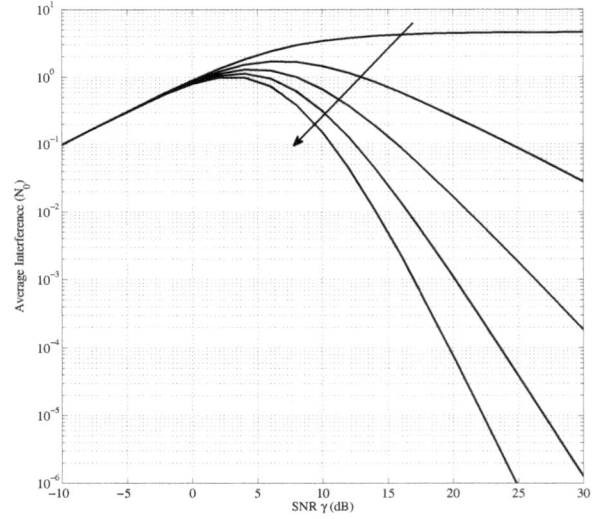

Figure 7. The efficiency–oriented scheme for HCoS with $I \leq N_0$. $k = N_0$. Along the direction of the arrow, $N = 1, 2, 3, 4, 5$.

When $\gamma \geq \frac{I_0}{k}$, by setting $P_{md,H}(N) = \frac{I_0}{k\gamma}$, together with Eqs. (23) and (26), the local decision threshold is obtained as

$$\theta_{E,l}(N) = -(\gamma + 1) \ln \left(1 - \left(\frac{I_0}{k\gamma} \right)^{\frac{1}{N}} \right) . \qquad (28)$$

Then, according to Eqs. (22) and (25), as $\gamma \to +\infty$, the spectrum loss factor is

$$\begin{aligned}
\eta_{E,H}(N) &= P_{f,H}(N) = 1 - \left(1 - e^{-\theta_{E,l}(N)} \right)^N \\
&\approx N e^{-\theta_{E,l}(N)} \\
&= N \left(\frac{1}{1 - \left(\frac{I_0}{k\gamma} \right)^{\frac{1}{N}}} \right)^{\gamma+1} \\
&\sim o(\gamma^{-p}) ,
\end{aligned} \qquad (29)$$

where p is an arbitrary positive real number. Therefore, similar to SCoS, the system efficiency for HCoS also has a diversity order of infinity. The numerical plot is given in Fig. 7.

In this section, we see that although local binary quantizations are made at local sensing secondary users, cooperative sensing with hard information fusion can also greatly improve the figures of merit for both efficiency- and reliability-oriented schemes, similar as SCoS. This implies that even a very simple cooperative scheme can help overcome the fundamental efficiency-reliability tradeoff incurred by NCoS.

6. Conclusions

In this paper, we analyzed the spectrum sensing performance of non-cooperative sensing (NCoS) and cooperative sensing with soft (SCoS) and hard information fusion (HCoS) in the context of a cognitive radio system, and gained some insightful observations on the role of cooperation. To capture the system tradeoff between efficiency and reliability, we introduced the spectrum loss factor η and the average interference I as the performance metrics. With these metrics, we analyzed both reliability-oriented and efficiency-oriented schemes for NCoS, SCoS, and HCoS. Results show that the seemingly unavoidable efficiency-reliability tradeoff in NCoS can be largely avoided by exploiting diversity via cooperating sensing. This is the case not only with SCoS collecting perfect information from local sensors but also with HCoS collecting only quantized information. In a nutshell, cooperation in spectrum sensing is not just a luxury but a necessity.

Acknowledgement. This paper was in part supported by National Science Foundation under awards CNS-1265227, CNS-1343155, ECCS-1305979 and ECCS-1232305.

References

[1] HAYKIN, S. (2005) Cognitive radio: brain-empowered wireless communications. *IEEE Journal on Selected Areas in Communications* 23(2): 201–220.

[2] MITOLA III, J. and MAGUIRE JR., G. (1999) Cognitive radio: making software radios more personal. *IEEE Personal Communications* 6(4): 13–18.

[3] STAPLE, G. and WERBACH, K. (2004) The end of spectrum scarcity. *IEEE Spectrum* 41(3): 48–52.

[4] CABRIC, D., TKACHENKO, A. and BRODERSEN, R.W. (2006) Experimental study of spectrum sensing based on energy detection and network cooperation. In *Proc. ACM 1st Int. Workshop on Technology and Policy for Accessing Spectrum (TAPAS)* (Boston, MA): 1–8.

[5] HONG, S., VU, M.H. and TAROKH, V. (2008) Cognitive sensing based on side information. In *IEEE Sarnoff Symposium* (Princeton, NJ): 1–6.

[6] QUAN, Z., CUI, S. and SAYED, A.H. (2008) Optimal linear cooperation for spectrum sensing in cognitive radio networks. *IEEE Journal of Selected Topics in Singal Processing* 2(1): 28–40.

[7] ZHANG, W. and LETAIEF, K. (2008) Cooperative spectrum sensing with transmit and relay diversity in cognitive radio networks. *IEEE Trans. on Wireless Communications* 7(12): 4761–4766.

[8] QUAN, Z., CUI, S., POOR, H. and SAYED, A. (2008) Collaborative wideband sensing for cognitive radios. *IEEE Signal Processing Magazine* 25(6): 60–73.

[9] UNNIKRISHNAN, J. and VEERAVALLI, V.V. (2008) Cooperative sensing for primary detection in cognitive radio. *IEEE Journal of Selected Topics in Singal Processing* 2(1): 18–27.

[10] WANG, H., hui YANG, E., ZHAO, Z. and ZHANG, W. (2009) Spectrum sensing in cognitive radio using goodness of fit testing. *IEEE Trans. on Wireless Communications* 8(11): 5427–5430.

[11] QUAN, Z., CUI, S., SAYED, A. and POOR, H. (2009) Optimal multiband joint detection for spectrum sensing in cognitive radio networks. *IEEE Trans. on Signal Processing* 57(3): 1128–1140.

[12] ZHANG, W., MALLIK, R. and LETAIEF, K. (2009) Optimization of cooperative spectrum sensing with energy detection in cognitive radio networks. *IEEE Trans. on Wireless Communications* 8(12): 5761–5766.

[13] CHAUDHARI, S., LUNDEN, J., KOIVUNEN, V. and POOR, H.V. (2010) Cooperative sensing with imperfect reporting channels: Hard decisions or soft decisions? *IEEE Trans. on Signal Processing* 60(1): 18–28.

[14] DUAN, D., YANG, L. and PRINCIPE, J. (2010) Cooperative diversity of spectrum sensing for cognitive radio

systems. *IEEE Trans. on Signal Processing* 58(6): 3218–3227.

[15] PEH, E., LIANG, Y.C., GUAN, Y.L. and ZENG, Y. (2010) Cooperative spectrum sensing in cognitive radio networks with weighted decision fusion schemes. *IEEE Trans. on Wireless Communications* 9(12): 3838–3847.

[16] KIM, S.J., DALL'ANESE, E. and GIANNAKIS, G.B. (2011) Cooperative spectrum sensing for cognitive radios using kriged kalman filtering. *IEEE Journal of Selected Topics in Signal Processing* 5(1): 24–36.

[17] SHEN, L., WANG, H., ZHANG, W. and ZHAO, Z. (2011) Blind spectrum sensing for cognitive radio channels with noise uncertainty. *IEEE Trans. on Wireless Communications* 10(6): 1721–1724.

[18] DUAN, D. and YANG, L. (2012) Cooperative spectrum sensing with ternary local decisions. *IEEE Communications Letters* 16(9): 1512–1515.

[19] DUAN, D., YANG, L. and SCHARF, L.L. (2012) Optimal detection fusion by large deviation analysis. In *Proceedings of European Signal Processing Conference (EUSIPCO)* (Bucharest, Romania): 744–748.

[20] DUAN, D., YANG, L. and SCHARF, L.L. (2012) The optimal fusion rule for cooperative spectrum sensing from a diversity perspective. In *Proc. of Asilomar Conf. on Signals, Systems, and Computers* (Pacific Grove, CA): 1056–1060.

[21] YUCEK, T. and ARSLAN, H. (2009) A survey of spectrum sensing algorithms for cognitive radio applications. *IEEE Communications Surveys Tutorials* 11(1): 116–130.

[22] DUAN, D., YANG, L. and PRINCIPE, J. (2009) Detection-interference dilemma for spectrum sensing in cognitive radio systems. In *Proc. of MILCOM Conf.* (Boston, MA): 1–7.

[23] LIANG, Y.C., ZHENG, Y., PEH, E.C.Y. and HOANG, A.T. (2008) Sensing-throughput tradeoff for cognitive radio networks. *IEEE Trans. on Wireless Communications* 7(4): 1326–1337.

[24] RABBACHIN, A., QUEK, T.Q.S. and WIN, M.Z. (2011) Cognitive network interference. *IEEE Journal on Selected Areas in Communications* 29(2): 480–493.

[25] POOR, H.V. (1994) *An Introduction to Signal Detection and Estimation* (Springer-Verlag), 2nd ed.

5

Distributed Spectrum Sharing Games Via Congestion Advertisement

Mahdi Azarafrooz * and R. Chandramouli

Department of Electrical and Computer Engineering, Stevens Institute of Technology, Hoboken,NJ

Abstract

Distributed spectrum sharing via congestion advertisement is modelled and studied as a game theoretic problem. A related graphical anti-coordination game problem and a suitable logit-response learning mechanism is proposed and studied. It has been shown that introducing an arbitrary small congestion advertisement term into the users utility can improve the convergence rate of the spectrum sharing game exponentially. Finally, simulation results are presented to evaluate the price of anarchy, convergence rate and phase transition properties.

Keywords: Spectrum Sharing, Game Theory, Learning

1. Introduction

Cognitive radio (CR) nodes learn to configure their transmission and reception parameters based on different cognitive processes. These cognitive processes vary from sensing an existing wireless channel, configuring a radio's parameters to accommodate the perceived wireless channel, evaluating the current situation and taking the best possible action based on this available knowledge [2], etc. Therefore, every action of a specific cognitive radio or user has an effect on the other nodes' payoffs.

In a recent work [3], we addressed the problem of distributed optimization of secondary user sharing of primary user spectrum, considering spatial re-use. This was modelled as a spatial or graphical game theoretic problem considering the radio interference induced by communication in a local neighborhood in a specific band. However as it will be shown later in this paper, this spectrum sharing game suffers from a high price of anarchy. Also, the distributed iterative algorithm to compute the equilibrium strategies for all the users is slow to converge. Therefore, we investigate the feasibility of computing a *good* (to be made precise later) equilibrium solution in polynomial steps (in the number of secondary nodes n).

Mechanism design is a tool that can be used to align incentives of the users with the system's objective. In systems where there are multiple Nash equilibria, using mechanism design, a central authority could move the system's behavior from a less efficient equilibrium to a more efficient one by promoting better user behavior. The objective of this paper is to investigate such a mechanism design and an iterative technique to compute an efficient Nash equilibrium solution with fast convergence properties. In this regard, we propose the idea of congestion advertisement by base stations as one of the mechanism design approaches.

In [5], spectrum sharing and spatial reuse in a wireless network is posed as an extended form of the congestion game where users' payoffs for using a spectrum band or channel is a function of the number of its interfering users sharing that channel. In [6], spectrum management is studied in CR networks by defining a secondary user specific utility as a function of the spectrum opportunity, congestion and bandwidth. The behavior of selfish nodes that dynamically switch their channels using broadcasted random public signal is presented in [7]. In [8], dynamic spectrum access is modelled as a minority game where the CR nodes try to minimize their cost in finding a clear band. A graphical game model for competitive spectrum access is discussed in [9].

---

*Corresponding author. Email: mazarafr@stevens.edu

This paper is directly related to [10-12]. In [10], the convergence rates of congestion games towards a good equilibrium is studied. Convergence of coordination games in a social networking context is presented in [11]. In [12], a framework for graphical games with global interactions is considered.

The main contributions of the paper are:

- The dynamics of the graphical spectrum sharing game is mapped to that of anti-ferromagnetic Ising model using a MECE logit-response learning mechanism.

- The effect of the graph of interaction on the convergence rate of the spectrum sharing games is studied both in terms of social welfare optimization and the convergence rate of the game.

- It has been shown that introducing an arbitrary small congestion advertisement term into the secondary users utility can improve the convergence rate exponentially.

This paper is organized as follows. In Section II a graphical anti-coordination spectrum sharing game model is proposed. A maximum entropy correlated logit response mechanism is discussed in Section III. Analysis of convergence rates of the response mechanism for specific graphs are presented in Section IV. Congestion advertisement for spectrum sharing is discussed in Section V. Simulation results are given in Section VI and conclusions in Section VII.

2. Spectrum Sharing as a Graphical Anti-Coordination Game

Consider a CR network scenario where n secondary users are placed in an undirected graph $G = (V, E)$, where $|V| = n$ and E is the set of edges. Let $\mathcal{N}_i$ denote the neighbor set of node i. We will interchangeably use the terms node and user throughout the paper. Users are assumed to have access to B primary user bands. Let $\mathbf{A}_{1 \times n}$ be the users' action vector where the ith element $a_i \in \{1, ..., B\}$ denotes the index of the spectrum band that user i is active in. Users can follow different approaches for evaluating the spectrum quality, e.g., based on whether data or video application needs to be supported. We assume that the evaluation approach is the same for all the users. Let $\mathbf{\Theta}_{1 \times B}$ represent the spectrum quality vector, i.e, θ_l, $l \in \{1, 2, \ldots, B\}$ denotes the quality of the lth spectral band. For example, this could be a function of the primary user activity, required data rate, etc. The higher the value of θ_l the more desirable that band is. let $\mathcal{I}_{a_i}$ be the set of interfering transmissions with user i scheduled in band a_i.

The secondary users compete for spectrum opportunities in a decentralized non-cooperative manner. The utility obtained by secondary user i is $U^i(|\mathcal{I}_{a_i}|, \theta_{a_i})$. That is, the utility function depends on the interference level as well as the quality of the operating band. However from the perspective of the designer of a wireless cognitive network it is important that the system as a whole entity can achieve a good operation point. A simple metric for example is social welfare which is defined as the accumulation of all users utility in the network i.e, .

$$U(\mathbf{A}, \mathbf{\Theta}) = \sum_i U^i(|\mathcal{I}_{a_i}|, \theta_{a_i}) \tag{1}$$

The optimal solution $\mathbf{A}^*$ to the spectrum sharing problem is then given by:

$$\mathbf{A}^* = \underset{\mathbf{A}}{\operatorname{argmax}} \, U(\mathbf{A}, \mathbf{\Theta}) \tag{2}$$

We first address the issue of solving this problem when users play the non-cooperative decentralized spectrum sharing game.

Consider a simple scenario seen in Fig. 1 where users 1, 2 and 3 are playing a graphical anti-coordination game as follows. Each user selects a color (spectrum band) white (W) or black (B) as their strategy based on the output of the evaluation function. Based on the color of their neighbors their utility is realized according to the payoff matrix shown in Fig.1. If two neighbors select the same color they incur a cost of -1 otherwise they get a reward 1. Moreover each user plays the game with each neighbor separately and its final decision is based on the realization of the composite game. For example, assume the users' strategy vector is $\mathbf{A} = (a_1 = W, a_2 = B, a_3 = W)$. Then user 3 obtains a cost -1 for choosing the same band as user 1. Also it obtains 1 from playing the game with user 2 since they have chosen different bands. Therefore user 3 obtains a total utility of $U^3 = 0$ from playing the composite game with it's neighbors. We can also define a potential function for this game, an example is shown in Fig. 1. For example given $a_1 = B, a_2 = B$, user 3 can improve it's utility from -2 to 2 by changing it's strategy from B to W which corresponds to the same change in it's potential function value.

We can generalize this example by defining the elements of a spectrum sharing graphical game $\mathcal{G}$:

1. Players are the secondary users $i \in V$.

2. Set of pure strategies for user (vertex) i is the set $a_i = \{1, ..., B\}$. Then the joint action strategy space for the entire network is $\mathcal{A} = \{1, ..., B\}^n$. Let us denote the joint-action by $\mathbf{A} \in \mathcal{A}$ and let $\mathbf{A}(i : a_i') \equiv (a_1, ..., a_{i-1}, a_i', ..., a_n)$ denote the vector resulting from setting the strategy of user i in vector $\mathbf{A}$ to a_i' while keeping all other

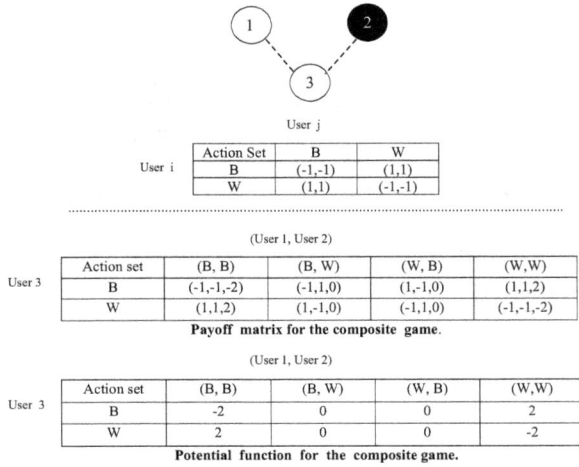

User j		
Action Set	B	W

User i		
B	(-1,-1)	(1,1)
W	(1,1)	(-1,-1)

(User 1, User 2)

Action set	(B, B)	(B, W)	(W, B)	(W,W)
User 3				
B	(-1,-1,-2)	(-1,1,0)	(1,-1,0)	(1,1,2)
W	(1,1,2)	(1,-1,0)	(-1,1,0)	(-1,-1,-2)

Payoff matrix for the composite game.

(User 1, User 2)

Action set	(B, B)	(B, W)	(W, B)	(W,W)
User 3				
B	-2	0	0	2
W	2	0	0	-2

Potential function for the composite game.

Figure 1. Simple example of a spectrum sharing graphical anti-coordination game.

strategies the same. The mixed strategy set for user i is the probability mass function $\{p(a_i) \equiv p_{ia_i}\}$, $\sum_{a_i=1}^{B} p_{ia_i} = 1$.

3. The utility user i receives, U^i, is given by the following linear function:

$$U^i(\mathbf{A}(i : a_i)) = \theta_{a_i} + \sum_{j \in \mathcal{N}_i} M_i(a_i, a_j) \quad (3)$$

where $j \in \mathcal{N}_i$ if there is an edge $e(i, j) \in E$ and $M_i(a_i, a_j)$ denotes the symmetric payoff user i obtains by playing the strategy a_i against the strategy a_j of user j. M_i is the following anti-coordination payoff matrix:

$$M_1 = M_2 = ... = M = \begin{pmatrix} -1,-1 & 0,0 & ... & 0,0 \\ 0,0 & -1,-1 & ... & 0,0 \\ ... & ... & ... & ... \\ 0,0 & 0,0 & ... & -1,-1 \end{pmatrix} \quad (4)$$

From M_i we see that if two users choose the same band then their respective payoff is -1 otherwise it is 0 for each. Also we assume the payoff Matrix to be symmetric i.e, $M_1 = M_2 = ... = M$.

Note 1 The linear selection of users utility is for the the ease of modelling of the distributed spectrum sharing games and does not reflect the actual performance of a practical communication network. However in the Simulation Results section it is shown that the result of the theoretical parts remains valid for the more generalized model where users utility follow a non-linear function of the interference.

Definition 2.1 A joint action strategy $\mathbf{A} \in \mathcal{A}$ is called a Nash equilibrium (NE) if no user $i, \forall i \in V$ has an incentive to deviate from the equilibrium strategy.

Definition 2.2 [4] A correlated equilibrium (CE) for game $\mathcal{G}$ is a joint-probability distribution Q over the

joint action space $\mathcal{A}$ such that for every user i, and every action pair $(j, a_i') \in \mathbf{A}^2, j \neq a_i'$,

$$\sum_{\mathbf{A} \in \mathcal{A}} Q(\mathbf{A}) U^i(\mathbf{A}(i, a_i)) \geq \sum_{\mathbf{A} \in \mathcal{A}} Q(\mathbf{A}) U^i(\mathbf{A}(i : a_i')) \quad (5)$$

A NE is a CE such that Q is a product distribution; that is $Q = \prod_{i=1}^{n} q_i$.

Definition 2.3 [20] Given a joint mixed strategy Q, let $H(Q) \equiv -Q(\mathbf{A}) \ln(Q(\mathbf{A}))$ denote the Shannon entropy. Then a maximum entropy correlated equilibrium (MECE) is the joint mixed strategy

$$Q^* = \underset{Q \in CE}{\operatorname{argmax}} H(Q) \quad (6)$$

Theorem 2.1. The spectrum sharing game $\mathcal{G}$ has a potential function $\Phi : \mathcal{A} \rightarrow \mathbb{R}$ given by:

$$\Phi(\mathbf{A}) = \sum_i \theta_{a_i} + \sum_i \sum_{j \in \mathcal{N}_i} H(a_i, a_j) \quad (7)$$

where H is :

$$H = \begin{pmatrix} -1 & 0 & ... & 0 \\ 0 & -1 & ... & 0 \\ ... & ... & ... & ... \\ 0 & 0 & ... & -1 \end{pmatrix} \quad (8)$$

Proof. We observe that if a matrix game M has a potential function H, then so does the associated graphical game with the following potential function Φ':

$$\Phi'(\mathbf{A}) = \sum_i \sum_{j \in \mathcal{N}_i} H(a_i, a_j) \quad (9)$$

To see this suppose that user i deviates, say by choosing strategy a_i'. Then,

$$U^i(\mathbf{A}(i : a_i)) - U^i(\mathbf{A}(i : a_i')) =$$
$$\sum_{j \in \mathcal{N}_i} [M(a_i, a_j) - M(a_i', a_j)] =$$
$$\sum_{j \in \mathcal{N}_i} [H(a_i, a_j) - H(a_i', a_j)] \quad (10)$$

From this it is now easy to see that matrix H characterizes a potential function:

$$H = \begin{pmatrix} -1 & 0 & ... & 0 \\ 0 & -1 & ... & 0 \\ ... & ... & ... & ... \\ 0 & 0 & ... & -1 \end{pmatrix} \quad (11)$$

Therefore it follows that:

$$\Phi(\mathbf{A}) = \sum_i \theta_{a_i} + \sum_i \sum_{j \in \mathcal{N}_i} H(a_i, a_j) \quad (12)$$

$\square$

The existence of the potential function then shows the existence of pure Nash strategies for $\mathcal{G}$ [1]. Let $\mathcal{E}(\mathcal{G}) \subseteq \mathcal{A}$ denote the set of pure NE equilibria.

Definition 2.4 [13]. The price of anarchy $PoA(\mathcal{G})$ is:

$$PoA(\mathcal{G}) := \frac{\max_{\mathbf{A} \in \mathcal{A}} U(\mathbf{A})}{\min_{\mathbf{A} \in \mathcal{E}} U(\mathbf{A})} \geq 1 \qquad (13)$$

$PoS(\mathcal{G})$ denotes the price of stability defined as:

$$PoS(\mathcal{G}) := \frac{\max_{\mathbf{A} \in \mathcal{A}} U(\mathbf{A})}{\max_{\mathbf{A} \in \mathcal{E}} U(\mathbf{A})} \geq 1 \qquad (14)$$

Consider a special case of the spectrum sharing game $\mathcal{G}$ when there are $B = 2$ available channels and $\Theta_{1 \times 2} = \mathbf{0}$ [14]. It can then be shown that $PoA(\mathcal{G})$ can be $\Omega(n^2)$ worse than $PoS(\mathcal{G})$ [15]. For example consider G to be the complete bipartite graph $G = K_{\frac{n}{2}, \frac{n}{2}}$. Since G is bipartite $PoS(\mathcal{G}) = 1$. To show that PoA can be $\Omega(n^2)$ worse it is enough to notice that one Nash equilibrium can be realized when half of the users on the left side of this bipartite graph occupy a same channel and the other half occupy the other channel. This implies that there are both good and bad Nash equilibria in spectrum sharing games. Let us call an equilibrium $\mathbf{A}$ *good* if $PoA(\mathbf{A})$ is small and bad otherwise. In this situation a central authority can be employed to move the system behavior from a bad to a good equilibrium. For example in [15], a central authority advertises the optimal equilibria. It has been demonstrated that in a general graph G, if users employ the advertisement strategy in their best response learning mechanism, with probability more than half, the game converges to the optimal equilibrium in polynomial time. In this work we consider using a distributed learning method such as Log-linear mechanism [17-18] modified by a congestion advertisement for two reasons. First, because finding the optimal configuration (for a centralized approach) even for a simplified game model is a NP-hard problem [14]. Second, transient properties of the available spectrum opportunities in CR makes methods such as [15] not applicable for this problem. Transient properties could be of several types. Primary users evacuate and occupy their band continually. Autonomous secondary users join or leave the network. Moreover the network structure can also be unknown.

2.1. Logit-Response Dynamic

We proposed a synchronous logit-response in [3] for spectrum sharing games. In asynchronous logit-response [17], it is assumed that players are equipped with independent and identically distributed (i.i.d.) rate 1 "Poisson alarm clocks" and when their alarm goes off they revise their strategy according to a noisy

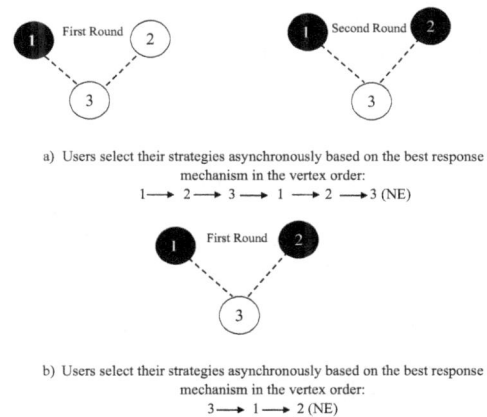

a) Users select their strategies asynchronously based on the best response mechanism in the vertex order:
$$1 \longrightarrow 2 \longrightarrow 3 \longrightarrow 1 \longrightarrow 2 \longrightarrow 3 \text{ (NE)}$$

b) Users select their strategies asynchronously based on the best response mechanism in the vertex order:
$$3 \longrightarrow 1 \longrightarrow 2 \text{ (NE)}$$

Figure 2. In graphical games how users learn is as much important as who learns first.

best response. Poisson distribution assumption implies that exactly one player at a time is allowed to update its strategy (asynchronous). Therefore the time between consecutive revision opportunities are independent and distributed with an exponential distribution of mean 1. When the user i alarm goes off it selects the strategy a_i with probability $p(a_i)$ according to a noisy best response mechanism given below::

$$p(a_i) = \frac{\exp(\beta U^i(\mathbf{A}(i : a_i)))}{\sum\limits_{a'_i = 1}^{B} \exp(\beta U^i(\mathbf{A}(i : a'_i)))} \qquad (15)$$

where β represents the inverse temperature parameter. $\beta \to \infty$ is equivalent to the best response mechanism. For $\beta \to 0$ the dynamics are totally random.

Proposition 2.1 [17] If the game has the potential function $\Phi(\mathbf{A})$ the logit-response mechanism leads to a reversible and irreducible Markov process on the state space $\mathcal{A}$ with the following stationary distribution:

$$\pi(\mathbf{A}) = \frac{\exp(\beta \Phi(\mathbf{A}))}{Z} \qquad (16)$$

where $Z = \sum\limits_{\mathbf{A} \in \mathcal{A}} \exp(\beta \Phi(\mathbf{A}))$ and as $\beta \to \infty$, $\pi(\mathbf{A})$ is concentrated on a Nash equilibrium. Moreover it turns out that the achieved equilibrium $\mathbf{A}^*$ for $\beta \to \infty$ using the logit-response is a good equilibrium. That is, the price of anarchy is small for the achieved equilibrium $\mathbf{A}^*$.

However the main problem with this mechanism is its slow convergence rate. Therefore, in the next section we introduce the fastest logit-response mechanism.

3. MECE Logit-Response Mechanism

In the standard logit-response mechanism, users find the opportunity to update their strategies with a fixed rate which is independent of the their positions in the graph and dynamics of the system. However using a simple example we can show that the order in which users update their strategy affects the speed of convergence to a NE.

Consider Fig.2a where user 1 as the first player selects strategy B and then user 2 selects W. There is no payoff dominant strategy for user 3 at this stage. Suppose user 3 randomly selects the strategy W. The same process is then repeated for another round in order to reach the Nash equilibrium. However if user 3 selects its strategy before others as in Fig.2b the game ends up in a NE after the first round. This implies that the standard logit-response should be modified with respect to parameters such as the position of the users in the graph and the system dynamics.

Consider the maximum entropy correlated equilibrium MECE logit learning as is shown in Algorithm 1. In the MECE learning mechanism the clock alarms of the users go off according a time varying probability distribution of Q^* as is described in (6). When a user gets the chance it updates according to (15).

As we showed in the previous example, the order of learning may cause negative effects in the convergence speed. This effect in fact can be explained via term Z in the stationary distribution $\pi(\mathbf{A})$ of (16) [10]. [1]

In the next lemma we show that MECE logit learning mechanism removes the term Z from the stationary distribution which can make the dynamics exponentially faster.

Lemma 3.1 The stationary distribution of $\mathcal{G}$, with the potential function Φ under the modified logit-response is $\pi(\mathbf{A}) \propto exp(\beta\Phi(\mathbf{A}))$.

Proof. A correlated equilibrium can be explained conceptually by introducing a mediator who has access to a randomization device. The "alarm clocks" described in the standard logit-response mechanism is one such randomization device. The i.i.d assumption on the alarm distribution in the standard logit-response implements the NE with $Q = \prod_{i=1}^{n} q_i$ with $q_i = \frac{1}{n}$. The corollary 4.1 in [20] shows that there exists a joint probability distribution Q^* which removes the term Z from stationary distribution $\pi(\mathbf{A})$. □

The modified learning approach can be thought of as a *Stackelberg* learning approach in which leaders and followers change roles along with the dynamics of the system.

The main importance of the modified mechanism is that it maps the dynamic of the game $\mathcal{G}$ to that of

Ising models (as described in the next theorem).This demonstrates how hard it is to achieve a good equilibrium for spectrum sharing game $\mathcal{G}$ in polynomial time.

For ease of analysis lets consider $B = 2$ and $\Theta = \mathbf{0}$. Assume if user i is transmitting in channel 1, $a_i = -1$ and if it is transmitting in channel 2, $a_i = 1$.

Theorem 3.1 The modified learning mechanism of the game coincides with the *Glauber dynamics* for the *anti-ferromagnetic Ising* models.

Proof. Consider the strategy set $a = \{-1, 1\}$ to be the set of spins. Glauber dynamics for Ising models are defined as algorithms that sample random assignments of spins to vertices V, according to a target distribution $\pi(\mathbf{A})$ using the following procedure: starting from any initial condition, repeatedly choose a site $i \in V$ *uniformly* at random, replace the spin of the site i with one sampled from $\pi(\mathbf{A})$ conditioned on the spins of $\mathcal{N}_i$. The Ising model is called a anti-ferromagnetic for $\pi(\mathbf{A}) \propto -\sum_i \sum_{j \in \mathcal{N}_i} a_i a_j, \forall i \in V$. Ising and ferromagnetic for $\pi(\mathbf{A}) \propto \sum_i \sum_{j \in \mathcal{N}_i} a_i a_j, \forall i \in V$ [21].

Using the simplified assumptions and Theorem 2.1 we can rewrite $\Phi(\mathbf{A}) = -\sum_i \sum_{j \in \mathcal{N}_i} a_i a_j, \forall i \in V$. Then the proof is complete by using Lemma 3.1. □

Theorem 3.1 establishes the connection between anti-ferromagnetic Ising models and dynamics of spectrum sharing games. Propositions 4.1 and 4.2. in the next section are the direct results of this theorem.

4. Convergence Rate for Specific Graphs

It is known that the lower bound for *mixing time* (defined in appendix) of the Glauber dynamics for the Ising models is $O(n \log n)$ if the $\beta < \beta_T$ where β_T is dependent on the graph G and the model of interaction (in our case an Anti-ferromagnetic Ising model) [21]. For $\beta > \beta_T$ the mixing time is exponential.

Proposition 4.1 For spectrum sharing game $\mathcal{G}$ there is a graph G where it is impossible to achieve a good equilibrium in polynomial steps.

Proof. As explained in section II-A in order to achieve a good equilibrium under logit-learning $\beta \to \infty$ is required. However Theorem 3.1 shows the game borrows the dynamic characteristic of the Ising model and for $\beta > \beta_T$ exponential steps in n is needed for convergence. □

The previous theorem raises the question of what kind of graphs show better behavior in terms of the convergence speed. This can be answered using the next Proposition.

---

[1] We have avoided the formal discussion on the effect of term Z in the convergence rate of the logit response to keep the context as consistent as possible. In order to understand the relation between the term Z and learning dynamics please refer to example 2 in [10].

Initialization: $t = 0$.

1. A central authority solves the alarm clock probability distribution $Q^* = \underset{Q \in CE}{\text{argmax }} H(Q)$.

2. When the alarm goes off for a player i, $t = t + 1$

   Update strategy a_i with probability $p(a_i) = \dfrac{\exp(\beta U^i(\mathbf{A}(i:a_i)))}{\sum\limits_{a'_i=1}^{B} \exp(\beta U^i(\mathbf{A}(i:a'_i)))}$.

3. If $\forall i, |U^i_{t+1} - U^i_t| \leq \epsilon_{\text{stop}}$, Stop.
   Otherwise go back to step 1.

Algorithm 1: MECE Logit Response Learning

Consider all the possible subdivisions of the graph in two disjointed subsets of vertices: S and its corresponding complement $V \backslash S$. The Isoperimetric function of graph $\mathcal{C}(G)$ is defined as the minimum value over all possible partitions of the number of edges connecting S with $V \backslash S$ divided by the number of sites in the smallest of the two subsets. That is, $\mathcal{C}(G) = \underset{S \subset V: |S| \leq \frac{n}{2}}{\min} \dfrac{\text{cut}(S, V \backslash S)}{|S|}$.

Proposition 4.2 The smaller the value of Isoperimetric function of a graph $\mathcal{C}_G$ the faster anti-ferromagnetic Ising model dynamics converges.

Proof. Let $\mathbf{W}$ represent the adjacency matrix of G defined as the $n \times n$ matrix $\mathbf{W} = (W_{ij})$ in which $W_{ij} = \begin{cases} 1 & j \in \mathcal{N}_i, \\ 0 & \text{Otherwise.} \end{cases}$. Moreover assume $\mathbf{I}$ to be the adjacency matrix of complete graph. Then let write $\Phi(\mathbf{A})$ as:

$$\Phi(\mathbf{A}) = -\sum_i \sum_j W_{ij} a_i a_j, \forall i \in V \quad (17)$$

Let G^c be the complementary graph of G with adjacency matrix $\mathbf{W}^c$, i.e $\forall i \neq j, W_{ij} + W^c_{ij} = 1$ then we can rewrite $\Phi(\mathbf{A})$ as:

$$\Phi(\mathbf{A}) = \sum_i \sum_j W^c_{ij} a_i a_j - \sum_i \sum_j I_{ij} a_i a_j, \forall i \in V \quad (18)$$

The first and second term of the right hand side of (18) can be recognized respectively as the (ferromagnetic) Ising model over the complementary graph G and anti-ferromagnetic Ising over the complete graph. The second term is independent of the phase transition analysis. This is because we can look at it as a symmetric congestion game which shows no phase transition behavior [10]. Therefore the phase transition behavior of the anti-ferromagnetic Ising can be shown by using Ising model on complementary graph G^c. The phase transition behavior of first term has been studied via Isoperimetric function of graphs in [11-27]. Using the

result of [11-27] over complementary graph G the proof is complete. $\qquad \square$

The previous proposition states that the more connected the graph of interaction G is, the faster the spectrum sharing game reaches an equilibrium.

4.1. Normalized Social welfare Optimization

In the previous discussions we addressed under what kinds of graphs, games converge faster to good equilibrium. This should not be confused with the problem of finding graph G whose purpose is to achieve the maximal social welfare $\sum_i U^i$.

Consider the optimization problem of (2). Our objective here is to find the graphs G which achieve high social welfare $U(\mathbf{A}) = \sum_{i \in V} U^i(\mathbf{A}(i, a_i)$. Let's modify the optimization problem with a new notion of normalized social welfare defined as:

$$\underset{\mathbf{A}: \mathbf{A} \in \{-1,1\}^n}{\max} \frac{U(\mathbf{A})}{2|E|} \quad (19)$$

This is because we are looking for the graphs that have a high capacity for achieving the optimal social welfare. Therefore it should be normalized with respect to the number of edges $|E|$.

Instead of solving the linear optimization problem of (19) let's consider solving the quadratic problem by rewriting $U(\mathbf{A}) = \sum_i U^i$ as $U(\mathbf{A}) = -\sum_i \sum_{j \in \mathcal{N}_i} (a_i - a_j)^2$.

We can rewrite in the graphical format

$$U(\mathbf{A}) = -\sum_i \sum_{j \in \mathcal{N}_i} (a_i - a_j)^2 = \quad (20)$$

$$-\mathbf{A}^T \mathcal{L}_G \mathbf{A}$$

where $\mathcal{L}_G$ is the Laplacian graph of G (refer to Appendix for definition). Moreover notice the term $2|E| = \mathbf{A}^T \mathbf{A}$. Then the normalized social welfare maximization problem can be written as minimization problem of

$$\underset{\mathbf{A}: \mathbf{A} \in \{-1,1\}^n}{\min} \frac{\mathbf{A}^T \mathcal{L}_G \mathbf{A}}{\mathbf{A}^T \mathbf{A}} \quad (21)$$

We assume G to be any arbitrary connected graph.

Lemma 4.1 Let λ be the smallest nonzero eigenvalue of $\mathcal{L}_G$, then

$$\lambda = \min_{\mathbf{A} \in \mathbb{R}^n} \frac{\mathbf{A}^T \mathcal{L}_G \mathbf{A}}{\mathbf{A}^T \mathbf{A}} \leq \min_{\mathbf{A}: \mathbf{A} \in \{-1,1\}^n} \frac{\mathbf{A}^T \mathcal{L}_G \mathbf{A}}{\mathbf{A}^T \mathbf{A}} \quad (22)$$

That is the optimal normalized social welfare is bounded below by λ.

Then to show the theorem now concentrate on the family of graphs that have high values of $C(G)$ and therefore fast convergence properties. A graph G is called ζ-expander for every subset S with $|S| \leq \frac{n}{2}$, $C(G) \geq \zeta$.

Lemma 4.2 (Cheeger inequality [29]) Let λ be the smallest nonzero Laplacian eigenvalue of graph ζ-expander graph G then

$$\zeta^2 \leq \lambda \leq 2\zeta \quad (23)$$

That is $C(G) = \zeta$ is upper bounded with λ.

(22) states that the graphs with lower value of λ provide larger normalized social welfare. However (23) shows larger λ are needed for faster convergence. *This shows a trade off between the normalized social welfare and the convergence speed.*

5. Spectrum Sharing Via Congestion Advertisement

We saw in the previous section that it is impossible to achieve a good equilibrium with polynomial mixing time on particular graphs. Therefore we are investigating a method to reach a good equilibrium in any graph.

Theorem 5.1 Assume that each user i evaluates it's utility as:

$$U_h^i = -\sum_{j \in \mathcal{N}_i} a_i a_j + a_i \epsilon \quad (24)$$

The game $\mathcal{G}$ under the MECE-logit response converges within polynomial steps to the good equilibrium if at each stage of the game $\epsilon = \text{sign}(\mathcal{E}_2 - \mathcal{E}_1)h$ where $h > 0$ is a small value, $\mathcal{E}_1 = \sum_i \sum_{j \in \mathcal{N}_i} \delta_{a_i,-1} \delta_{a_j,-1}$, $\mathcal{E}_2 = \sum_i \sum_{j \in \mathcal{N}_i} \delta_{a_i,1} \delta_{a_j,1}$ and δ is the Kronecker delta .

Proof. With the utility function of (24) the best response strategy of user i can be written as $\text{sign}(-\sum_{j \in \mathcal{N}_i} a_j + \epsilon)$.

By doing this the term ϵ makes one specific strategy for user i "risk dominant". The risk dominant strategy for user i is the one yielding the highest payoff and that is when user i have no information about its neighbors $\mathcal{N}_i$. If half of the $\mathcal{N}_i$ are active in a channel

strategy and the other half are active in the other, user i will select the risk dominant one. Theorem 3.1 shows that dynamic of $\mathcal{G}$ can be analysed by studying the dynamics of Glauber algorithms for anti ferromagnetic Ising model. [11] implies that polynomial mixing time for ferromagnetic Ising model over any graph G can be achieved by introducing a risk dominant term in the favor of one of the strategies. Let's show this risk dominant strategy by $a_d \in \{-1, 1\}$. Then in the stationary state with probability converging to 1 every user select strategy a_d. This is the key to polynomial mixing time.

However having a fixed risk dominant in the spectrum sharing games is disastrous as it provides wrong incentive for the users to occupy the same channel and therefore cause a large interference. Let's write the social welfare as $U^i(\mathbf{A}) = -(n_{11} + n_{22})$ where n_{11} and n_{22} represent the number of edges $e(i, j)$ in which $a_i = a_j = 1$ and $a_i = a_j = 2$ respectively. This format is clear since the other edges that don not experience interference add zero value to the social welfare. The logit-response leads to good equilibrium for $\beta \to \infty$ when the term $n_{11} + n_{22}$ will be minimized. The risk dominant strategy should be selected in favor of the appropriate strategy to minimize this term. In order to select the appropriate term, lets define the energy of a strategy in the system, by the value of concentration of users on that specific strategy, $\mathcal{E}_1 = \sum_i \sum_{j \in \mathcal{N}_i} \delta_{a_i,-1} \delta_{a_j,-1}$, $\mathcal{E}_2 = \sum_i \sum_{j \in \mathcal{N}_i} \delta_{a_i,1} \delta_{a_j,1}$. For example in a complete graph due to the symmetry of strategy configuration, the channel with more active users has higher energy. Energy also depends on the graphical characteristics of the users that have occupied a strategy. Then it would be enough to flip the sign of ϵ in favour of the strategy with less energy. Assume that the energy of strategy -1 (channel 2) is more than +1 (channel 1). Then by making the $\epsilon > 0$ we actually balance the energy by making the strategy +1 risk dominant (channel 1). This reduces the term $(n_{11} + n_{22})$ as it prevents it from any existing permanent risk dominant strategy in the network. Notice that if one strategy becomes risk dominant permanently without considering the energy difference, every user with probability converging to 1 chooses the same strategy and therefore increases the existing interfering links. $\square$

The previous theorem states that to have an exponentially faster spectrum sharing, users need to introduce an *arbitrary small* risk dominant term ϵ in their utility. Producing these suitable risk-dominant terms can be based upon an advertisement entity such as a base station. Finding the exact risk dominant terms is a difficult problem but it can be approximated by a congestion announcement. A simple scenario of spectrum sharing via congestion advertisement can be described as follows:

Base stations announce the number of active users in different spectrum bands during each time slot. When users experience the same signal interference for both available channels, they select the one with less congestion. Let's $\rho_{a_i} = \frac{1}{|V|} \sum_{j \in V} \delta_{a_i, a_j}$ be associated with action profile $\mathbf{A}$ and advertisement control parameter $h > 0$. Then user's i utility can be displayed in the following form:

$$U_h^i(\mathbf{A}(i, a_i)) = U^i(\mathbf{A}(i, a_i)) - h\rho_{a_i} \qquad (25)$$

This generalizes (24) by presenting the energy difference as the congestion term ρ. The congestion advertisement method has been applied to the spectrum sharing game with utility format of (25). The results are explained in the next section.

6. Simulation Results

Our simulations have been conducted for a more generalized version than the theoretical part. It shows that many of our theoretical results maintain their validity even with the change of some assumptions. These generalized assumptions are:

- Learning method is a Responsive Learning Automata (RLA) [24] with the learning parameter of $\alpha \in (0, 1)$. The description on this learning algorithm is given in Appendix.

- Users update their strategies simultaneously.

- In the simulations there are four channel strategies $B = 4$ and arbitrary Θ.

- Utility format is of the form $U^i(\mathbf{A}(i, a_i)) = \frac{\theta_{a_i}}{\rho_{a_i}(1 + |\mathcal{I}_i|)}$.

- G is a geometric graph.

6.1. Price of Anarchy

In a distributed cognitive system when there are multiple Nash equilibria it is important to understand the gap between worst and best possible equilibria of the network. As it is described in Section 2 a good way for understanding this is the comparison between the price of anarchy and stability. These metrics then imply how bad or good the network output can be from its best achievable one. When the price of anarchy is much worst than the price of stability it becomes crucial for the network designer to come up with mechanisms that could move the system's behaviour from a less efficient equilibrium to a more efficient one by promoting better user behavior. This paper

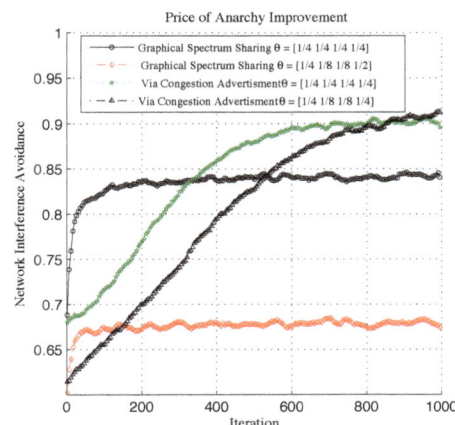

Figure 3. Price of Anarchy Improvement, $B = 4$ and the simulation has averaged over 100 realizations, best α for RLA learning for each case has selected to make the comparison independent of learning process, the network interference avoidance is: $\frac{1}{|V|} \sum_{i \in V} \frac{1}{1 + |\mathcal{I}_i|}$.

introduced congestion advertisement for the spectrum sharing games as one of these mechanisms.

Fig. 3 shows the improvement of spectrum sharing with congestion advertisement mechanism in terms of network interference avoidance. As well as demonstrates that by injecting congestion incentive into users utility, there is less probability that users will herd to a spectrum with higher quality. This in turn reduces the interference.

6.2. Convergence

Network volatility has been plotted in oder to show the convergence rate has improved. Volatility is defined as the variance of alternating between different strategies. Fig. 4 shows that with the increase in the communication range, the convergence rate improves since it increases the graph connectivity. This validates the result of Proposition 4.2. Fig. 5 also shows congestion advertisement method enhances the dynamics of $\mathcal{G}$.

6.3. Phase transition

We have run several simulations for different values of learning parameter α. We have also selected the best α which will bring the highest social welfare of $\frac{1}{1 + |\mathcal{I}_i|}$ for different values of h. These results indicate a transition point in Fig 6. You can see at the beginning, with the increase in h, the exploration rate α required to find the good equilibrium reduces. This improves the convergence rate. However by continually increasing h, the system arrives at a transition point h_T. This is where users herd on the channel with less congestion which

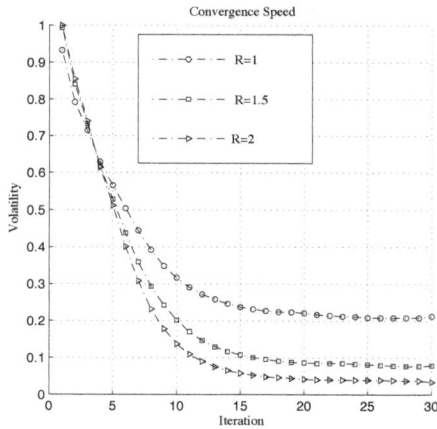

Figure 4. The simulation has run for the geometric graph. We assume a square area 100 units uniformly distributed random configuration of $n = 100$ secondary users. We consider different communication ranges R. For example $R = 1$ means that for a secondary node i all other nodes within an Euclidean distance of $R = 1$ are considered to be the neighbors $\mathcal{N}_i$. It shows well connected graphs have higher convergence speed.

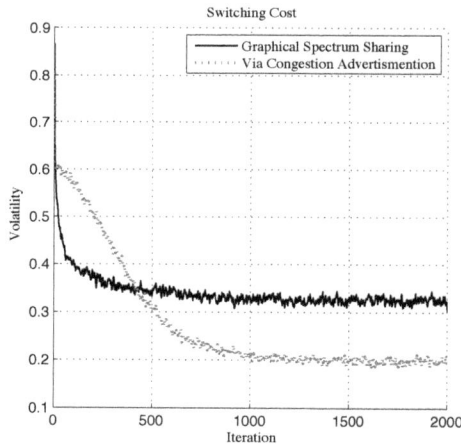

Figure 5. Network volatility improvement (as a convergence criteria) comparing with graphical spectrum sharing, $B = 4$, homogeneous spectrum quality $\theta = [1/4 \ 1/4 \ 1/4 \ 1/4]$, simulation has averaged over 15 realization, α in RLA learning for each case has selected so that it shows the best performance possible in the shown region, utility format is $U^i(\mathbf{A}(i, a_i)) = \frac{\theta_{a_i}}{\rho_{a_i}(1+|\mathcal{I}_i|)}$.

starts to increase the signal interference. Therefore in order to reduce the interference, a higher level of irrationality becomes necessary. This is similar to the behavior of the Glauber dynamics for $\beta > \beta_T$

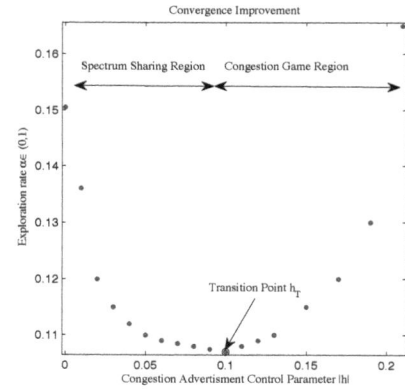

Figure 6. We have used 20 realizations for a observation window size of 500 for a network size of $n = 100$ and $B = 4$ numbers of channel. Also the simulation has run for a constant range of interaction on a random geometric graph to make the analysis independent of range of interaction.

7. Conclusion

We addressed the spectrum sharing games using graphical anti-coordination games. We showed how a modified logit learning mechanism establishes the connection between the simplified spectrum sharing games and anti-ferromagnetic Ising models. We studied the convergence rate of these spectrum sharing games and discovered the trade offs between achieving a good social welfare and convergence rate. This demonstrated spectrum sharing games under graphs which have lower isoperimetric values, tend to converge faster to equilibrium. We also showed how introducing an arbitrary small advertisement parameter into equations can enhance the convergence significantly.

Appendix A. Laplacian Graph

Definition A.1 The Laplacian of the graph G is defined as the $n \times n$ matrix $\mathcal{L}_G = (L_{ij})$ in which $L_{ij} =$
$$\begin{cases} d_{ij} & i = j, \\ -w_{ij} & i \neq j \end{cases}.$$

Appendix B. Markov

Definition A.2 The total variation distance between two probability distributions μ and v on $\mathcal{A}$ is defined by

$$\|\mu - v\|_{TV} = \frac{1}{2} \sum_{i \in \mathcal{A}} |\mu_i - v_i| \tag{B.1}$$

Definition A.3 The time it takes for a process to reach to its stationary distribution v is known as mixing times $\tau(\epsilon)$:

$$\tau(\epsilon) = \min_t \{\|\mu(t) - v\|_{TV} \leq \epsilon\} \tag{B.2}$$

Appendix C. Responsive Learning Automata

Let r_i^t represents the payoff obtained at time t by playing strategy i. The update rules for responsive learning automata are:

$$p_i^{t+1} = p_i^t + \alpha r_i^t \sum_{j \neq i} s_j^t p_j^t$$

$$\forall j \neq i, p_i^{t+1} = p_i^t - \alpha r_i^t s_j^t p_j^t$$

$$s_j^t = \min[1, \frac{p_j^t - \alpha/2}{\alpha p_j^t r_i^t}]$$

where α is the learning parameter and p_i^t is the probability of playing the strategy i at time t.

References

[1] D. Monderer and L. Shapley, "Potential Games", *Games and Economic Behavior 14*, pp.124-143, 1996.

[2] S. Haykin, "Cognitive radio: brain-empowered wireless communications", *IEEE J. Select. Areas Commun.*, vol. 3, no. 2, pp. 201-220, 2005.

[3] M. Azarafrooz, R. Chandramouli, "Distributed Learning in Secondary Spectrum Sharing Graphical Game". *IEEE GLOBECOM*, 2011.

[4] R.J. Aumann, "Subjectivity and correlation in randomized strategies", *Journal of Mathematical Economics*, 1974.

[5] M. Liu and Y. Wu, "Spectum sharing as congestion games", *Annual Allerton Conf. Commun., Control, and Compu.*, pp. 1146 - 1153, 2008.

[6] I. Malanchini,M. Cesana and N. Gatti, "On Spectrum Selection Games in Cognitive Radio Networks", *IEEE GLOBECOM*, pp. 1 - 7, 2009.

[7] P. Mertikopoulos and A. Moustakas, "Correlated Anarchy in Overlapping Wireless Networks", *IEEE J. Select. Areas Commun.*, vol. 26, no. 7, pp. 1160-1169, 2008.

[8] S. Sengupta, R. Chandramouli, S. Brahma and M. Chatterjee, "A game theoretic framework for distributed self-coexistence among IEEE 802.22 networks ", *IEEE GLOBECOM*, pp. 1-6, 2008.

[9] H. Li and Z. Han, "Competitive Spectrum Access in Cognitive Radio Networks: Graphical Game and Learning". *Wireless Communications and Networking Conference (WCNC)*, 2010.

[10] D. Shah and J. Shin, "Dynamics in congestion games", *SIGMETRICS*, pp-107-118, 2010.

[11] A. Montanari and A. Saberi, "The Spread of Innovations in Social Networks", *Proc. Natl. Acad. Sci.*, 2010.

[12] A. Ramezanpour, J. Realpe-Gomez and R. Zecchina "Statistical physics approach to graphical games: local and global interactions". *The European Physical Journal on Condensed Matter and Complex Systems*, 2011.

[13] C. Papadimitriou, *Algorithms, games, and the internet, In STOC*, 2001.

[14] G. Christodoulou, V. S. Mirrokni, A. Sidiropoulos, "Convergence and Approximation in Potential Games" . *In: Durand, B., Thomas, W. (eds.) STACS 2006. LNCS, vol. 3884, pp. 3490. Springer, Heidelberg* 2006.

[15] M. Balcan, A. Blum and Y. Mansour, "Improved equilibria via public service advertising", *SODA Proceedings of the twentieth Annual ACM-SIAM Symposium on Discrete Algorithms*, 2009.

[16] S. Kirkpatrick, Jr. Gelatt and M. Vecchi,"Optimization by Simulated Annealing" *Science, 220, 671-680.* 1983.

[17] L. Blume, "The statistical mechanics of strategic interaction.","*Games and economic behavior, Vol. 5, No. 3., pp. 387-424*, 1993.

[18] C. A. Ferrer and N. Netzer, "The logit-response dynamics", *TWI Research Paper Series 28, Thurgauer Wirtschafts institut, Universitat Konstanz*, 2008.

[19] J. R. Marden and J. S. Shamma, "Revisiting log-linear learning: asynchrony, completeness, and payoff based implementation", *Games and Economic Behavior*, 2011.

[20] L. E. Ortiz, R. E. Schapire and S. M. Kakade, "Maximum Entropy Correlated Equilibria", *In Eleventh International Conference on Artificial Intelligence and Statistics (AISTATS)*, 2007.

[21] T. P. Hayes and A. Sinclair, "A general lower bound for mixing of single site dynamics on graphs", *In FOCS: Proceedings of the 46th Annual IEEE Symposium on Foundations of Computer Science*, 2005.

[22] A. Montanari and A. Saberi,"Supplementary Information for: On the Spread of Innovations in Social Networks", 2010.

[23] K. Satyen and C. Seshadhri, "Combinatorial Approximation Algorithms for MaxCut using Random Walks", *In 2nd Symposium on Innovations in Computer Science (ICS , 2011.*

[24] E. J. Friedman and S. Shankar "Decentralized learning and the design of Internet". *Mimeo*, 1996.

[25] H. Young "Evolution of Conventions" *Econometrica, Vol. 61, No. 1., pp. 57-84*, 1993.

[26] A. Galeotti. "Network Games." *The Review of Economic Studies. 77(1), 218-244.* 2010.

[27] N. Berger, C. Kenyon, E. Mossel and Y. Peres,"Glauber dynamics on trees and hyperbolic graphs", *Probab. Theory Rel. 131:311-340.*

[28] S. Kirkpatrick, Jr. Gelatt and M. Vecchi, "Optimization by Simulated Annealing", *Science, 220, 671-680*, 1983.

[29] N. Alon, "Eigenvalues and expanders", *Combinatorica 6, no. 2, 83-96*, 1986.

6

Towards Effective Intra-flow Network Coding in Software Defined Wireless Mesh Networks

Donghai Zhu[1], Xinyu Yang[1], Peng Zhao[1,*], and Wei Yu[2]

[1]Department of Computer Science and Technology, Xi'an Jiaotong University, Xi'an, China
[2]Department of Computer and Information Sciences, Towson University, Towson, MD, USA

Abstract

Wireless Mesh Networks (WMNs) have potential to provide convenient broadband wireless Internet access to mobile users. With the support of Software-Defined Networking (SDN) paradigm that separates control plane and data plane, WMNs can be easily deployed and managed. In addition, by exploiting the broadcast nature of the wireless medium and the spatial diversity of multi-hop wireless networks, intra-flow network coding has shown a greater benefit in comparison with traditional routing paradigms in data transmission for WMNs. In this paper, we develop a novel OpenCoding protocol, which combines the SDN technique with intra-flow network coding for WMNs. Our developed protocol can simplify the deployment and management of the network and improve network performance. In OpenCoding, a controller that works on the control plane makes routing decisions for mesh routers and the hop-by-hop forwarding function is replaced by network coding functions in data plane. We analyze the overhead of OpenCoding. Through a simulation study, we show the effectiveness of the OpenCoding protocol in comparison with existing schemes. Our data shows that OpenCoding outperforms both traditional routing and intra-flow network coding schemes.

Keywords: wireless mesh network, software defined networking, intra-flow network coding

1. Introduction

Wireless mesh networks (WMNs) [1] have been considered a promising solution for providing convenient broadband wireless Internet access to mobile clients (e.g., smartphones and notebooks). In a WMN, stationary wireless mesh routers can establish the backbone for mobile clients, controlling and monitoring traffic flows in the network and managing routing paths according to algorithms and protocols implemented in routers. After being connected to one of mesh routers, a mobile client can communicate with other mobile clients in the same WMN or can get access to the Internet via the hop-by-hop forwarding way.

Nonetheless, because of the heterogeneous infrastructure of WMNs, deploying and managing WMNs is a challenging problem. The network administrator will have to setup and maintain each mesh router individually as network devices are vendor-specific. To address the issue of network deployment and management, Software-Defined Networking (SDN) [2] has been proposed. The essential idea of SDN is to exploit the ability of decoupling the data plane and the control plane in routers or switches, and keeping only data forwarding functions in network devices. By sending configuration commands from the control plane that consists of one or multiple controllers down to the data plane, the network administrator then has the ability to control and configure the data plane globally rather than configuring each device individually. As we can see, by using SDN, the network can be easily deployed and managed. In addition, with the view of the entire network, the network performance can be improved by deploying global optimization strategies through a powerful control plane [3]. For example, existing research efforts [4–6] have shown the great benefit of integrating SDN with WMNs.

---

*Please ensure that you use the most up to date class file, available from EAI at http://doc.eai.eu/publications/transactions/latex/
*Corresponding author. Email: p.zhao@mail.xjtu.edu.cn

Meanwhile, another challenge in establishing WMNs is to achieve a high network performance despite the unreliable and lossy wireless communication channels. The inherent broadcast nature of wireless communication channels can be exploited to improve network performance. Network coding [7] is one mechanism, which makes use of the broadcast nature in wireless networks. For example, by integrating intra-flow random linear network coding that enables relay nodes to combine information content in packets before forwarding and performing opportunistic routing [8], the MORE protocol [9] has demonstrated that network coding can improve network throughput significantly and can be extended to support applications, including data distribution in peer-to-peer networks [10], etc.

As we can see, the SDN technology and network coding scheme are two different techniques to address different challenges in WMNs. Again, the key idea of SDN is to keep routers or switches simple, while moving the control function into the control server deployed in the network. The intra-flow network coding scheme can achieve much better performance than hop-by-hop forwarding schemes because of leveraging the nature of the wireless broadcast medium. Therefore, one research problem is raised: *How can we integrate the SDN and intra-flow network coding schemes to make that the integrated scheme performs better than the individual ones?*

To answer this question, in this paper we make the following contributions.

- We designed the OpenCoding protocol to address the challenge of easily deploying WMNs and further improving the performance of WMNs. The key idea of OpenCoding is to use SDN in the network where the intra-flow network coding is deployed. That is, a controller that works on the control plane makes routing decisions for mesh routers and the hop-by-hop forwarding function is replaced by the network coding function in data plane. By decoupling control plane and data plane, mesh routers can be simple, leading to a much easier ability to manage a network. More importantly, through the aid of viewing the entire network, a global optimization strategy can also be easily applied. Based on this idea, we developed a novel scheduling scheme for intra-flow network coding, which can achieve a high network performance gain, as well as the fairness of multiple network flows.

- We conducted extensive simulations to demonstrate the effectiveness of OpenCoding in comparison with existing schemes. Our evaluation data shows that our proposed OpenCoding outperforms OLSR [11], OpenFlow [12], and MORE [9] protocols. On average, OpenCoding achieves 68.32%, 12.46% and 9.88% gain over OLSR, OpenFlow and MORE in terms of data transmission throughput, respectively. Meanwhile, the throughput of OpenCoding drops little when the number of concurrent flows increases. The results also show that OpenCoding outperforms OLSR, OpenFlow and MORE protocols in terms of end-to-end delay. To our best knowledge, there has been no previous work, which integrates the SDN technique into the intra-flow network coding in WMNs and evaluates performance gain.

An earlier version of this work was published in [13]. Based on the conference version, we have made substantial extensions and revisions. The rest of the paper is organized as follows. In Section 2, we give the background and review existing research efforts that are closely relevant to our research. In Section 3, we present our proposed OpenCoding scheme in detail, including the architecture, the detailed design, and how to achieve fairness optimization. In Section 4, we analyze the overhead of OpenCoding. In Section 5, we show the performance evaluation of OpenCoding in comparison with several representative schemes. In Section 6, we discuss some open issues related to OpenCoding. Finally, we conclude the paper in Section 7.

2. Background and Related Work

In this section, we give the background and literature review of SDN and OpenFlow, as well as intra-flow network coding.

2.1. Wireless Mesh Networks

Wireless Mesh Networks (WMNs) are developed for a new broadband Internet access technology with growing interests in both research and industry [1]. Because of highly self-organized and self-configured features of WMNs, they are used to extend last-mile access to the Internet and have become a promising technology for providing easy wireless access for mobile users.

The infrastructure of WMNs is shown in Figure 1. As we can see from the figure, WMNs consist of two types of nodes: mesh routers and mesh clients. Mesh routers are almost stationary and establish the mesh backbone for mesh clients, thus there are no power limits on mesh routers. Mesh routers can equip multiple wireless interfaces and serve as the Internet gateway or bridge and integrate heterogeneous networks, including both wired and wireless networks. To this end, the deployment and management of such a heterogeneous infrastructure will be a big challenge. Meanwhile, because of the unreliable and lossy wireless communication channels of WMNs, another challenge in establishing WMNs is to achieve a high enough network throughput performance.

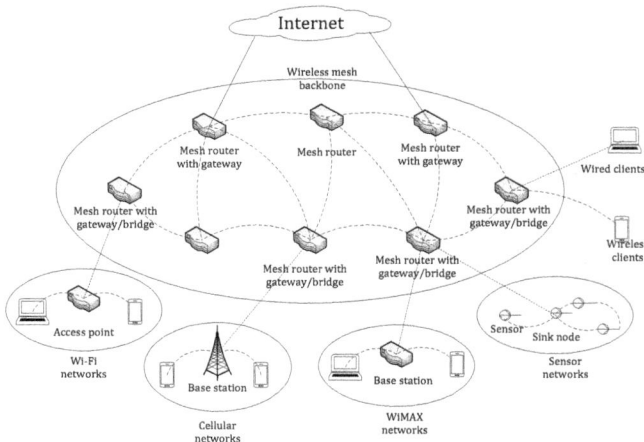

Figure 1. Infrastructure/backbone WMNs

With the rapid development of radio techniques, a number of approaches have been proposed in order to increase the capacity and flexibility of WMNs. Typical examples include cognitive radios [14], cooperative communications [15], Multi-Input Multi-Output (MIMO) systems [16] and Multi-radio Multi-channel (MRMC) systems [17]. On the other hand, by exploiting the broadcast nature of wireless medium, innovative transmission schemes, including Opportunistic Routing (OR) [8] and network coding [7], have also been developed to increase the performance of WMNs.

2.2. SDN and OpenFlow

Generally speaking, SDN is an emerging network architecture, which can decouple control plane and data plane, with goals of achieving a higher transmission speed, greater scalability, and greater flexibility. The key idea of SDN is to allow software developers and network administrators to allocate and manage the network resources in an easy way via an open interface. Its concept was originally developed by Nicira Networks based on their earlier development at UCB, Stanford, CMU, Princeton [2]. Figure 2 illustrates the standard architecture of SDN, which is defined by Open Networking Foundation (ONF) [18]. As we can see from the figure, the control plane can send commands down to the data planes of the hardware (e.g., routers or switches) [19]. The hardware in the data plane can only simply deliver data among them by checking the flow tables, which are distributed by the controller in the control panel. In this way, the hardware devices can be greatly simplified.

One of well-known SDN protocol standards is the OpenFlow [12], which was originally proposed for wired networks in order to make Internet switches intelligent and programmable through a standardized interface. OpenFlow allows network devices from many different companies to utilize the abstraction of the control planes and data planes [20–22], thus OpenFlow receives a considerable amount of industry attention. Besides OpenFlow, there are other SDN implementations and standards, such as IEEE P1520 standards [23], ForCES [24] and SoftRouter [25]. Also, there are a number of earlier work related to programming networks and active networks [26, 27].

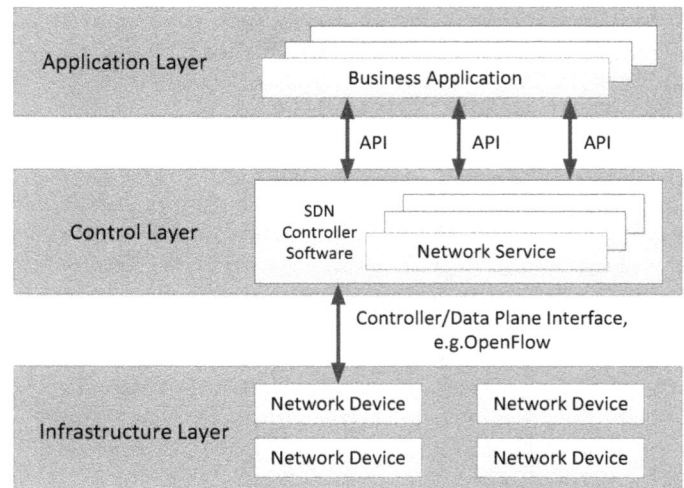

Figure 2. SDN architecture defined by ONF[18]

There have also been several research efforts on applying SDN/OpenFlow paradigm in wireless mobile networks. For example, Figure 3 shows an generic software defined wireless network-based architecture of a mobile network operator [28]. In Figure 3, we can see that SDN technology can be applied to various types of wireless networks. In this paper, we focus on applying SDN technology to wireless mesh networks [1]. In [4], an architecture that allows the flexible and efficient use of OpenFlow in WMNs was demonstrated. As the first attempt to apply OpenFlow in WMNs, the evaluation of this proposed architecture on a real test-bed called KAUMesh showed that the SDN architecture could achieve better performance than traditional schemes in terms of transmission throughout, control traffic overhead, and rule activation time. Riggio et al. in [29] proposed a set of high-level programming abstractions for WiFi networks, which enables new features and services to be implemented as software modules.

In order to obtain the link information and controlling link behavior, the OpenFlow protocol extensions were proposed to allow controllers to remotely control and manage wireless links through a set of Media-Independent Management (MIM) mechanisms [30]. In addition, one OpenFlow-based

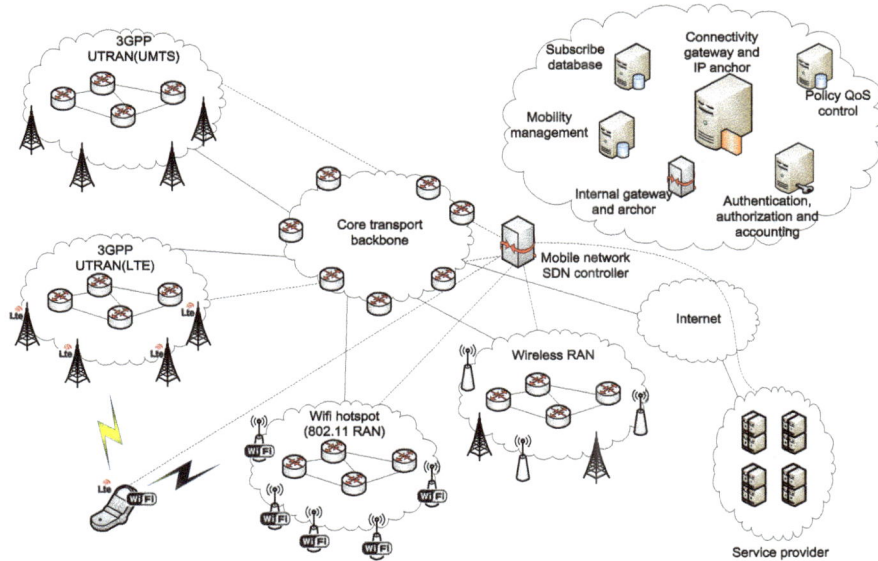

Figure 3. SDN-based wireless mobile network architecture

communication protocol was proposed to provide the access to the data plane of network routers [31], which can be used to balance traffic loads in the network. These existing research efforts confirm that OpenFlow can be a useful technique to improve the performance of WMNs.

There have been other research efforts that show the benefits of using the SDN/OpenFlow paradigm in wireless networks. The examples include the minimum-energy reprogramming in software-defined sensor networks [32], capacity sharing in hybrid networked environments [33], etc.

2.3. Intra-flow Network Coding

Network coding, which was initially introduced for improving the performance of wired networks in [7], has been considered as a promising communication paradigm to improve network performance in terms of throughput and energy efficiency. In the seminal paper [7], Ahlswede et al. proved that in a directed graph with lossless links, if operations are allowed at the intermediate nodes, the theoretical maximum multicast rate can be achieved and it is equal to the min-cut from the source to each receiver. Later, Li et al. showed that the linear coding operation is sufficient to achieve that maximum multicast rate [34]. Also, Ho et al. further proposed a random linear coding and proved that if the linear coefficients are random selected over a large enough finite field at each intermediate nodes, the destination could decode the packets with a high probability [35]. What's more, Chou et al. developed the random linear network coding infrastructure, which makes network coding

practical [36]. As the network coding shows its significant advantage over the traditional routing in terms of bandwidth efficiency, there has been a growing interest to applying network coding to improve the performance of wireless networks. For example, MORE [9] is one protocol implemented on a real-world test-bed and achieves a significant performance gain than the traditional routing in WMNs by integrating network coding with opportunistic routing [8].

It is worth noting that MORE protocol does not take multiple flows, fairness, and scheduling into account so that its performance drops as the number of flows increases. To tackle this issue, a cross-layer optimization framework [37] was proposed to optimize the rate of packet transmissions between source and destination pairs. In this framework, a distributed heuristic algorithm was developed to solve the problem. In addition, the problem for the network coding-based opportunistic multicast routing was studied and a duality-based distributed algorithm was proposed to solve the problem [38]. In this proposed scheme, each node only needs to maintain the local information.

Although these aforementioned distributed solutions for optimizing network performance is indeed helpful and can improve performance in terms of throughput, energy efficiency and fairness, they are very costly to be deployed in real-world applications. Fortunately, with the support of the SDN technique, the controller in software-defined networks could acquire the network states dynamically by receiving regular reports from deployed network devices. With the collected information, the controller is able to acquire a global view of the network and can deliver this important information to network devices. Therefore, the global optimization can be achieved by exploiting the SDN technique.

3. OpenCoding

In this section, we present our approach. Particularly, we first brief the architecture, and then introduce the detailed design of our approach as well as how to conduct fairness optimization.

3.1. Architecture

OpenCoding is a novel transmission protocol based on the concept of SDN for WMNs. In our study, we consider a WMN, which consists of one control server, one or more mesh gateways, and multiple mesh routers that are connected to some mobile clients (e.g., smartphones, notebooks, etc.). In such a software-defined network, the complexity of managing the network is low and the network administrator can manage the entire network. In addition, the network can be controlled in real-time based on the intelligence of network controllers.

Similar to the OpenFlow protocol, in OpenCoding-enabled wireless mesh routers, packet forwarding intelligence are moved to the control server, which is referred to as the OpenCoding controller. As there is no control functionality residing at the OpenCoding-enabled routers, the data forwarding functionality can be kept simple by only containing packet coding, recoding, and decoding functions, which are basic primitives defined in the random linear network coding. Particularly, when a new data flow arrives at an OpenCoding-enabled router, the router will perform a *PACKET_IN* function, to ask for the routing for this flow. Then, the controller makes the best decision and computes the path for this flow, and distributes flow entries, which contain actions to perform on individual routers on the selected routing. After receiving these flow entries, those designated routers will carry out corresponding actions (e.g., the packet coding function in source router, recoding function in forwarding router, and decoding function in destination router).

The OpenCoding controller, which runs a network-wide network operating system (e.g., NOX [39], etc.), is the network element that is responsible for managing routers and making decision to meet requirements. All routers are connected to the controller through the secure channel via one hop or multiple hops. This secure channel can be provided by using the SSIDs described in [4] or the use of multi-radio multi-channel (MRMC) technique [17]. Notice that it is easier and cheaper to use the different SSIDs to separate the controller-to-router communication from router-to-router communication, while the communication quality may be worse due to the interference in the wireless channel. Also notice that using the multi-radio

multi-channel technique can address the interference problem, but it requires extra hardware and also needs to deal with the channel assignment issue.

3.2. The Design of OpenCoding

In the following, we first introduce the basic idea of our proposed approach and then present the detailed design of our approach.

Basic Idea. In a conventional WMN, once a packet is received and needs to be forwarded by a mesh router, it first stores this packet into the packet buffer pool and then uses a set of rules to find the next-hop of the packet according to routing protocols. When there is a transmission opportunity, the router sends the packet to the corresponding next-hop node. In such a hop-by-hop way, packets can be reliably delivered to destinations.

As stated, in a conventional mesh router, the function that handles the forwarding of packets (i.e., data plane) and the control of forwarding rules (i.e., control plane) are closely coupled. This makes it is hard for routers to extend with new functions. SDN/OpenFlow addresses this issue by decoupling the control and data plane. To be specific, the control plane is implemented partially in a server rather that resides on the router only. This leads to an easier network management. More importantly, in the view of entire network, the global optimization can be achieved, leading to an efficient use of network resources. By leveraging SDN, we can develop an effective scheme for carrying out intra-flow network coding to achieve a higher network performance gain than the network, in which the SDN technique is not used, in terms of both throughput and fairness.

Network State Measurement and Management. In order obtain a global view of the entire network, the controller in OpenCoding uses the Link Layer Discovery Protocol (LLDP), which is also adopted in Open-Flow protocol. To do so, the controller first sends a *PACKET_OUT* message to all mesh routers. As soon as the mesh router receives the *PACKET_OUT* message, it will send LLDP messages to all of its neighbor routers. Because there is no corresponding flow tables to process these LLDP messages, the neighbor routers will send *PACKET_IN* messages to the controller to report the link quality measured by itself. In this way, after receiving the *PACKET_IN* messages, the controller can acquire the topology and current network state of the global network.

In addition, to maintain the network topology and state, the controller will periodically send *PACKET_OUT* messages and the LLDP packets to mesh routers. The controller will also obtain the network information from the feedback *PACKET_IN* messages generated from mesh routers, which can be used to update network topology dramatically.

Controller. In the following, we describe the detailed protocol in the perspective of the components of Open-Coding, including controller, source router, forwarding router, and destination router. In a WMN, mobile mesh clients can access the network through mesh routers. If a mobile client wants to access the Internet or needs to communicate with other clients in the same WMN, the client shall first connects to one of mesh routers, and that mesh router will be responsible for forwarding all traffic flows. Denote S as the source mesh router that connects to the mobile client that has data to transmit and denote D as the destination mesh router that is the mesh router with gateway or connecting to the destination mobile client. When a source mobile client initiates a traffic flow, S will send *PACKET_IN* message to the controller to request for a new traffic flow.

After receiving a *PACKET_IN* message from S, the controller computes the best routing with the global view. As we know, in a traditional hop-by-hop routing, after the routing path is established by some routing protocols, a node keeps sending a packet until the next hop receives it or the number of transmissions exceeds a particular threshold. Nonetheless, in an intra-flow network coding that is similar with an opportunistic routing, there is no particular next hop. The nodes closer to the destination than the current transmitter can participate in forwarding the packet. Therefore, in an intra-flow network coding, the routing of each node consists of one or multiple next-hop forwarding routers. What is more, the threshold of transmission count for the current batch shall also be assigned to ensure the efficiency of bandwidth use.

The detail of how to derive threshold of the number of transmissions can be referred to Equations (1) and (2) in [9]. Here, we use a simple topology to demonstrate the computation of the transmission count. Consider the scenario shown in Figure 4, where the value under the arrow is the packet delivery probability of corresponding link. For every packet from S to D, in expectation, only 1.67 transmissions rather than 2 transmissions are needed when the network coding is used, because R only needs to forward 67% of linear combination of received packets. That is enough for D to recover all packets as D receives the other 33% through link "$S \rightarrow D$".

Further, notice that the number of transmissions computed in [9] is only based on the average loss rate of links and can be optimized. After the routing is computed, the controller installs these routing rules to all potential transmitting routers through *MODIFY_STATE* messages. The rules contain: (i) the flow ID that is used to identify each traffic flow, (ii) the batch number in each flow, (iii) source IP, (iv) destination IP, (v) forwarding router list for each router, and (vi) forwarding transmission count. In the simple scenario shown in Figure 4, C will tell S and

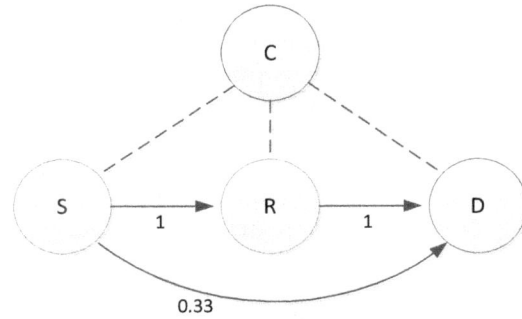

Figure 4. A simple topology

R to transmit once and 0.67 times for each encoded packet to R and D, respectively. In addition, as soon as the *BATCH_FINISHED* message that is sent from the destination router is received by the controller, the controller will immediately send *BATCH_NEXT* message to all transmitting nodes. The source router will stop forwarding packets in the current batch and activate the transmission of the next batch, while forwarding routers will also stop forwarding packets in the current batch as well and wait for the next batch.

Some key message used in the OpenCoding is summarized in Table 1.

Source. When a source mobile client initiates a traffic flow, S performs the *PACKET_IN* function, to request the controller for a new traffic. After receiving the request, the controller makes a decision whether to accept the request or not. If so, the controller computes the best routing for this traffic and installs the forwarding rules on the routers, which will participate in this traffic transmission. As soon as the rule is installed on the S, S will inform source mobile clients to start sending packets to S.

On receiving the stream of packets, S stores the packets first and divides them into batches. Each batch consists of m packets

$$\mathbf{x} = \underline{\mathbf{x}}_1, \underline{\mathbf{x}}_2, \ldots, \underline{\mathbf{x}}_m,$$

where $\underline{\mathbf{x}}_i$ is represented as a vector of

$$\underline{\mathbf{x}}_i = (\underline{x}_{i,1}, \underline{x}_{i,2}, \ldots, \underline{x}_{i,n}).$$

Also, each $\underline{x}_{i,j}$ is a symbol of finite field $\mathbb{F}_q^n$. Here, m is the batch size, q is a prime of a proper size, and all the arithmetic operations are done over $\mathbb{F}_q$. In addition, S generates an augmented packet $\mathbf{x}_i$ by prefixing $\underline{\mathbf{x}}_i$ with the ith unit vector of dimension m:

$$\mathbf{x}_i = (\overbrace{0, \ldots, 0, 1, 0, \ldots, 0}^{m}, \underline{x}_{i,1}, \underline{x}_{i,2}, \ldots, \underline{x}_{i,n})$$
$$\underbrace{\qquad}_{i-1}$$

Table 1. Message type in the OpenCoding

Types of Message:	Description
MODIFY_STATE:	The controller send MODIFY_STATE messages to routers to add, delete and modify routing rules on routers.
PACKET_OUT:	The controller sends PACKET_OUT message to routers that tell routers to send corresponding messages out.
PACKET_IN:	The source mesh router sends the PACKET_IN message to the controller to inform that there is a new packet received in routers and ask for corresponding instructions.
BATCH_FINISHED:	The destination mesh router sends the BATCH_FINISHED message to inform the controller that the transmission of this batch should be completed.
BATCH_NEXT:	The controller sends the NEXT_BATCH to inform the source and forwarding router to stop the transmission of this batch, and another batch of packets should be activated
FLOW_REMOVED:	Routers sends Flow_REMOVED messages to the controller to inform that one of the flow tables is removed

For a random linear network coding, S randomly picks the coefficients from the finite field $\mathbb{F}_q$ and creates the combinations of packet $\mathbf{x}_i$. This means that for packets

$$\mathbf{x} = \mathbf{x}_1, \mathbf{x}_2, \ldots, \mathbf{x}_m,$$

and coefficients

$$\alpha = \alpha_1, \alpha_2, \ldots, \alpha_m,$$

an output packet is represented as

$$\mathbf{y} = \sum_{i=1}^{m} \alpha_i \mathbf{x}_i = (\sum_{i=1}^{m} \alpha_i x_{i,1}, \ldots, \sum_{i=1}^{m} \alpha_i x_{i,m+n})$$

The first m symbol of $\mathbf{y}$ is called the global coding coefficient. Subsequently, S sends encoded packets to the forwarding routers defined in installed rules and waits for the *BATCH_NEXT* message from the controller.

Forwarders. After rules are installed, mesh routers work in promiscuous mode, listening all transmissions. When the router receives a packet, it first checks the packet to confirm whether the packet matches the rules that are installed. More specifically, it checks whether the flow ID and the batch ID extracted from the packet's header matches any entry in the flow table, and whether it is in the forwarder list of packets. If so, it further checks whether the packet is an innovate packet by *Guassian eliminations*. An innovative packet is linearly independent from the packets that the node has previously received in this batch. A packet that does not match any flow table or is not innovative will be dropped. Otherwise, the packet will be stored for encoding.

After the forwarding router accepts an innovate packet, it then randomly selects the coefficients and linearly combines the all received packets in the same batch in a similar way with the source. This can be

formulated as follows:

$$\mathbf{z} = \sum_{i=1}^{m} \gamma_i \mathbf{y}_i.$$

Here, $\mathbf{z}$ is the new packet created by the forwarding router, γ_i is the random coefficient for the ith packet and all the $\mathbf{y}$ are packets received in the same batch, including the newly received one. Notice that in a fault-free execution of the random linear network coding, the packets transmitted in the network are all linear combinations of the original augmented messages: $x = \mathbf{x}_1, \mathbf{x}_2, \ldots, \mathbf{x}_m$.

Subsequently, a packet header, which contains the source and destination IP, flow ID, batch ID and forwarding list, will be appended to the newly created packet. When a transmission opportunity arises, the packet will be send to its next-hop forwarding routers.

Destination. The destination router D receives packets in a similar way as the forwarders. After receiving m linearly independent packets $\mathbf{z}_j, j = 1 \ldots m$, D can recover the original packets $\mathbf{x}_i, i = 1, \ldots, m$ using the global coding matrix $\mathbf{G}$, where i^{th} row is the corresponding global coding vector through *Guassian eliminations* with a high probability:

$$[\mathbf{I}, \mathbf{X}] = \mathbf{G}^{-1} \cdot \mathbf{Z}. \tag{1}$$

Here, $\mathbf{I}$ is the identity matrix, $\mathbf{X}$ is the matrix, in which i^{th} row is the corresponding $\mathbf{x}_i$, $\mathbf{G}^{-1}$ is the inverse of the matrix $\mathbf{G}$, and $\mathbf{Z}$ is the matrix, in which i^{th} row is the corresponding $\mathbf{z}_i$.

As soon as the current batch can be decoded, D performs the *BATCH_FINISHED* function, and sends a message to inform the controller that the transmission of this batch should be complete and the next batch transmission can be activated. The controller then performs the *NEXT_BATCH* function and sends a message to the source and all the forwarders.

In addition, D forwards the decoded packets to destination mobile clients or to the Internet.

The control flow of OpenCoding is summarized in Figure 5. On the sender's side, before sending a packet, the router will add the header to the packet, and check whether the transmission count for this batch has reached the pre-determined value. If so, the router will turn to transmit the next packet; otherwise, this packet will be sent to its next-hop forwarding routers. On the receiver's side, the router checks the header of packet to determine whether to accept or not. If so, the router will store the packet for recoding (i.e., if it is forwarding router) or decoding (i.e., if it is destination router).

3.3. Fairness Optimization for OpenCoding

Benefited from the centralized control logic and the global network view, OpenCoding can address other issues, which are not easily solved in a distributed manner. We now introduce a scheme to ensure the fairness of OpenCoding.

Considering a scenario in which there are two traffic flows f_1 and f_2 showed in Figure 6, which are transmitted with intra-flow random linear network coding. If the node R is the common forwarding router of both flows, it will participate in forwarding packets in these two flows. In the following, we present our idea using an example. We assume that R just begins to transmit a new batch from f_1 and p_1 is the first packet in this batch, while another batch from f_2 is nearly at the end and p_2 is the last packet in that batch. In Figure 6, packets belong to different flows are marked with different colors. In the original MORE protocol, these two packets will not be differentiated and they will be transmitted in First-In-First-Out (FIFO) manner. Nonetheless, it is more important to transmit p_2 than p_1 because as soon as the destination router receives p_2, it can decode all packets in this batch while the reception of p_1 will not bring any benefit at that moment. To address this issue, we develop a novel scheme to ensure the fairness of OpenCoding by exploiting the centralized control of SDN.

When there are multiple flows in the network, the controller periodically requests the destination routers to report their states of the current batch. The reports from the destination routers contain the current batch ID and the progress of the corresponding batch. In our proposed scheme, destination routers only reports whether they receives $m/2$ innovative packets. Meanwhile, forwarding routers maintain two virtual queues, one for low priority packets (e.g., the packet is in the first half of the batch), and the other for high priority packets (e.g., the packet is in the second half of the batch). When a new packet is added to the output queue, an entry is added to the appropriate virtual queue based on the progress of packet's batch.

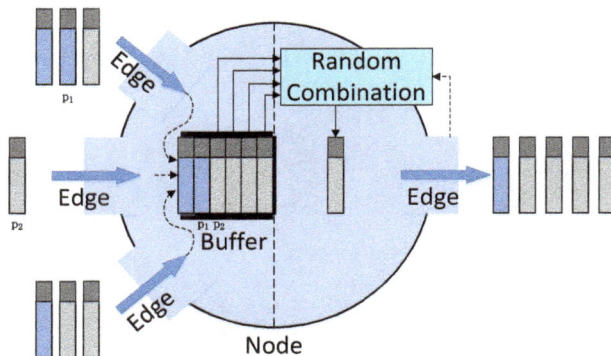

Figure 6. An Example of fairness optimization

In addition, when the last packet in the high priority queue is sent, the low priority queue will be the high priority queue and the high priority queue will be the low priority queue. The controller periodically updates rules, which are installed in the forwarding routers based on the report from destination routers. The rules inform forwarding routers the progress of the current transmitting batch. To be specific, when a new packet is received, if more than half of packets in that batch are received by the destination router, the new encoded packet will be added to the high priority queue waiting to be forwarded, and vice versa. When there is a transmission opportunity, it will first look up the high priority queue. If it is not empty, the head packet in the high priority queue will be transmitted.

Notice that the progress of corresponding batch can be further subdivided and more corresponding virtual queues can be used to further improve the fairness. Nonetheless, from the simulation results described in the performance evaluation section, dividing the batch into two halves and two virtual queues are sufficient to ensure the fairness. Again, it is worth noting that developing such a special scheduler is benefit from a global view enabled by SDN, while in the traditional system with network coding, such a scheduler is hard to achieve.

4. Overhead Analysis

In this section, we analyze the routing overhead of our OpenCoding. As we described in Section 3, there are three types of control messages needed for the controller to obtain a global view of the entire network: (i) *PACKET_OUT* message, (ii) LLDP message and (iii) *PACKET_IN* message. In addition, the flow table request and distribution incur extra routing overhead. Therefore, the routing overhead of OpenCoding consists of following two components and can be represented by

$$O_{OpenCoding} = O_{TO} + O_{FT}, \qquad (2)$$

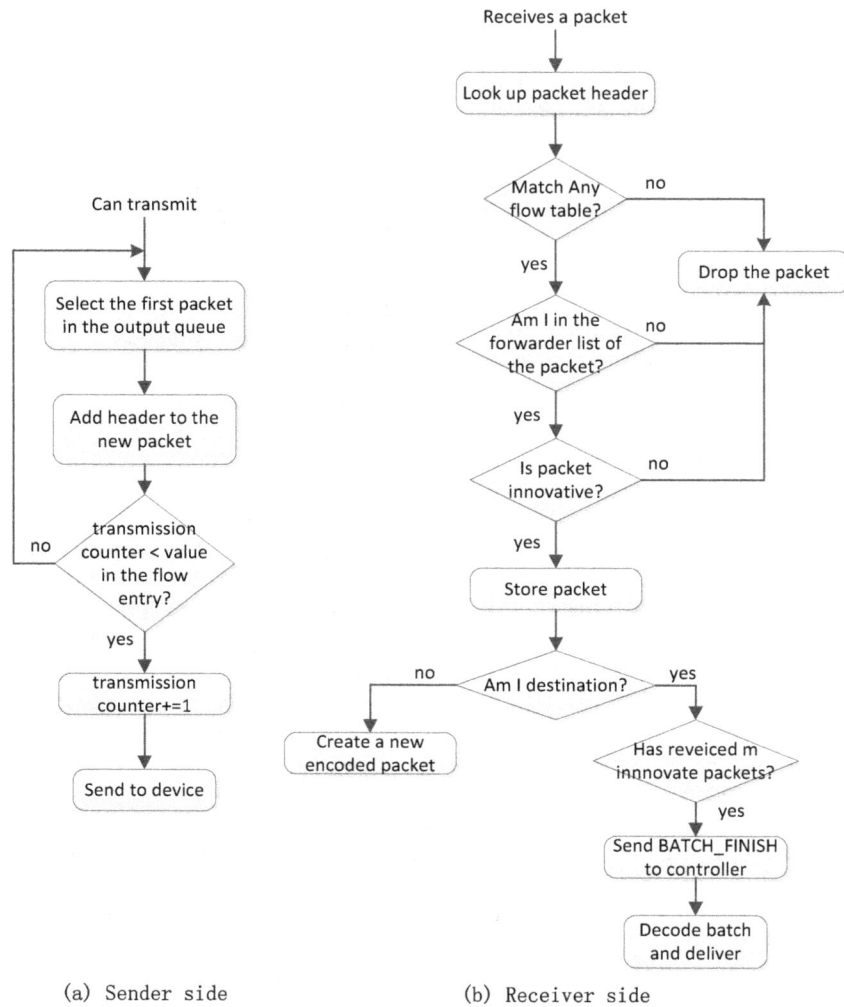

(a) Sender side   (b) Receiver side

Figure 5. Flow chart of the OpenCoding

where O_{TO} is the overhead caused by the routing discovery and maintenance, and O_{FT} is the overhead caused by the flow table request and distribution. In the following, we derive the closed formulae for the two components. All main notations used in this section are listed in Table 2.

4.1. Overhead for Routing Discovery and Maintenance

The total overhead of routing discovery and maintenance O_{TO} is defined as the total amount of information needed by control messages. Assume that a wireless mesh network consists of N wireless nodes that are uniformly distributed in a square field with a size of $a \times b$. To simply the analysis, assume the transmission range of the routers is equal and is set to r. Then, the ratio of the node's coverage area to the whole area p is provided by $\frac{\pi r^2}{a \times b}$.

Table 2. Notation

N:	Number of nodes in the network
a, b:	Size of the square field of network
r:	Transmission range of a node
p:	Ratio of node's coverage area to the whole network area
τ:	Time step that is also referred to as the interval for network topology update
η:	Minimum average number of transmission to broadcast a message to all the other nodes
h	Random variable of number of hops between two arbitrary nodes
O_{TO}:	Overhead caused by the routing discovery and maintenance
O_{FT}:	Overhead caused by the flow table request and distribution

Denote η as the minimum number of transmissions to broadcast a message to every node in the network. According to [40], if the average number of neighbors for any node is at least 2, η can be determined as

$$\eta = \left\lceil \frac{N-2}{pN-2} \right\rceil. \tag{3}$$

Denote I_{PO} as the amount of information needed for the controller to broadcast the *PACKET_OUT* message within a time step τ. The controller periodically broadcast the *PACKET_OUT* message of length $\log N$ to all the mesh routers in the network. Here, $\log N$ is the amount of information required to identify the controller. Therefore, the amount of information that controller broadcasts the *PACKET_OUT* message can be derived by

$$I_{PO} \geq \eta \log N. \tag{4}$$

Denote I_L as the amount of information needed for mesh routers to advertise LLDP messages to their neighbor routers within time step τ. As described, when the mesh router receives the *PACKET_OUT* message from the controller, it will advertise LLDP messages of length $\log N$ to all of its neighbor routers to inform its neighbor about the existence. Here, $\log N$ is the amount of information required to differentiate one node from the others. There are $N-1$ mesh routers in network. Then, I_L can be derived by

$$I_L = (N-1)\log N. \tag{5}$$

Denote I_{PI} as the amount of information needed for mesh routers to report the link state through *PACKET_IN* messages within time step τ. As described, when the mesh router receives LLDP messages from other routers, it is informed that the sender of LLDP message is one of its neighbors. Then, it will send *PACKET_IN* message to the controller to report the link states measured by itself. That is to say, the message contains the information about itself and its pN neighbor routers. Therefore, the message length is $(pN+1)\log N$. In addition, according to [41], the expect number of hops between two randomly located nodes in a $a \times b$ square field can be determined by Equations 6 and 7.

In Equation 6, $f_L(x)$ is the probability distribution function (PDF) of the random variable x, which is the Euclidean distance between two arbitrary nodes. Then,

$$E[h] = \left\lceil \frac{\int x f_L(x)dx}{r} \right\rceil. \tag{7}$$

Therefore, the amount of information needed for mesh routers to report the link state through *PACKET_IN* messages I_{PI} can be derived by

$$I_{PI} = E[h] \cdot N(pN+1)\log N. \tag{8}$$

Overall, given a time step τ, the lower bound on the total overhead caused by the LLDP protocol for the whole network per second is given by

$$O_{TO} = \frac{1}{\tau}(I_{PO} + I_L + I_{PI}), \tag{9}$$

where I_{PO}, I_L, and I_{PI} are given in Equations 4, 5 and 8.

From the result of Equation 9, we can see that the overall overhead of routing discovery and maintenance O_{TO} is $O(N^2 \log N)$ and it depends on the overhead of sending *PACKET_IN* messages to the controller. According to the result of [40], the routing overhead of proactive routing protocols is $O(N \log N)$. Therefore, the routing overhead in this part is larger. That is because in both our OpenCoding and OpenFlow, the controller periodically asks routers for the topology information of all of their neighbors. In order to reduce the overhead of sending *PACKET_IN* messages, the controller can increase the time interval and the routers can send the topology message only when there are changes in the states of their links between themselves and their neighbors.

4.2. Overhead of Flow Table Request and Distribution

The total overhead of flow table request and distribution O_{FT} is defined as the total amount of information needed in transmitting the request and flow table itself.

Denote $I_{S \to C}$ be the amount of information of routing request from a source router S to the controller C. Denote $I_{C \to \mathbb{F}}$ as the amount of information of the flow table distribution from the controller C to all the forwarding mesh routers $\mathbb{F}$.

A source router S sends a routing request message to the controller C when there is a new traffic to be transmitted. The message contains the information of source and destination. Then, the message length is $2 \log N$. In addition, if the source and the controller are randomly located in the network, the expect number of hops between them is $E[h]$. Therefore, $I_{S \to C}$ can be derived by

$$I_{S \to C} = E[h] \cdot 2\log N. \tag{10}$$

After receiving the request message, the controller computes the best routing path for this traffic flow and installs the flow table on the forwarding routers, which will participate in the transmission of this traffic flow. The flow table contains the information of all the forwarding routers. Notice that as $E[h]$ is the expect number of hops between two random located routers, it can also represent the number of routers in the shortest path from source to destination. In a random linear network, the number of nodes participating in the transmission will be larger than that of traditional routing. Therefore, the length if flow table message

$$f_L(x) = \begin{cases} 2x\left[\frac{\pi}{ab} - \frac{2x}{a^2b} - \frac{2x}{ab^2} + \frac{x^2}{a^2b^2}\right] & 0 \le x < b \\ \frac{2x}{ab}\left[\frac{\pi}{2} - \arcsin(1 - \frac{2b^2}{x^2})\right] - \frac{4x}{ab^2}\left[x - \sqrt{x^2 - b^2}\right] - \frac{2x}{a^2} & b \le x < a \\ \frac{2x}{ab}\left[\arcsin(\frac{2a^2}{x^2} - 1) - \arcsin(1 - \frac{2b^2}{x^2})\right] + \frac{2x}{a^2b^2}[a^2 + b^2 - x^2] - \frac{4x}{ab^2}[a - \sqrt{x^2 - b^2}] + \frac{4x}{a^2b}\left[\sqrt{x^2 - a^2} - b\right] & a \le x < \sqrt{a^2 + b^2} \end{cases} \tag{6}$$

is at least $E[h]\log N$. The flow table will be installed to at least $E[h]$ routers, and every flow table will be transmitted through $E[h]$ hops. Thus, $I_{C\to\mathbb{F}}$ can be estimated by

$$I_{C\to\mathbb{F}} \ge E[h]^3 \log N. \tag{11}$$

Then, the overhead caused by the flow table request and distribution can be estimated by

$$O_{FT} = \left(E[h]^3 + 2E[h]\right)\log N. \tag{12}$$

From the result of Equation 12, we can see that the overall overhead of each flow table distribution O_{FT} is $O(\log N)$. According to the result of [41], the routing overhead of reactive routing protocols is $O(N\log N)$. Therefore, routing overhead of this part is smaller. That is because in the flow table distribution procedure of OpenCoding and OpenFlow, the source is aware of the route to the controller and the controller also knows the routes to all the forwarding routers so that the broadcast RREQ messages are not needed.

5. Performance Evaluation

We perform simulation experiments to show the effectiveness of our proposed protocol. In the following, we first present the simulation methodology and then show the simulation results.

5.1. Methodology

We evaluate the performance of our proposed scheme in comparison with existing schemes in terms of throughput, average end-to-end delay, and rule activation time. In our simulations, the end-to-end delay is defined as the duration from a packet is sent from the source router until the packet is received by the destination router. In MORE and OpenCoding, the end-to-end delay is averaged across batches. The rule activation time of both OpenFlow and OpenCoding is defined as the duration from the time when the first packet of a new flow arrives at the source router and the source router performs *PACKET_IN* function until all the flow-tables are distributed to the corresponding forwarding routers.

We use ns-3 simulator [42] and compare the performance of our proposed OpenCoding protocol in comparison with the following protocols: OLSR [11], OpenFlow [12], and MORE [9]. In the simulations to evaluate the throughput, we use the topology of Roofnet

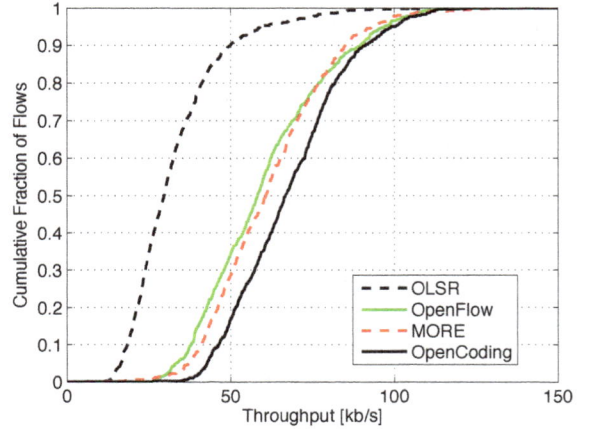

Figure 7. Throughput CDF of OLSR, OpenFlow, MORE and OpenCoding

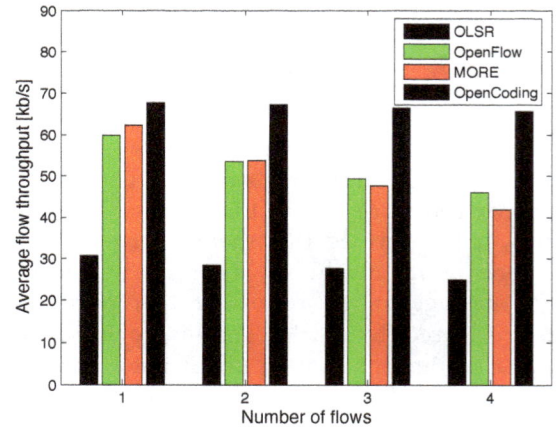

Figure 8. The throughput of OLSR, OpenFlow, MORE and OpenCoding with different numbers of concurrent flows

[43], which is a known experimental 802.11b/g wireless mesh network developed by MIT (Massachusetts Institute of Technology). While in the simulations to evaluate end-to-end delay and rule activation time, we use the random topology that consists of 20, 35, and 50 nodes to evaluate the scalability of network, respectively. To keep the same node density, nodes are randomly placed in a 377*377, 500*500 and 600*600 area, respectively. In all simulations, we use User Datagram Protocol (UDP) traffic and the packet size for

all protocols is set to 1500B. In addition, to implement the network visualization as described in OpenFlow, the nodes in a software-defined network are equipped with two network cards, one for the control path and the other for the data path. In each simulation, we randomly select the source-destination pairs in the corresponding topology. In addition, the nodes works in 802.11b and the data transmission rate is 1 Mbps. For both MORE and our OpenCoding, the batch size is set to 32.

5.2. Results

In the following, we show the evaluation results in terms of throughput, end-to-end delay and rule activation time.

Throughput. Figure 7 shows the Cumulative Distribution Function (CDF) of the transmission throughput of OLSR, OpenFlow, MORE and our proposed Open-Coding, measured over 2000 different runs. In each run, there is only one traffic flow in the network and the source-destination pairs are randomly selected. Our data shows that OpenCoding outperforms the other three protocols in terms of throughput. Particularly, as we can see from the figure, OpenCoding achieves a significant larger throughput gain than the traditional routing OLSR, and outperforms the OpenFlow and MORE as well. In the median case, OpenCoding has a 68.32%, 12.46%, and 9.88% throughput gain over OLSR, OpenFlow and MORE, respectively. We also observe that there is a performance gap between our results and the results of [4] which is based on a real test-bed. In our simulation, the data communication and control communication is separated by using multi-radio multi-channel technique to avoid the channel interference, while they used different SSIDs that may degrade the throughput performance to some extent.

Figure 8 shows the average per-flow throughput as a function of the number of concurrent flows for these protocols. In all cases, OpenCoding still outperforms the other three protocols. In addition, the throughput gain increases as the number of flows increases. As we can see from the figure, the average per-flow throughout of OpenCoding drops little, while the others decline much more than that of OpenCoding. The evaluation data confirms that OpenCoding can ensure the fairness with the global optimization.

Because of the intelligent management of the controller introduced by the concept of SDN, the best routing can be optimally selected by the controller, which could have a global view of the network. To this extent, OpenFlow outperforms traditional routing mechanisms. Meanwhile, because of the utilization of the broadcast nature of wireless communication medium, MORE also achieves a significant throughput

gain over the traditional routing mechanisms. Exploiting the benefits of both characteristics, the throughput performance of OpenCoding is the best. In addition, the developed scheduler for OpenCoding can effectively ensure the fairness of multiple flows.

End-to-end Delay. Figure 9 illustrates the relationship between the average per-packet end-to-end delay and the number of flows for different numbers of nodes in the network for OLSR, OpenFlow, MORE, as well as our OpenCoding, respectively. Each simulation is repeated 1000 times and the results are averaged as illustrated. As we can see from the figure, all the average delays increase as the number of flows and number of nodes in the network increases. In particular, when the number of nodes is 20, as the number of concurrent flows increases from 1 to 4, the end-to-end delay of OpenCoding increases 18.1% only, while those of OLSR, OpenFlow and MORE increase 75.4%, 53.5%, and 124%, respectively. The result shows that OpenCoding achieves a better performance as the number of network flows increases. In addition, when the number of flows is 1, as the number of nodes in the network increases from 20 to 50, the end-to-end delay increases 43.9%, 41.7%, 37.2%, and 43.1% for OLSR, OpenFlow, MORE and OpenCoding, respectively. This result shows that the performance of OpenCoding is almost same as the network size increases.

Rule Activation Time. In order to compare the rule activation time of OpenFlow and OpenCoding, we change the number of concurrent flows, as well as the number of network nodes, and plot the average value as shown in Figure 10. As we can see from the figure, the rule activation time of OpenFlow is a little shorter than that of OpenCoding. That is because in a network with network coding, there are more routers that participate in the transmission to fully exploit the broadcast nature in wireless communication. In particular, when the number of flows increases, the rule activation time will grows as well. That is because the controller needs to compute the best routing for more flows. Therefore, when there are a large number of traffic flows concurrently transmitted over the network, the computation ability of the controller needs to be powerful enough to avoid being performance bottleneck.

6. Discussion

In this section, we discuss some open issues related to OpenCoding.

6.1. Controller Design

When the network size becomes larger, the SDN controllers could become the performance bottleneck due to a large number of requests from routers. For

(a) node number:20

(b) node number:35

(c) node number:50

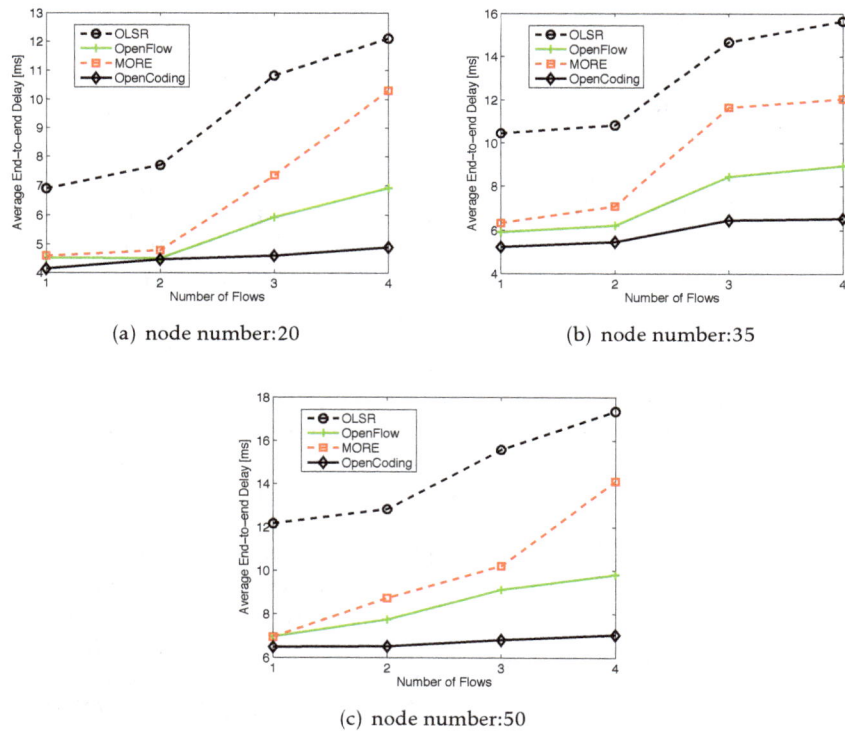

Figure 9. Average Delay of OLSR, OpenFlow, MORE and OpenCoding versus number of flows with different node numbers

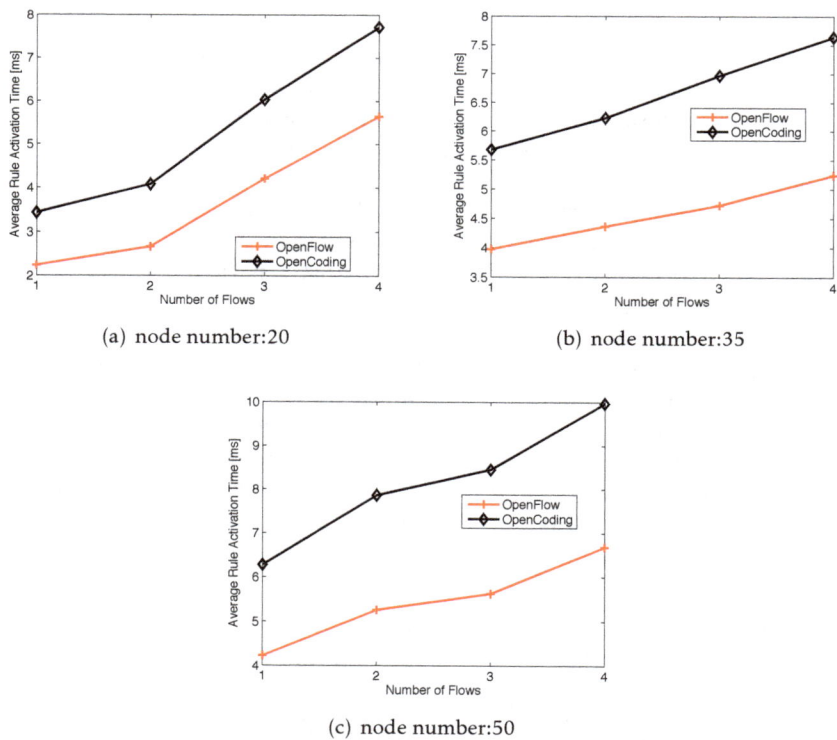

(a) node number:20

(b) node number:35

(c) node number:50

Figure 10. Average Rule Activation Time of OLSR, OpenFlow, MORE and OpenCoding versus number of flows with different node numbers

the sake of scalability, reliability and robustness, it has been realized that the logically-centralized controller must be physically distributed [44] in Software-Defined Networks. Because of the lossy and unreliable wireless channels, the transmission delay will be unacceptable if the distance between controller and router is long. Therefore, it is more necessary to distribute multiple controllers in software-defined wireless mesh networks.

Nonetheless, the controllers located in a distributed way may compete for common resources (such as communication channels). To this end, an optimal controller placement strategy and a carefully designed scheduling strategy to avoid collision should be developed. On designing such a strategy, two essential questions should be concerned. First, given a wireless network topology, how many controllers are needed in order to achieve a desirable performance? Second, where should the controllers be located? When we design the optimal deployment strategy, we need to consider the dynamics of traffic, the density distribution of mobile customers and routers, and realistic constraints (e.g., geo-restriction and information availability), deployment costs, the coverage and quality of service for mobile customers, and applications. We will develop scenarios for validate the optimal deployment of controllers. In addition, there are other questions to be solved: How to synchronize the information among the controllers and how do the mesh routers connect to the controllers? We will answer these questions in the future work.

6.2. Security

There has been limited research effort to date on the security issues associated with SDN. Nonetheless, as SDN is a new network architecture, it also brings some new targets for potential attacks. Our developed OpenCoding could operate in hostile environments and all modules and devices could increase the possibility of being compromised. Therefore, this calls for the framework to systematically explore possible attacks against OpenCoding and develop mitigation schemes [45]. Based on security objectives (e.g., availability, integrity, etc.), the adversary can launch attacks against various against various components (e.g., mesh routers, communication networks, and controller) associated with data and control planes.

Some vulnerabilities of SDN are listed below. At the controller level, authentication and authorization is a critical issue. When there is multiple organizations and applications accessing the network resources, if the resources are not enough to be allocated to all the applications, the demand of application with higher privilege should be satisfied with less delay. That is to say, a security model should be designed to isolate the applications with different privileges.

Meanwhile, privacy preservation is another critical issue. Since the controller obtains the global view of network traffic information, adversaries may try to compromise the controller or overhear and analyze the traffic through the controller to get these critical information. Therefore, a privacy preservation strategy is also necessary.

At the data transmission level, there are numerous attacks that can be launched to disrupt the effectiveness of networks. As an example, one potential attack is denial-of-service or jamming attacks. The adversary launches an attack by originating a large number of new traffic flows. As the computation and bandwidth resources of controller is limited, this type of attack may be devastating. Moreover, as we know, because of the mixing nature, network coding is vulnerable to network attacks such as pollution attacks [46]. The security problem of OpenCoding should also be considered in future work.

7. Conclusions

In this paper, we proposed a novel OpenCoding protocol, which integrates the software-defined networking technique with intra-flow network coding technique for WMNs. Our developed protocol makes a full use of the broadcast nature of the wireless transmission medium and the global intelligence of SDN controller, which enables the ability of improving the network performance (e.g., throughput, end-to-end delay, and fairness). Similar to the known OpenFlow protocols for wired software-defined networks, OpenCoding decouples the data plane and the control plane in wireless mesh routers, and leaves only network coding functions in each router for easy deployment and management of WMNs. Benefited from a global view of the controlled network, OpenCoding can ensure the fairness of flows. Through a simulation study, our evaluation data shows that OpenCoding outperforms intra-flow network coding protocols such as MORE, and OpenFlow. As ongoing work, we are implementing OpenCoding in a real-world test-bed to further evaluate its performance gain.

Acknowledgement. This work is sponsored in part by the following funds: the National Natural Science Foundation of China under Grant 61373115, Grant 61402356 and Grant 61502381, the Fundamental Research Funds for the Central Universities under Grant xjj2015065, and the China Post Doctoral Science Foundation under Grant 2015M570836. This work was also supported in part by US National Science Foundation (NSF) under grant: CNS 1350145. Any opinions, findings and conclusions or recommendations expressed in this material are those of the authors and do not necessarily reflect the views of the funding agencies.

References

[1] AKYILDIZ, I. and WANG, X. (2005) A survey on wireless mesh networks. *IEEE Communications Magazine* **43**(9): S23–S30.

[2] ORTIZ, S. (2013) Software-defined networking: On the verge of a breakthrough? *IEEE Computer* **46**(7): 10–12.

[3] HU, F., HAO, Q. and BAO, K. (2014) A survey on software-defined network and openflow: From concept to implementation. *Communications Surveys Tutorials, IEEE* **16**(4): 2181–2206.

[4] DELY, P., KASSLER, A. and BAYER, N. (2011) Openflow for wireless mesh networks. In *2011 Proceedings of 20th International Conference on Computer Communications and Networks (ICCCN)* (IEEE): 1–6.

[5] ABOLHASAN, M., LIPMAN, J., NI, W. and HAGELSTEIN, B. (2015) Software-defined wireless networking: centralized, distributed, or hybrid? *IEEE Network* **29**(4): 32–38.

[6] HUANG, H., LI, P., GUO, S. and ZHUANG, W. (2015) Software-defined wireless mesh networks: architecture and traffic orchestration. *IEEE Network* **29**(4): 24–30.

[7] AHLSWEDE, R., CAI, N., LI, S.Y. and YEUNG, R. (2000) Network information flow. *IEEE Transactions on Information Theory* **46**(4): 1204–1216.

[8] BISWAS, S. and MORRIS, R. (2005) Exor: Opportunistic multi-hop routing for wireless networks. *SIGCOMM Computer Communication Review* **35**(4): 133–144.

[9] CHACHULSKI, S., JENNINGS, M., KATTI, S. and KATABI, D. (August 2007) Trading structure for randomness in wireless opportunistic routing. In *Proceedings of ACM SIGCOMM Conference*.

[10] BIOGLIO, V., GRANGETTO, M., GAETA, R. and SERENO, M. (2013) A practical random network coding scheme for data distribution on peer-to-peer networks using rateless codes. *Performance Evaluation* **70**(1): 1–13.

[11] CLAUSEN, T., JACQUET, P., ADJIH, C., LAOUITI, A., MINET, P., MUHLETHALER, P., QAYYUM, A. *et al.* (2003) Optimized link state routing protocol (olsr) .

[12] MCKEOWN, N., ANDERSON, T., BALAKRISHNAN, H., PARULKAR, G., PETERSON, L., REXFORD, J., SHENKER, S. *et al.* (2008) Openflow: Enabling innovation in campus networks. *SIGCOMM Computer Communication Review* **38**(2): 69–74.

[13] ZHU, D., YANG, X., ZHAO, P. and YU, W. (2015) Towards effective intra-flow network coding in software defined wireless mesh networks. In *Proceedings of IEEE International Conference on Computer Communication and Networks (ICCCN)* (IEEE): 1–8.

[14] CHOWDHURY, K.R. and AKYILDIZ, I.F. (2008) Cognitive wireless mesh networks with dynamic spectrum access. *IEEE Journal on Selected Areas in Communications (JSAC)* **26**(1): 168–181.

[15] HONG, Y.W., HUANG, W.J., CHIU, F.H. and KUO, C.C.J. (2007) Cooperative communications in resource-constrained wireless networks. *IEEE Signal Processing Magazine* **24**(3): 47–57.

[16] BHATIA, R. and LI, L.E. (2007) Throughput optimization of wireless mesh networks with mimo links. In *Proceedings of 26th IEEE International Conference on Computer Communications (INFOCOM)* (IEEE): 2326–2330.

[17] KODIALAM, M. and NANDAGOPAL, T. (2005) Characterizing the capacity region in multi-radio multi-channel wireless mesh networks. In *Proceedings of the 11th Annual International Conference on Mobile Computing and Networking* (ACM): 73–87.

[18] ONF (2012) Software-defined networking: The new norm for networks. *ONF White Paper* .

[19] YEGANEH, S., TOOTOONCHIAN, A. and GANJALI, Y. (2013) On scalability of software-defined networking. *IEEE Communications Magazine* **51**(2): 136–141.

[20] HU, Y., WANG, W., GONG, X., QUE, X. and CHENG, S. (2014) On reliability-optimized controller placement for software-defined networks. *Communications, China* **11**(2): 38–54.

[21] CONGDON, P.T., MOHAPATRA, P., FARRENS, M. and AKELLA, V. (2014) Simultaneously reducing latency and power consumption in openflow switches. *IEEE/ACM Transactions on Networking (ToN)* **22**(3): 1007–1020.

[22] KHAN, A. and DAVE, N. (2013) Enabling hardware exploration in software-defined networking: A flexible, portable openflow switch. In *Proceedings of 2013 IEEE 21st Annual International Symposium on Field-Programmable Custom Computing Machines (FCCM)*: 145–148.

[23] BISWAS, J., LAZAR, A., HUARD, J.F., LIM, K., MAHJOUB, S., PAU, L.F., SUZUKI, M. *et al.* (1998) The ieee p1520 standards initiative for programmable network interfaces. *IEEE Communications Magazine* **36**(10): 64–70.

[24] DORIA, A., SALIM, J.H., HAAS, R., KHOSRAVI, H., WANG, W., DONG, L., GOPAL, R. *et al.* (2010) *Forwarding and control element separation (ForCES) protocol specification*. Tech. rep.

[25] LAKSHMAN, T., NANDAGOPAL, T., RAMJEE, R., SABNANI, K. and WOO, T. (2004) The softrouter architecture. In *Proceedings of ACM SIGCOMM Workshop on Hot Topics in Networking*, **2004**.

[26] YU, W., CHELLAPPAN, S., XUAN, D. and ZHAO, W. (2004) Distributed policy processing in active-service based infrastructures. *International Journal of Communication Systems (IJCS)* **19**(7): 727–750.

[27] GALIS, A., PLATTNER, B., SMITH, J.M., DENAZIS, S., MOELLER, E., GUO, H., KLEIN, C. *et al.* (2000) A flexible ip active networks architecture. In *Proceedings of Second International Working Conference on Active Networks, ed. H. Yasuda* (Springer-Verlag): 1–15.

[28] BERNARDOS, C., DE LA OLIVA, A., SERRANO, P., BANCHS, A., CONTRERAS, L., JIN, H. and ZU?NIGA, J. (2014) An architecture for software defined wireless networking. *IEEE Wireless Communications* **21**(3): 52–61.

[29] RIGGIO, R., GOMEZ, K.M., RASHEED, T., SCHULZ-ZANDER, J., KUKLINSKI, S. and MARINA, M.K. (2014) Programming software-defined wireless networks. In *Proceedings of 2014 10th International Conference on Network and Service Management (CNSM)*: 118–126.

[30] GUIMARAES, C., CORUJO, D. and AGUIAR, R. (2014) Enhancing openflow with media independent management capabilities. In *Proceedings of 2014 IEEE International Conference on Communications (ICC)*: 2995–3000.

[31] YANG, F., GONDI, V., HALLSTROM, J., WANG, K.C. and EIDSON, G. (2014) Openflow-based load balancing for

wireless mesh infrastructure. In *Proceedings of IEEE International Conference on Consumer Communications and Networking Conference (CCNC)*: 444–449.

[32] ZENG, D., LI, P., GUO, S. and MIYAZAKI, T. (2014) Minimum-energy reprogramming with guaranteed quality-of-sensing in software-defined sensor networks. In *Proceedings of 2014 IEEE International Conference on Communications (ICC)*: 288–293.

[33] SANTOS, M., DE OLIVEIRA, B., MARGI, C., NUNES, B., TURLETTI, T. and OBRACZKA, K. (2013) Software-defined networking based capacity sharing in hybrid networks. In *Proceedings of 2013 21st IEEE International Conference on Network Protocols (ICNP)*: 1–6.

[34] LI, S.Y., YEUNG, R. and CAI, N. (2003) Linear network coding. *IEEE Transactions on Information Theory* **49**(2): 371–381.

[35] HO, T., MEDARD, M., KOETTER, R., KARGER, D., EFFROS, M., SHI, J. and LEONG, B. (2006) A random linear network coding approach to multicast. *IEEE Transactions on Information Theory* **52**(10): 4413–4430.

[36] CHOU, P.A., WU, Y. *et al.* (2007) Network coding for the internet and wireless networks. *IEEE Signal Processing Magazine* **24**(5): 77.

[37] RADUNOVIĆ, B., GKANTSIDIS, C., KEY, P. and RODRIGUEZ, P. (2010) Toward practical opportunistic routing with intra-session network coding for mesh networks. *IEEE/ACM Transactions Networking (ToN)* **18**(2): 420–433.

[38] KHREISHAH, A., KHALIL, I.M. and WU, J. (2012) Distributed network coding-based opportunistic routing for multicast. In *Proceedings of the Thirteenth ACM International Symposium on Mobile Ad Hoc Networking and Computing*, MobiHoc '12 (New York, NY, USA: ACM): 115–124.

[39] GUDE, N., KOPONEN, T., PETTIT, J., PFAFF, B., CASADO, M., MCKEOWN, N. and SHENKER, S. (2008) Nox: towards an operating system for networks. *ACM SIGCOMM Computer Communication Review* **38**(3): 105–110.

[40] TRAN, Q.M. and DADEJ, A. (2014) Optimizing topology update interval in mobile ad-hoc networks. In *Proceeding of 2014 IEEE 79th Vehicular Technology Conference (VTC Spring)*: 1–5.

[41] NASERIAN, M., TEPE, K. and TARIQUE, M. (2005) Routing overhead analysis for reactive routing protocols in wireless ad hoc networks. In *Proceedings of IEEE International Conference on Wireless And Mobile Computing, Networking And Communications (WiMob)*, **3**: 87–92 Vol. 3.

[42] HENDERSON, T.R., LACAGE, M., RILEY, G.F., DOWELL, C. and KOPENA, J. (2008) Network simulations with the ns-3 simulator. *SIGCOMM demonstration* **15**: 17.

[43] *MIT Roofnet* (http://pdos.csail.mit.edu/roofnet/doku.php).

[44] NUNES, B., MENDONCA, M., NGUYEN, X.N., OBRACZKA, K. and TURLETTI, T. (2014) A survey of software-defined networking: Past, present, and future of programmable networks. *IEEE Communications Surveys Tutorials* **16**(3): 1617–1634.

[45] BHATTARAI, S., ROOK, S., GE, L., WEI, S., YU, W. and FU, X. (2014) On simulation studies of cyber attacks against lte networks. In *Proceedings of IEEE International Conference on Computer Communication and Networks (ICCCN)* (IEEE): 1–8.

[46] ZHU, D., YANG, X. and YU, W. (2015) Spais: A novel self-checking pollution attackers identification scheme in network coding-based wireless mesh networks. *Computer Networks* **91**: 376 – 389.

7

Popular Content Distribution in CR-VANETs with Joint Spectrum Sensing and Channel Access using Coalitional Games

Tianyu Wang[1], Lingyang Song[1,*], Zhu Han[2]

[1]School of Electrical Engineering and Computer Science, Peking University, Beijing, China.
[2]Electrical and Computer Engineering Department, University of Houston, Houston, USA.

Abstract

Driven by both safety concerns and commercial interests, popular content distribution (PCD), as one of the key services offered by vehicular networks, has recently received considerable attention. In this paper, we address the PCD problem in highway scenarios, in which a popular file is distributed to a group of on-board units (OBUs) driving through an area of interest (AoI). Due to high speeds of vehicles and deep fading of vehicle-to-roadside (V2R) channels, the OBUs may not finish downloading the entire file. Consequently, a peer-to-peer (p2p) network should be constructed among the OBUs for completing the file delivery process. Here, we apply the cognitive radio technique for vehicle-to-vehicle communications and propose a cooperative approach based on coalition formation games, which jointly considers the spectrum sensing and channel access performance. Simulation results show that our approach presents a considerable performance improvement compared with the non-cooperative case.

Keywords: VANET, cognitive radio, coalitional games

1. Introduction

Vehicular ad hoc networks (VANETs) have been envisioned to provide increased convenience and efficiency to drivers, with numerous applications ranging from traffic safety, traffic efficiency to infotainment [1, 2]. One particular type of service, popular content distribution (PCD) has attracted a lot of attentions recently, in which multimedia contents are distributed from roadside units (RSUs) to on-board units (OBUs) driving through an area of interest (AoI) [3–5]. Examples of PCD may include: a local hotel periodically broadcasts multimedia advertisements to the vehicles entering the city on suburban highway; and a traffic authority delivers real-time traffic information ahead, or disseminates an update version of local GPS map [3].

Unlike downloading services on the Internet or their direct extensions in VANETs where various vehicles are interested in downloading different files [6, 7], the PCD is different in the following characteristics:

1. The content for dissemination is a single large file, such as an emergency video.

2. The only interested users are the OBUs driving through an AoI.

3. The OBUs are moving at high speeds with unstable network topology.

4. The wireless links are unreliable and may suffer interferences from each other.

Due to the high speeds of vehicles and the large file size of popular content, the OBUs may fail to download the entire popular file within the limited time for vehicle-to-roadside (V2R) communications, and each OBU receives only a portion depending on the location and speed of passing the AoI. However, in most cases, the OBUs as whole has already obtained all the segments of the popular file when they sequentially pass the AoI. For completing the file delivery process, a peer-to-peer (p2p) network can be constructed among the OBUs, in which popular segments can be transmitted by vehicle-to-vehicle (V2V) communications. Considering the pressing demand of spectrum, the cognitive radio technique can be applied in V2V communication and the p2p network therefore turns out to be a cognitive radio vehicular ad hoc network (CR-VANET), where the OBUs as secondary users (SUs) sense the licensed spectrum for primary users (PUs) and access the PU channels at vacant time. Actually, in the proposed highway scenario where most portions are in rural areas, the spectrum is generally quite clean and cognitive radio is always a suitable and efficient technique.

Peer-to-peer (p2p) networks, which are distributed systems without any hierarchical organization or centralized control, have become immensely popular on the Internet in both application field and academic field [8]. Based on the existence of prior knowledge of network topology, p2p networks can be classified into structured P2P systems and unstructured P2P systems [8]. In our scenario, the network topology is unknown due to the randomness of OBUs. Thus, the p2p system in this paper is an unstructured p2p system as the BitTorrent and eDonkey. However, those p2p techniques used on the Internet should be carefully inspected before applying in our problem. First, the network topology is not only unknown but also ever-changing due to the mobility of OBUs. Second, the wireless links are very unreliable and may even interfere each other. Thus, the existing methods for Internet services can be inefficient and novel approaches should be proposed. Moreover, by introducing the cognitive radio technique, the performance of spectrum sensing and the throughput of channel access should be jointly considered in order to maximize the efficiency of the network, which also increases the complexity of networking.

In this paper, we propose a distributed approach based on coalition formation games to address the PCD problem in CR-VANETs. Coalitional game as a game theory model, in which players adopt cooperative strategies to form a coalition for improving individual profits, has recently been for modeling the cooperative behaviors of communication nodes. In [9], the cooperation between RSUs for V2R communication has been modeled as a coalition formation game with transferable utility, and the RSUs have been partitioned into many coalitions each of which applies a cooperative protocol to maximize their profits. In [10], coalitional games in partition form have been used for modeling the cooperation between SUs in cognitive radio networks, and the SUs in each coalition jointly sense the spectrum and coordinately access the PU channel for improving the total performance. In this paper, the considered p2p network not only has an ever-changing topology but also is strongly affected by the initial content distribution and the corresponding content request, which is different from the RSUs network in [9] or the SUs network in [10]. We propose a coalition formation game with non-transferable utility for modeling the cooperation between OBUs, which jointly considers the performance of cooperative sensing and the throughput of channel access.

The rest of this paper is organized as follows. Section II provides the system model and the non-cooperative spectrum sensing and access. In Section III, we present a coalition formation game with non-transferable

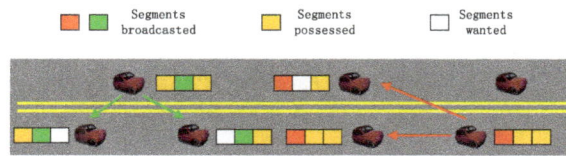

Figure 1. System model of the popular content distribution problem in cognitive radio ad hoc vehicular networks.

utility for modeling the cooperative manner, while in Section IV the proposed approach based on a coalition formation algorithm is presented for the entire PCD problem in CR-VANETs. In Section V simulation results and analysis are presented, and in Section VI we conclude the paper.

2. System Model

2.1. Network Model

Consider a CR-VANET with N OBUs (SUs) engaged in the sensing of K PU channels in order to access the spectrum and transmit data for PCD applications. A popular file, which is equally divided into M segments, has been requested by all the OBUs. Let $\mathcal{N}$, $\mathcal{K}$ and $\mathcal{M}$ denote the set of OBUs, the set of PU channels and the set of segments, respectively. Due to the fast speeds of OBUs and deep fading of V2R channels, each OBU in $\mathcal{N}$ only receives a part of all the M segments directly from RSUs, and the rest are gradually achieved by p2p transmissions using V2V communications. Let $\mathcal{M}_i$ denote the set of segments possessed by OBU i, the initial elements of which are randomly distributed in $\mathcal{M}$.

We suppose the OBUs of the p2p network periodically sense and access the PU channels, cooperatively or individually. In each cycle, or we say time slot, any OBU $i \in \mathcal{N}$ can decide to access one of the empty PU channels, the set of which is denoted by $\mathcal{K}_i$, or not to access. If OBU i decides to access PU channel $k_i \in \mathcal{K}_i$, then OBU i broadcasts one and only one segment $m_i \in \mathcal{M}_i$ in PU channel k_i for the current slot. We suppose the data traffic of PU channels can be modeled as K independent Poisson processes with the same arriving rate λ per time slot. The PU channel is empty only when no package of PU arrives at the current slot.

2.2. Channel Model

For any OBU i broadcasting at any PU channel k_i, we suppose only the "neighbors" (the OBUs with a line of sight to OBU i), denoted by $\mathcal{N}_i$, can receive the signal [12]. For a given group of OBUs S broadcasting in the current slot, we suppose only the non-interfered neighbors of any OBU $i \in S$, the set of which is denoted by $\mathcal{N}_i^*$, have the possibility of successfully achieving a segment. Thus, we have

$$\mathcal{N}_i^* = \{j \in \mathcal{N}_i \mid \forall l \in \left(\mathcal{N}_j \backslash \{i\}\right) \cap S, k_l \neq k_i,$$
$$\text{and } k_j \neq k_i, \text{if } j \in S\}. \quad (1)$$

The set of non-interfered neighbors of any OBU i, denoted by $\mathcal{N}_i^*$, represents the OBUs that are within the range of OBU i but not within the range of any other OBU broadcasting via channel k_i.

We suppose the possibility of successful delivery between any two non-interfered OBUs is proportional to the corresponding V2V channel capacity. Here, we only consider the pathloss without any small-scale fading, and the capacity of PU channel k between OBU i and OBU j is given by

$$c_{i,j}^k = \begin{cases} W \log_2\left(1 + \alpha d_{i,j}^{-n}\right), & \text{LOS exits,} \\ 0, & \text{otherwise,} \end{cases} \quad (2)$$

where $d_{i,j}$ is the distance between OBU i and OBU j, n is the pathloss exponent, α is a scale factor representing signal-to-noise ratio (SNR) at transmitters, and W is the bandwidth of any PU channel. Thus, the possibility of successful delivery between OBU i and OBU j through PU channel k is given by

$$p_{i,j}^k = \begin{cases} 0, & c_{i,j}^k < c, \\ \frac{c_{i,j}^k - c}{4c}, & c \leq c_{i,j}^k \leq 5c, \\ 1, & c_{i,j}^k > 5c, \end{cases} \quad (3)$$

where c is the size of any segment in $\mathcal{M}$. In (3), we assume the probability of success is zero if the capacity is less than the data rate, i.e., $c_{i,j}^k < c$, and the probability of success is 1 if the capacity is five times greater than the data rate, i.e., $c_{i,j}^k > 5c$. If $c < c_{i,j}^k < 5c$, we assume the probability of success is proportional to the capacity.

2.3. Mobility Model

The mobility model we use is similar to the Freeway Mobility Model (FMM) proposed in [13], which is well accepted for modeling the traffic in highway scenarios. In FFM, the simulation area includes many multiple lane freeways without intersections. At the beginning of the simulation, the vehicles are randomly placed in the lanes, and move at history-based speeds. The vehicles randomly accelerate or decelerate with the security distance $d_s > 0$ maintained between two subsequent vehicles in a lane and no change of lanes is allowed.

In our scenario, the map has been simplified to a one-way traffic road with double lanes as shown in Fig. 1. To better reflect the changing topology of CR-VANETs, we allow the change of lanes when a vehicle is overtaking the vehicle ahead. We assume all the OBUs accelerate and decelerate with acceleration $a > 0$ by probability

p, and the velocity of any OBU $i \in \mathcal{N}$ is limited by $v_{min} \leq v_i(t) \leq v_{max}$ at any time. The mobility constraints are listed as follows:

1. The OBUs are randomly placed on both lanes within length L when the simulation begins.

2. The initial speed of OBU $i \in \mathcal{N}$, denoted by $v_i(0)$, is randomly given in $[v_{min}, v_{max}]$.

3. The speed of OBU $i \in \mathcal{N}$ satisfies:

$$v_i(t+1) = \begin{cases} v_i(t), & 1 - 2p \\ \min\left(v_i(t) + a, v_{max}\right), & p \\ \max\left(v_i(t) - a, v_{min}\right), & p \end{cases} \quad (4)$$

where p is the probability of acceleration or deceleration.

4. For any OBU $i \in \mathcal{N}$ with OBU j_1 ahead in the same lane and OBU j_2 ahead in the other lane, OBU i switches to the other lane if $d_{i,j_1}(t) \leq d_s$ and $d_{i,j_2}(t) > d_s$, or OBU i decelerates to $v_i(t) = v_{j_1}(t)$ if $d_{i,j_k}(t) \leq d_s, k = 1, 2$.

2.4. Non–cooperative Spectrum Sensing and Access

In non-cooperative manner, the OBUs in $\mathcal{N}$ individually sense the PU channels. For any PU channel $k \in \mathcal{K}$ and any OBU $i \in \mathcal{N}$, we suppose the probability of miss (i.e., probability of missing the detection of the PU) and false alarm (i.e., probability of the false detection of the PU) are denoted by P_m^{ik} and P_f^{ik}, respectively. For simplicity without loss of of generality, we assume the sensing devices in all OBUs have the same performance for any PU channels, which implies $P_m^{ik} = P_m, P_f^{ik} = P_f, \forall i \in \mathcal{N}, k \in \mathcal{K}$.

After sensing all the PU channels, OBU i randomly selects a PU channel in $\mathcal{K}_i$ to access, as long as $\mathcal{K}_i \neq \emptyset$. To avoid potential collisions among OBUs, the carrier sense multiple access with collision avoidance (CSMA/CA) protocol has been adopted by the OBUs broadcasting at the current slot. Each OBU $i \in \mathcal{N}$ can only access the PU channel when no other OBUs are using it. If a data collision is detected by OBU i, OBU i will stop broadcasting and randomly choose another time point to access.

3. Cooperative Spectrum Sensing and Access

For the PCD problem in the considered p2p network, it is natural for the OBUs to cooperate with each other by sharing the sensing result or avoiding potential data collisions. In this section, we introduce the coalition game to model the cooperation of OBUs, in which we suppose it is beneficial for OBUs to contribute for the file delivery process with each successfully delivered segment corresponding to unit profit. By proposing a suitable utility function which jointly considers the sensing and access performance, we model the PCD problem as a coalitional game with non-transferable utility.

3.1. Utility Function

For any given PU channel $k \in \mathcal{K}$, the OBUs broadcasting at the current slot, the set of which is denoted by S^k, cooperatively sense the spectrum and simultaneously broadcast at PU channel k. Using OR-rule [14] (the PU channel is decided to be empty only when the sensing results of all the SUs are empty), the miss-detection and the false-alarm probabilities of the cooperative spectrum sensing for any OBU in S^k are, respectively, given by

$$Q_m^k(S^k) = P_m^{|S^k|}, \qquad (5)$$

and

$$Q_f^k(S^k) = 1 - \left(1 - P_f\right)^{|S^k|}. \qquad (6)$$

We denote by $\mathcal{H}_1$ the event that PU channel k is actually occupied by a primary user, and by $\mathcal{H}_0$ the complementary event of $\mathcal{H}_1$. Also, we denote by $\mathcal{H}_1'$ the hypothesis that the sensing result shows PU channel k is occupied, and $\mathcal{H}_0'$ the alternative hypothesis. Thus, the possibility that the empty decision of PU channel k is false can be expressed by:

$$
\begin{aligned}
&P^k\left(H_1 | H_0'\right) \\
&= \frac{P^k(H_1) P^k(H_0' | H_1)}{P^k(H_0')} \\
&= \frac{P^k(H_1) P^k(H_0' | H_1)}{P^k(H_0' | H_0) P^k(H_0) + P^k(H_0' | H_1) P^k(H_1)},
\end{aligned}
\qquad (7)
$$

and the possibility that the empty decision is correct is given by

$$P^k\left(H_0 | H_0'\right) = 1 - P^k\left(H_1 | H_0'\right). \qquad (8)$$

As we noted, the data traffic of PU channel k is modeled as a Poisson process with parameter λ. Thus, the possibility of no PU data arriving at PU channel k in the current slot is given by $P^k(H_0) = e^{-\lambda}$, and the alternative $P^k(H_1) = 1 - P^k(H_0)$. For any given group of OBUs S^k broadcasting at the current slot, we directly have $P^k(H_0' | H_1) = Q_m^k(S), P^k(H_0' | H_0) = 1 - Q_f^k(S)$. Thus, $P^k\left(H_1 | H_0'\right)$ and $P^k\left(H_0 | H_0'\right)$ can be expressed by P_m, P_f, λ and S^k, where $P^k\left(H_0 | H_0'\right)$ represents the probability of successful channel access and $P^k\left(H_1 | H_0'\right)$ represents the probability of data collision with PUs.

For quantifying the contribution made by each OBU in S^k broadcasting at PU channel k, we need to determine the only segment broadcasted by each member of S^k. For maximizing individual profits, we

suppose each OBU in S^k chooses the "most valuable" segment $x_i^* \in \mathcal{M}_i$ to broadcast, which satisfies

$$\sum_{j \in \mathcal{N}_i^*(x_i^*)} p_{i,j}^k \geq \sum_{j \in \mathcal{N}_i^*(x_i)} p_{i,j}^k, \ \forall x_i \neq x_i^*, \ x_i, x_i^* \in \mathcal{M}_i, \qquad (9)$$

where $\mathcal{N}_i^*(x)$ represents the set of OBUs in $\mathcal{N}_i^*$ lacking x, which is given by

$$\mathcal{N}_i^*(x) = \{j \in \mathcal{N}_i^* | x \notin \mathcal{M}_j\}. \qquad (10)$$

As we see, the proposed greedy algorithm for choosing the broadcasted segments selects the segments that can be successfully delivered to the most interested OBUs, which corresponds to the largest profits. As we noted, each successful delivery of one segment to an OBU will bring unit profit to the transmitter. Thus, the expecting profits of OBU $i \in S^k$ can be given as

$$R_i(S^k) = \sum_{j \in \mathcal{N}_i^*(x_i^*)} p_{i,j}^k, \qquad (11)$$

where x_i^* is the "most valuable" segment for OBU i satisfying (9). Additionally, as the p2p network is also a cognitive radio network, the accuracy of spectrum sensing should also be considered. By introducing the probability that the empty decision is correct in (8), we have the throughput of OBU $i \in S^k$ given by

$$U_i(S^k) = P^k(H_0 | H_0') R_i(S^k). \qquad (12)$$

As we see, for calculating the value of $U_i(S^k)$, certain information needs to be exchanged among the OBUs. This cost in V2V transmissions is necessary and should be considered in the utility function. We consider the following linear cost function with a pricing factor denoted by β:

$$C_i(S^k) = \begin{cases} \beta |S^k|, & \text{if } |S^k| > 1, \\ 0, & \text{otherwise,} \end{cases} \qquad (13)$$

Thus, given the payment formulated in (12) and the cost in (13), the utility function of any OBU $i \in S^k$ is given by

$$V_i(S^k) = U_i(S^k) - C_i(S^k). \qquad (14)$$

This utility function represents the expecting contribution that OBU $i \in S^k$ makes in the current slot and equally the payoff OBU i achieves. Based on this utility function, we consider the cooperation and competition among OBUs by coalitional game model in the following subsection.

3.2. Coalitional Game

We suppose the group of OBUs $\mathcal{N}^k \subset \mathcal{N}$ are competing for the opportunity of broadcasting at PU channel k

(the rest OBUs, the set of which is $\mathcal{N} \backslash \mathcal{N}^k$, have already got the chance of broadcasting at other PU channels). As we noted, for maximizing their individual profits given by (12), the OBUs may choose to form coalitions to compete against each other. For mathematically modeling the competition and cooperation among OBUs, we introduce the coalitional game with non-transferable utility [15–17]:

Definition 1: A coalitional game with *non-transferable utility* is defined by a pair $(\mathcal{N}, V)$, where $\mathcal{N}$ is the set of players and V is a mapping such that for each coalition of players $S \subseteq \mathcal{N}$, $V(S)$ is the payoff vector that players in S can achieve.

The concept of non-transferable utility means that each member in a given coalition has a specific profit or performance, and their profits cannot are non-transferable even they are in the same coalition. By adopting the value function in (14) and substituting $S^k, \mathcal{N}^k$ for $S, \mathcal{N}$, we directly have that the proposed problem can be modeled as a coalitional game with non-transferable utility:

Remark 1: The proposed problem can be modeled as a $(\mathcal{N}^k, V)$ coalitional game with non-transferable utility where the mapping V is given by

$$V(S^k) = (V_{i_1}(S^k), V_{i_2}(S^k), \ldots, V_{i_{|S^k|}}(S^k)),$$
$$i_l \in S^k, \ \forall l = 1, \ldots, |S^k|. \tag{15}$$

In this coalitional game, the players (the OBUs) can choose to join or leave any coalition based on its own payoff given in (14). If the structure of the players can finally get to a stable situation (no players has the incentive to leave its current coalition), then each coalition in the final structure represents an interest group that competes for the broadcasting opportunity in PU channel k. If coalition S^k gets the chance of broadcasting in PU channel k, then the OBUs in S^k simultaneously broadcast their "most valuable" segments in PU channel k at the current slot. To maximize the profits of the entire network, we suppose the broadcasting coalition is the one with the largest total payoff, which is defined as

$$T(S^k) = \sum_{i \in S^k} V_i(S^k). \tag{16}$$

Remark 2: In the proposed $(\mathcal{N}^k, V)$ coalitional game with non-transferable utility, due to the changing network topology and the benefit-cost tradeoffs from cooperation as expressed in (15), any coalitional structure may form in the network and the grand coalition $(\mathcal{N}^k)$ is seldom beneficial considering the potential interference and collisions. Hence, the proposed game is classified as a coalition formation game with non-transferable utility.

By carefully inspecting the expression of the payoff in (14), we can see the OBUs with no interference to each other always have the incentive to form a coalition, as the transmission rate in (11) keeps the same while the accuracy of sensing in (8) can be improved in cooperative manner. On the other hand, the OBUs close to each other (always means interference from a line of sight) seldom form a coalition because of the severe interference. Thus, the grand coalition $\mathcal{N}^k$ is seldom the outcome of the coalitional game due to the potential data collisions. Therefore, the proposed coalitional game is a $(\mathcal{N}^k, V)$ coalition formation game [17]. We will devise a coalition formation algorithm to get these disjoint coalitions in the next section.

4. Coalition Formation Algorithm

Given the above analysis, the cooperative manner of OBUs for a single PU channel can be classified as a coalition formation game with non-transferable utility, where the OBUs form several disjoint coalitions to maximize their own payoffs. In this section, we devise a distributed coalition formation algorithm and additionally propose the entire approach for the PCD problem.

4.1. Coalition Formation Concepts

Before constructing a coalition formation algorithm, we first introduce some necessary concepts, taken from [18]. Here, the set of players is denoted by $\mathcal{N}^k$ representing the OBUs cooperatively competing for PU channel k. First, we give the concept of coalition partition for describing the structure of players.

Definition 2: A *coalition partition* is defined as a set of disjoint coalitions that cover the player set, i.e., for any coalitional partition $\Pi^k = \{S_1^k, \ldots, S_r^k\}$ we have $\forall l, S_l^k \subseteq \mathcal{N}^k$ and $\bigcup_{l=1}^{r} \{S_l^k\} = \mathcal{N}^k$. Also, we denote by $S_{\Pi^k}^k(i)$, the coalition $S_l^k \in \Pi^k$ including OBU i, i.e., $i \in S_l^k$.

One key approach in coalition formation is to enable the players to join or leave a coalition based on well-defined preferences. Therefore, we need to introduce the concept of preference relation, using which the players can decide which coalition they prefer more to be a member of.

Definition 3: For any player $i \in \mathcal{N}^k$, a *preference relation* or *order* $\succeq_i$ is defined as a complete, reflexive, and transitive binary relation over the set of all coalitions that player i can possibly form, i.e., the set $\{S_l^k \subseteq \mathcal{N} : i \in S_l^k\}$.

For any given player $i \in \mathcal{N}^k$, $S_1^k \succeq_i S_2^k$ implies that player i prefers being a member of coalition S_1^k with $i \in S_1^k$ over being a member of coalition S_2^k with $i \in S_2^k$, or at least, OBU i prefers both coalitions equally. In this paper, the preference of any OBU $i \in \mathcal{N}^k$ with $i \in S_1^k, S_2^k$ is quantified as follows:

$$S_1^k \succeq_i S_2^k \Leftrightarrow V_i(S_1^k) \geq V_i(S_2^k) \ \& \ S_1^k \in A^k(i), \tag{17}$$

where $A^k(i)$ is the set of OBU i's "friendly" coalitions defined by

$$A^k(i) = \{S_l^k \subseteq \mathcal{N}^k \mid i \in S_l^k \;\& \\ \forall j \in S_l^k \backslash \{i\}, V_j(S_l^k) \geq V_j(S_l^k \backslash \{i\})\}. \quad (18)$$

As we see, a coalition S_l^k is "friendly" to OBU i, only when OBU i's joining increases or at least maintains the payoffs of all the other OBUs in S_l^k. Thus, the definition of preference implies that OBU i prefers being a member of S_1^k over S_2^k only when OBU i gains an increase in individual profit and meanwhile no other OBUs in S_1^k suffers a decrease because of OBU i's joining. The asymmetric counterpart of $\geq_i$, denoted by $>_i$, is defined as

$$S_1^k >_i S_2^k \Leftrightarrow V_i(S_1^k) > V_i(S_2^k) \;\& \; S_1^k \in A^k(i). \quad (19)$$

Based on the defined preference relation, the basic operation in our proposed coalition formation algorithm, the switch operation, is defined as follows:

Definition 6: Given a partition $\Pi^k = \{S_1^k, \ldots, S_r^k\}$ of the players set $\mathcal{N}^k$, if player $i \in \mathcal{N}^k$ performs a switch operation from $S_{\Pi^k}^k(i) = S_m^k$ to $S_l^k \in \Pi^k \cup \{\emptyset\}, S_l^k \neq S_{\Pi^k}^k(i)$, then the current partition Π^k of $\mathcal{N}^k$ is modified into a new partition $\Pi^{k'}$ such that $\Pi^{k'} = (\Pi^k \backslash \{S_m^k, S_l^k\}) \cup \{S_m^k \backslash \{i\}, S_l^k \cup \{i\}\}$.

Definition 7: Given any player $i \in \mathcal{N}^k$, the history collection $H^k(i)$ is defined as the set of coalitions that player i visited and then left for the competition of PU channel k.

4.2. Distributed Coalition Formation Algorithm

For competing a given PU channel k, we propose a coalition formation algorithm that allows the OBUs in $\mathcal{N}^k$ to make distributed decisions as to which coalitions they decide to join at any time slot. The basic rule of the algorithm is as follows:

Basic Rule: Given a partition $\Pi^k = \{S_1^k, \ldots, S_r^k\}$ of the OBUs set $\mathcal{N}^k$, a switch operation from $S_{\Pi^k}^k(i)$ to $S_l^k \in \Pi^k \cup \{\emptyset\}, S_l^k \neq S_{\Pi^k}^k(i)$ is allowed for any OBU $i \in \mathcal{N}^k$, if and only if $S_l^k \cup \{i\} >_i S_{\Pi^k}^k(i)$ and $S_l^k \cup \{i\} \notin H^k(i)$.

For any partition Π^k, the basic rule provides a mechanism whereby any OBU can leave its current coalition S_m^k and join another coalition $S_l^k \in \Pi^k$, given that the new coalition is strictly preferred over S_m^k through the preference relation defined by (19). Thus, the basic rule can be seen as an individual rule abided by each member of $\mathcal{N}^k$, to move from its current coalition to a new coalition for improving its payoff, meanwhile maintains the profits of other members of this new coalition. Further, we suppose whenever an OBU decides to switch from its current coalition $S_m^k \in \Pi^k$ to join a different coalition, coalition S_m^k is stored

Table 1. The Coalition Formation Algorithm for On-board Units Competing for Primary User Channel k

Given any partition $\Pi_{initial}^k$ of the OBUs set $\mathcal{N}^k$ with the initialized history collections $H^k(i) = \emptyset, \forall i \in \mathcal{N}^k$, the OBUs engage in the coalition formation algorithm as follows:

* **repeat**

For a randomly chosen OBU $i \in \mathcal{N}^k$, with current partition $\Pi_{current}^k$ ($\Pi_{current}^k = \Pi_{initial}^k$ in the first round)

1. OBU i searches for a possible switch operation according to the basic rule in Section IV.

2. If such a switch exists, OBU i performs the following steps:

   (a) Updates the history collection $H^k(i)$ by adding coalition $S_{\Pi_{current}^k}^k(i)$ before leaving it.

   (b) Leaves coalition $S_{\Pi_{current}^k}^k(i)$.

   (c) Joins the new coalition $S_{\Pi_{next}^k}^k(i)$ that improves its payoff.

* **until** the partition converges to a final Nash-stable partition Π_{final}^k.

in its history collection $H^k(i)$ (if $|S_m^k| > 1$). With the order in which the OBUs make their switch operations considered to be random, the complete form of our distributed coalition formation algorithm is shown in Table 1.

4.3. Convergence and Stability

The convergence of the proposed coalition formation algorithm is guaranteed as follows:

Proposition 1: Starting from any initial coalitional partition $\Pi_{initial}^k$, the proposed algorithm maps to a sequence of switch operations, which always converges to a final partition Π_{final}^k composed of a number of disjoint coalitions.

Proof: Denote by $\Pi_{n,i}^k$ the partition formed at the time OBU $i \in \mathcal{N}^k$ needs to act after the n switch operations made by one or more OBUs. Given any initial starting partition $\Pi_{initial}^k = \Pi_{0,i}^k$, the coalition formation algorithm consists of a sequence of switch operations, e.g., $\Pi_{0,1}^k \rightarrow \Pi_{1,3}^k \rightarrow \ldots \rightarrow \Pi_{n,i}^k \ldots \rightarrow \Pi_{m,j}^k \ldots$. For any two partition $\Pi_{n,i}^k$ and $\Pi_{m,j}^k$, such that $n < m$, $\Pi_{m,j}^k$ is a result of the transformation of $\Pi_{n,i}^k$ after $m - n$ switch operations, we have have $\Pi_{n,i}^k \neq \Pi_{m,j}^k$ since each

OBU maintains a history collection to avoid joining a coalition repeatedly. As the number of partitions of a set is finite (given by the Bell number [19]), then the number of transformations in the sequence is finite. Hence, the algorithm will always terminate after finite iterations and converge to a final partition Π^k_{final}.

The stability of the final partition Π^k_{final} resulting from the proposed algorithm can be studied using the following concept from the hedonic games [18].

Definition 8: A partition $\Pi^k = \{S^k_1, \ldots, S^k_r\}$ is Nash-stable, if $\forall i \in \mathcal{N}^k$, $S^k_{\Pi^k}(i) \succeq_i S^k_l \cup \{i\}$ for all $S^k_l \in \Pi^k \cup \{\emptyset\}$.

The definition of Nash-stable implies that any coalition partition Π^k in which no OBU has an incentive to move from its current coalition to another coalition in Π^k or to deviate and act alone, is considered to be Nash-stable.

Proposition 2: Any final partition Π^k_{final} resulting from the proposed coalition formation algorithm is Nash-stable.

Proof: Suppose the final partition Π^k_{final} is not Nash-stable, which implies there exists an OBU $i \in \mathcal{N}^k$, and a coalition $S^k_l \in \Pi^k_{final}$ such that $S^k_l \cup \{i\} >_i S^k_{\Pi^k_{final}}(i)$. Based on the basic rule, OBU i can perform a switch operation from $S^k_{\Pi^k_{final}}(i)$ to S^k_l, which contradicts the fact that Π^k_{final} is the final partition. Thus, any final partition Π^k_{final} resulting from the proposed coalition formation algorithm is Nash-stable.

4.4. Cooperative Approach for Popular Content Distribution

The competition for PU channel k among OBUs $\mathcal{N}^k$ has been modeled as a coalition formation game with non-transferable utility and a convergent coalition formation algorithm with a Nash-stable outcome has been given in Table 1. Consider the considered scenario has multiple PU channels, we propose a sequential method for allocating the OBUs in $\mathcal{N}$ to the PU channels in $\mathcal{K}$. To be specific, the OBUs in $\mathcal{N}$ implement the coalition formation algorithm on each channel sequentially. After the calculation for a PU channel, the OBUs achieving the opportunity of broadcasting will quit the competition for the rest PU channels.

A summary of the entire approach for the proposed PCD problem is shown in Table 2. In the first stage, a distributed neighbor discovery method, e.g., those in [20, 21], is used to capture any changes in the network structure and record the necessary information for the following calculation. In the second stage, the sequential coalition formation algorithm has been used to allocate PU channels for the OBUs in the network. In the thrid stage, the OBUs access their corresponding PU channels by broadcasting their most valuable segments.

Table 2. The Approach for the Popular Content Distribution Problem

---

With the initial segments randomly distributed among $\mathcal{M}$, the OBUs periodically perform the following three stages:

Stage I: Network Discovery

Each OBU performs a neighbor discovery algorithm to obtain the necessary information for calculating individual payoff in (14), and the initial partition is set as $\Pi_{initial} = \mathcal{N}$.

Stage II: Sequential Coalition Formation

For PU channel $k = 1, 2, \ldots, |\mathcal{K}|$, we performs the following steps:

1. Determine the OBUs competing for the current PU channel $\mathcal{N}^k = \mathcal{N}^{k-1} \backslash S^{k-1}$ (For PU channel 1, $\mathcal{N}^1 = \mathcal{N}$) where S^{k-1} is the broadcasting coalition for PU channel $k-1$.

2. The OBUs in $\mathcal{N}^k$ perform the coalition formation algorithm in Table 1 to obtain the final partition Π^k_{final}.

3. The coalition with the largest total payoff (16) is decided to be the broadcasting coalition S^k.

Stage III: Spectrum Sensing and Broadcasting

For any broadcasting coalition S^k in PU channel k, the OBUs in S^k cooperatively sense PU channel k with OR-rule. If the sensing result is empty, the OBUs in S^k simultaneously access PU channel k by broadcasting their "most valuable" segments defined in (9).

To adapt to the changing network topology, the three stages are repeated periodically.

In the non-cooperative manner, each OBU has to sense all the PU channels (or at least enough PU channels to fine an empty one) before deciding which to access. In the proposed approach, the channel access decision is based on the expecting data throughput which needs no sensing result from any OBU. Thus, the potential PU channel for any OBU can be decided before any spectrum sensing. Consequently, in our proposed approach, each OBU only needs to sense the PU channel that has been allocated to it rather than sense the entire spectrum, which decreases the sensing cost or equally increases the transmission time.

Table 3. Parameters for Simulation

$N = 5 \sim 15$	number of OBUs in the network
$L = 500m$	initial length of the fleet of vehicles
$M = 5 \sim 15$	number of segments of the entire file
$K = 1 \sim 2$	number of PU channels
$Mc = 1Mbit$	the size of the entire popular file
$v_{min} = 20m/s$	the minimal speed
$v_{max} = 40m/s$	the maximal speed
$d_s = 100m$	the safe distance
$a = 2m/s^2$	the acceleration
$p = 0.1$	the probability for acceleration or deceleration
$\rho_c = 0.3 \sim 0.6$	the initial content density
$\beta = 0.01$	the pricing factor
$n = 4$	the exponent for pathloss
$W = 30MHz$	the bandwidth
$\alpha = 10^6$	the signal-to-noise rate at the transmitter
$\lambda = 0.1 \sim 0.3$	the arriving rate of PU traffic per slot
$P_m = 0.1$	the possibility of missing
$P_f = 0.1$	the possibility of false alarm

5. Simulation Results

In this section, the performance of the proposed approach in Table 2 is simulated in various environmental conditions and compared with the non-cooperative manner in many aspects. The parameters are taken from a general highway scenario as shown in Table 3. Here, the content density ρ_c is defined as the average number of initially possessed segments dividing the total number of segments M, which represents the level of initial content dissemination. Also, we assume each slot lasts 1 second.

In Fig. 2, we show the number of total possessed segments, given by $P(t) = \sum_{i \in \mathcal{N}} |\mathcal{M}_i|$, for networks with $N = 5, M = 5, K = 1, 2 \rho_c = 0.6, \lambda = 0.1$, where the vertical coordinate has been normalized by the total demand NM. The performance of the proposed algorithm is compared with the non-cooperative scheme in both conditions of single PU channel and multiple PU channels. In the non-cooperative scheme, each OBU individually senses the spectrum and makes an individual decision on the presence or absence of the PU for each PU channel. Then, each OBU randomly accesses one of the vacant channels where no neighbor OBUs are broadcasting. Fig. 2 shows that, as the transmitting time increases, the number of total possessed increases for both the methods. However, our proposed algorithm achieves a better performance in both the total number and the ascending rate. In the non-cooperative manner, the OBUs sense the

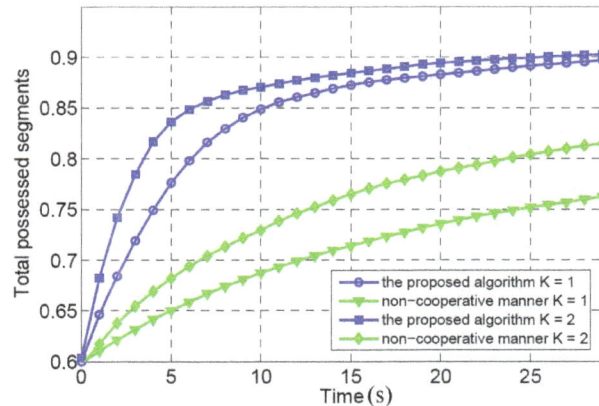

Figure 2. Total segments of all OBUs by the proposed approach and the non-cooperative manner as a function of time for networks with $N = 5, M = 5, K = 1, 2, \rho_c = 0.6, \lambda = 0.1$.

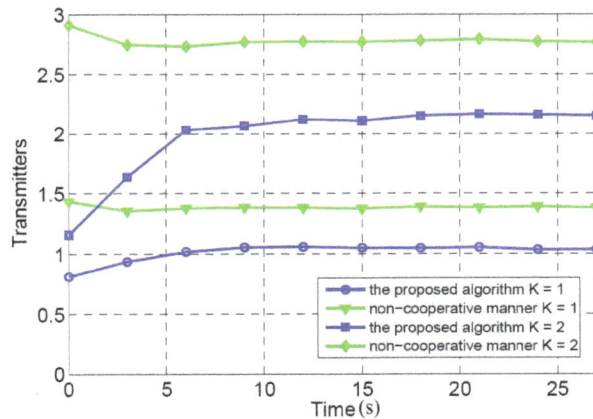

Figure 3. Total transmitters in the network by the proposed approach and the non-cooperative manner as a function of time for networks with $N = 5, M = 5, K = 1, 2, \rho_c = 0.6, \lambda = 0.1$.

spectrum individually and access by avoiding potential collisions (though still with hidden terminal problem). However, in our proposed algorithm, the spectrum sensing and channel access have been jointly in the value function of the coalition formation game, which represents the expecting throughput for each OBU. Aiming at maximizing the throughput rather than simply avoiding collisions makes our algorithm more efficient than the non-cooperative manner.

In Fig. 3, we show the number of transmitters for the networks with $N = 5, M = 5, K = 1, 2 \rho_c = 0.6, \lambda = 0.1$. As we can see, the number of transmitters of the proposed approach is lower than the non-cooperative manner in both single PU channels or multiple PU channels. In the non-cooperative manner, if OBU i detects that no other OBUs are using an empty PU channel, OBU i immediately accesses the spectrum

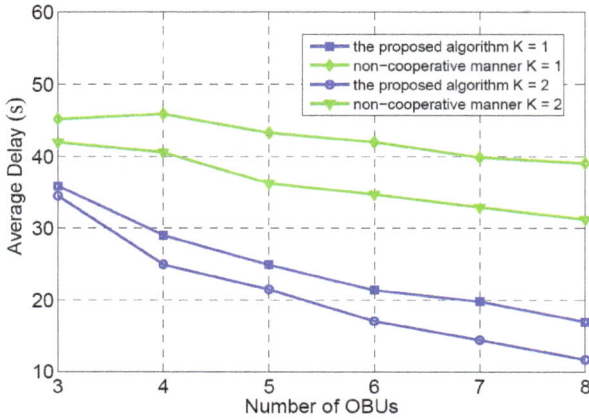

Figure 4. Average delay by the proposed approach and the non-cooperative manner as a function of N for networks with $M = 5, K = 1, 2, \rho_c = 0.6, \lambda = 0.1$.

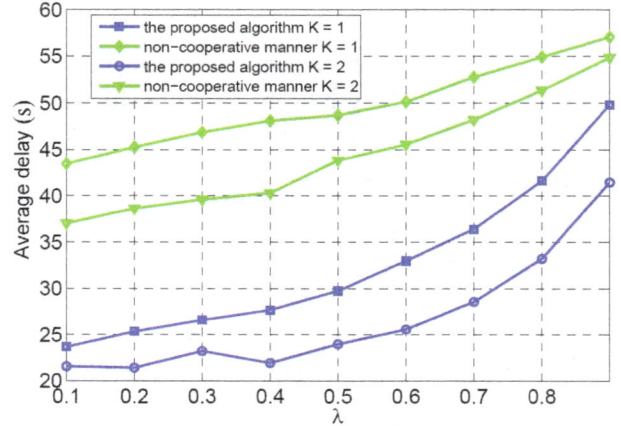

Figure 5. Average delay by the proposed approach and the non-cooperative manner as a function of λ for networks with $N = 5, M = 5, K = 1, 2, \rho_c = 0.6$.

without considering how many OBUs could actually receive the broadcasting segment. However, in our proposed approach, the channel conditions together with the wanted segments of each OBUs are considered, which highly reduces unnecessary broadcastings when the transmitter is far way from other OBUs, or the segment for broadcasting is not requested by neighborhood OBUs. Fig. 3 shows that, although we do not consider any energy model in our scenario, the proposed approach still achieves a low power consumption.

In Fig. 4, we show the average delay experienced by the OBU for achieving the entire file as a function of N for networks with $M = 5, K = 1, 2\rho_c = 0.6, \lambda = 0.1$. In the proposed PCD problem, the core parameter considered by the content owner is the average delay experienced by the OBUs, which is generally defined as $\tau_a = \tau_t/N$ with τ_t representing the total delay experienced by all OBUs. For any given content distribution scheme, τ_t is given by the area between the cumulative demand curve and cumulative service curve [22]. In our scenario, we assume all arrivals of demand occur instantaneously [23] at the beginning of p2p transmission and stay unchanged ever since. Thus, the demand curve is a constant NM and the average delay τ_a is given by

$$\tau_a = \frac{1}{N} \int_{t=0}^{\tau_m} [NM - P(t)] \, dt, \qquad (20)$$

where $P(t)$ is the cumulative service curve representing the number of segments possessed by the OBUs, and τ_m is the maximal delay defined by $P(\tau_m) = NM$ when the PCD completes. The performance of the proposed algorithm is compared with the non-cooperative manner in conditions of single PU channel

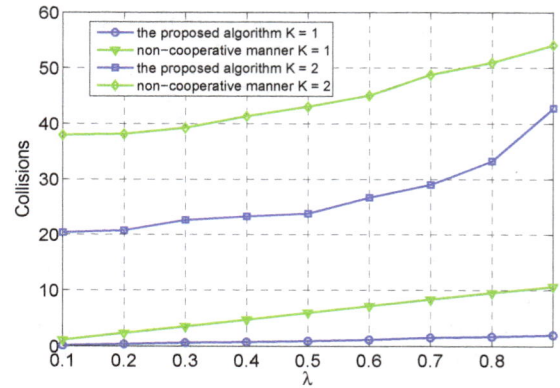

Figure 6. Collisions with primary users by the proposed approach and the non-cooperative manner as a function of λ for networks with $N = 5, M = 5, K = 1, 2$.

and multiple PU channels. Fig. 4 shows our proposed method enjoys an average delay approximately 40% smaller than the non-cooperative scheme. As we see, the average delay decreases as the number of OBUs increases by both methods, which can be explained by the increasing connectivity among OBUs.

In Fig. 5, we show the average delay as a function of λ for networks with $N = 5, M = 5, K = 1, 2, \rho_c = 0.6$. The performance of the proposed algorithm is compared with the non-cooperative manner in conditions of single PU channel and multiple PU channels. As the arriving rate of PU data λ increases, the potential channels and time slots for SUs (the OBUs) decrease, which increases the average delay of the PCD problem. Also, Fig. 5 shows that our method has a better performance than the non-cooperative manner with 40% decrease in average delay for any level of PU traffic.

In Fig. 6, we show the collisions with PU traffic as a function of λ for networks with $N = 5, M = 5, K = 1, 2, \rho_c = 0.6$. The performance of the proposed algorithm is compared with the non-cooperative manner in conditions of single PU channel and multiple PU channels. As the λ increases, the PU traffic becomes heavy and the probability of collisions also increases as shown in Fig. 6. As we see, our proposed algorithm has less collisions compared with the non-cooperative manner, which means our proposed approach has a smaller interference to the PUs.

6. Conclusions

In this paper, we have proposed a p2p scheme using V2V communications to solve the PCD problem in VANETs. Cognitive radio technique has been introduced for V2V communications which makes the network a CR-VANET. By carefully defining the utility function, we have proposed a coalition formation game with non-transferable utility for modeling the behavior of the OBUs. Further, we have introduced a coalition formation algorithm which could map into a sequence of switch operations and converge to a final Nash-stable partition. Based on the algorithm, we have proposed the overall approach for the PCD problem. In the simulation part, we have shown that our approach has a better performance with various conditions in average delay, power consumption, and interference to PUs, compared with the non-cooperative scheme.

References

[1] OLARIU, S. and WEIGLE, M. A. C. [eds.] (2009) *Vehicular Networks: From Theory to Practice.* (London, UK: Chapman & Hall/CRC).

[2] HARTENSTEIN, H. and LABERTEAUX, K. P. (2008) A tutorial survey on vehicular ad hoc networks. *IEEE Communications Magazine* 46(6): 164-171.

[3] LI, M., YANG, Z. and LOU, W. (2011) Codeon: cooperative popular content distribution for vehicular networks using symbol level network coding. *IEEE Journal on Selected Areas in Communications* 29(1): 223-235.

[4] WANG, T. SONG, L. and HAN, Z. (2013) Coalitional Graph Games for Popular Content Distribution in Cognitive Radio VANETs. *IEEE Transactions on Vehicular Technology* 31(9): 538-547.

[5] WANG, T. SONG, L. HAN, Z. and JIAO, B. (2013) Dynamic Popular Content Distribution in Vehicular Networks using Coalition Formation Games. *IEEE Journal on Selected Areas in Communications* 62(8): 4010-4019.

[6] NANDAN, A. DAS, S. PAU, G. GERLA M. and SANADIDI, M. Y. (2005) Co-operative downloading in vehicular ad-hoc wireless networks. In *Second Annual Conference on Wireless On-demand Network Systems and Services*(St. Moritz, Switzerland: IEEE), 32-41.

[7] FIORE, M. and BARCELO-ORDINAS, J. M. (2009) Cooperative download in urban vehicular networks. In *IEEE 6th International Conference on Mobile Adhoc and Sensor Systems* (IEEE), 20-29.

[8] LUA, K. CROWCROFT, J. PIAS, M. SHARMA, R. and LIM, S. (2005) A survey and comparison of peer-to-peer overlay network schemes. *IEEE Communications Surveys & Tutorials* 7(2): 72-93.

[9] SAAD, W. HAN, Z. HJORUNGNES, A. NIYATO, D. and HOSSAIN, E. (2011) Coalition formation games for distributed cooperation among roadside units in vehicular networks. *IEEE Journal on Selected Areas in Communications* 29(1): 48-60.

[10] SAAD, W. HAN, Z. ZHENG, R. HJORUNGNES, A. BASR, T. and VINCENT POOR, H. (2012) Coalitional games in partition form for joint spectrum sensing and access in cognitive radio networks. *IEEE Journal on Selected Areas in Communications* 6(2): 195-209.

[11] ZHANG, R. SONG, L. HAN, Z. and JIAO, B. (2011) Distributed coalition formation of relay and friendly jammers for secure cooperative networks. In *IEEE International Conference on Communications* (Kyoto, Japan: IEEE), 1-6.

[12] MECKLENBRAUKER, C. F. MOLISCH, A. F. KAREDAL, J. TUFVESSON, F. PAIER, A. BERNADO, L. ZEMEN, T. KLEMP, O. and CZINK, N. (2011) Vehicular channel characterization and its implications for wireless system design and performance. *Proc. the IEEE* 99(7): 1189-1212.

[13] MAHAJAN, A. POTNIS, N. GOPALAN, K. and WANG, A. I. A. (2006) Urban Mobility Models for VANETs. In *Proceedings of the 2nd IEEE International Workshop on Next Generation Wireless Networks* (IEEE).

[14] PEH, E. and LIANG, Y.-C. (2007) Optimization for cooperative sensing in cognitive radio networks. In *Proceedings of IEEE Wireless Communication and Networking Conference* (Hong Kong, China: IEEE), 27-32.

[15] MYERSON, R. B. (1991) *Game Theory, Analysis of Conflict* (Cambridge, MA, USA: Harvard University Press).

[16] OWEN, G. (1995) *Game theory* (London, UK: Academic Press), 3rd ed.

[17] SAAD, W. HAN, Z. DEBBAH, M. HJOUNGNES, A. and BASAR, T. (2009) Coalition game theory for communication networks: a tutorial. *IEEE Signal Processing Magzine, Special Issue on Game Theory* 26(5): 77-97.

[18] BOGOMONLAIA, A. and JACKSON, M. (2002) The stability of hedonic coalition structures. *Games and Economic Behavior* 38: 201-230.

[19] RAY, D. (2007) *A Game-Theoretic Perspective on Coalition Formation* (New York, USA: Oxford University Press).

[20] DAMLJANOVIC, Z. (2008) Cognitive radio access discovery strategies. In *International Symposium on Communication Systems, Networks, and Digital Signal Processing* (IEEE).

[21] ARACHCHIGE, C. J. L. VENKATESAN, S. MITTAL, N. (2008) An asynchronous neighbor discovery algorithm for cognitive radio networks. In *IEEE Symposium on New Frontiers in Dynamic Spectrum Access Networks* (IEEE).

[22] BERTSEKAS, D. and GALLAGER, R. [eds.] (1987) *Data Networks* (Boston, MA: Longman Higher Education).

[23] QIU, D. and SRIKANT, R. (2004) Modeling and performance analysis of bittorrent-like peer-to-peer networks. In *Proc. ACM SIGCOMM* (Karlsruhe, Germany), 367-378.

8

Alleviate Cellular Congestion Through Opportunistic Trough Filling

Yichuan Wang*, Xin Liu

Department of Computer Science, University of California, Davis, CA 95616, USA

Abstract

The demand for cellular data service has been skyrocketing since the debut of data-intensive smart phones and touchpads. However, *not all data are created equal.* Many popular applications on mobile devices, such as email synchronization and social network updates, are delay tolerant. In addition, cellular load varies significantly in both large and small time scales. To alleviate network congestion and improve network performance, we present a set of opportunistic trough filling schemes that leverage the time-variation of network congestion and delay-tolerance of certain traffic in this paper. We consider average delay, deadline, and clearance time as the performance metrics. Simulation results show promising performance improvement over the standard schemes. The work shed lights on addressing the pressing issue of cellular overload.

Keywords: opportunistic scheduling, traffic engineering, cellular operation

1. Introduction

Cellular operators are facing grand challenges in satisfying the fast increasing demand for wireless data service. Mobile Internet is experiencing unforeseen growth in the number of users, capability of devices, and most importantly, volume of traffic. There are 717 million 3G subscribers globally in 2009 [1], and the number will reach 2776 million in 2014 [2]. Data-intensive mobile devices such as smartphones and touchpads are the fastest selling consumer electronics ever. These devices enable game-changing applications, such as video streaming and social networks, on mobile Internet. As a result, mobile Internet has 240,000 terabytes of traffic in 2010, and are expected to grow to 6.3 exabytes (10^6 terabytes) in 2015 [3].

While mobile traffic imposes great pressure on service providers, *not all traffic are created equal.* Many popular applications on mobile devices, such as email synchronization and social network updates, are delay tolerant. In addition, data from a large cellular network shows that there exists a significant lag between content generation and user-initiated upload time, more that 55% uploaded content on mobile network is at least 1 day old [4]. We collected data from regular mobile users, and conservatively assess the delay requirement

of each application (details in §6). On average, 65% of the uplink traffic and 70% of the downlink traffic are delay tolerant. In addition to existing delay-tolerant data, service providers are also considering dynamic prices that incent certain applications and users to be more flexible when transmissions occur [5]. We call such data *delay-tolerant data*, which can be leveraged to improve network resource utilization, by opportunistically scheduling its transmission when the network congestion is low and the network condition is more favorable.

The delay tolerance of such jobs varies from subseconds to hours. For example, Emails can tolerate seconds to tens of seconds of delay. Content update, such as Facebook, Twitter, and RSS feeds, can tolerant hundreds of seconds of delay. Content precaching and uploading can tolerant minutes to hours of delay. OS/application updates can tolerant even longer delay. Current cellular networks more or less treat all traffic as equal, which results in performance degradation during high load period.

In this paper, we present a set of opportunistic trough filling schemes that leverage the time-variation of network congestion and delay-tolerance of certain traffic, in order to alleviate network congestion and improve network performance. The key idea is as follows: we classify cellular traffic into two categories: delay-sensitive and delay tolerant. Delay-sensitive

*Corresponding author. Email: yicwang@ucdavis.edu

traffic include voice calls, online games, and all user initiated applications, such as web surfing and texting. For delay sensitive jobs, we assume the BS schedules them with high priority and as usual. Then the remaining resource is smartly allocated among delay tolerant jobs using the opportunistic job scheduling policies we propose in the paper. This is referred to as trough filling, i.e., avoiding the high congestion period and filling the valley of network load. We note that network congestion varies in both large time scale (e.g., hours) and small time scale (e.g., seconds). The idea of trough filling is to shape and shift traffic from high congestion period to low congestion period, within the delay expectation range of the delay tolerant jobs.

In such trough filling schemes, intuitively, we want to schedule users in low congestion period and with relatively good channel condition so that resource utilization efficiency is high and congestion is low. At the same time, we need to consider the delay performance of the users. The desirable properties include low average delay, low probability of missing deadline (if the delay tolerant jobs have deadlines), low network congestion, and graceful degradation when the network is overloaded (i.e., not all jobs can be satisfied). The challenge is for the BS to schedule jobs judiciously to achieve such goals.

Compare to machine-job scheduling literature, the main difference of this work is that the channel condition of users is time-varying. In the machine-job scheduling literature, learning effect has been studied, e.g., in [6]. However, learning is a monotonic and deterministic process, while channel variation is a stochastic process. Therefore, their impacts on job scheduling are significantly different. In addition, in the cellular network job scheduling, we consider a pool of resource to be shared among all users in a single cell or cell sector. Therefore, it is a single-server scheduling instead of multiple servers. This property also impacts the selection of scheduling schemes.

This approach also shares features with opportunistic scheduling, also named multi-user diversity, that has been extensively studied and standardized in all 3G/4G cellular systems, e.g., in [7–10]. The basic idea of opportunistic scheduling is to schedule active users in relatively good channel conditions so that resource utilization efficiency is high. At a given time, channel condition information (CQI) is sent back to the BS, and the BS decides the user or the set of users to transmit in a time scale on the order of 1ms. Opportunistic scheduling exploits fast fading of multiple users to improve spectrum efficiency. The main difference is the time scale. The time scale considered in opportunistic trough filling in this paper is much larger, in the order of subseconds to tens of minutes. Because of the time scale difference, we schedule jobs; e.g., a Facebook synchronization or an album upload, while

the original multi-user scheduler schedules packets or even bits. In addition, a job has a fixed and known size during arrival, which is useful information in designing scheduling schemes. The time scale difference results in different policies, as well as different performance metrics. For example, job delay and deadlines, instead of (head-of-line) packet delay, are important metrics. In addition, even if all users have the same and static channel condition, intelligent scheduling schemes are still needed in our context for good performance, which is not the same in the multi-user diversity schemes.

The rest of the paper is organized as follows. In Section 2, we present the system model. We then present three problem formulations, minimizing deadline missing probability, average delay, and clearance time in Sections 3, 4, and 5, respectively. Simulations results are then reported, followed by conclusions and future work.

2. System Model

We consider a time-slotted system. The time slot length can be of a subsecond or a few minutes, depending on the time scale of the delay tolerance of the jobs. Note that this time-slot length is much larger than the time slot length in the existing opportunistic scheduling schemes that leverages small-time-scale fast fading, where each time slot is a few milliseconds. Because of this difference, it is possible that a job can be finished in one time slot. For notation convenience, we assume each time slot is one unit of time.

At time slot t, the available resource for DTJ (Delay Tolerant Jobs) is noted as $\eta(t)$. For example, $\eta(t)$ can be considered as the number of available resource blocks in LTE systems, the number of subcarriers in WiMAX systems, or available power in UMTS systems. We emphasize here $\eta(t)$ is available resource left after allocating resources to delay-sensitive jobs, not capacity. We do not make specific assumptions on the distributions of $\eta(t)$. At time t, the resource utilization efficiency of user i is denoted as $Z_i(t)$. We note that $Z_i(t)$ varies over different time slots due to fading environment change and user mobility. We assume that $Z_i(t)$ remains constant at time slot t. Therefore, if user i is given $x_i(t)$ units of resource at time slot t, it can transmit data of volume of $x_i(t) * Z_i(t)$ (recall that each time slot is one unit of time). Note that $x_i(t)$ is the decision variable. Let $F_i(t)$ be the remaining file size of user i at time t. We note that a rational allocation scheme will have $x_i(t) * Z_i(t) \leq F_i(t)$.

The job scheduling process proceeds as follows: at the beginning of the time slot t, the BS allocates (or estimate the need for) resource to delay-sensitive users first, and thus obtain the amount of available resource $\eta(t)$. The BS obtains channel condition and file size information of users (i.e., $Z_i(t)$s and $F_i(t)$s). Then the

BS decides the resource allocation $x_i(t)$. By the end of time slot t, each user has a remaining file size of $F_i(t+1) = \max(0, F_i(t) - x_i(t) * Z_i(t))$. For simplicity, we assume that $x_i(t)$ is a continuous variable, i.e., the BS can allocate resource in arbitrary slices. The assumption is reasonable because we are considering a relatively large time scale, at least subseconds, and each frame size is on the order of 1ms. Therefore, if needed, BS can share resource among users in a TMD fashion.

To simplify the presentation, we assume that each user has at most one job. If a user has multiple jobs, we treat them as a single job. When different jobs have different deadlines, we could handle them in an iterative manner. Alternatively, we can treat different jobs as different users. In addition, we note that in each time slot (which is relatively large), one could still perform small time scale opportunistic scheduling schemes with other DTJs and delay sensitive jobs, e.g., using a proportional fair scheduler in frames. We assume that the impact of such small time scale transmission scheduling can be approximated in the value of $Z_i(t)$, and thus ignored here for simplicity. We hope to better explore this issue in future research.

We summarize the notations used in the paper as follows.

- i: user index

- t: time index

- $Z_i(t)$: channel condition of user i at time t

- $F_i(t)$: remaining file size of user i at time t

- $\eta(t)$: the available resource for delay tolerant jobs at time t

- $x_i(t)$: the amount of resource allocated to user i at time t, the decision variable

- A_i: arrival time of job i

- D_i: deadline of job i

- K_i: time when job i is finished

- J: job completion time

In the following sessions we present three problem fomulations with different optimizing goals: missing deadline, average delay, and job completion time. We also propose a practical index policy in each section to address the complexity in the stardard solutions such as MDP (Markov Desicion Process).

3. Deadline

In this case, we assume each job, when generated, has a deadline attached. We consider the problem of minimizing the number of missed deadlines. Let A_i and

D_i denote the arrival time and the deadline of the ith job. Therefore, $F_i(A_i)$ is the original file size of job i. The problem can be formally stated as:

$$\min \quad \limsup_{t \to \infty} \frac{1}{t}\mathbf{I}\left(\sum_{t=A_i}^{D_i} x_i(t)Z_i(t) < F_i(A_i)\right) (1)$$
$$\text{subject to} \quad \sum_i x_i(t) \leq \eta(t).$$

The objective of the function is to minimize the time-average of the expected probability of missing a deadline. The expectation is taken over the distributions of random channel conditions ($Z_i(t)$s) and random available resource ($\eta(t)$). The constraint is the resource availability constraint at each time slot t. In the problem formulation, $\mathbf{I}(\cdot)$ is the indicator function and $\sum_{t=A_i}^{D_i} x_i(t)Z_i(t) < F_i(A_i)$ is the event that job i does not finish by its deadline. The left hand side ($\sum_{t=A_i}^{D_i} x_i(t)Z_i(t)$) is the total service received by job i between its arrival and its deadline, and the right hand side ($F_i(A_i)$) is the file size.

To solve this problem, one can use a standard MDP (Markov Decision Process) framework, in principle. However, in addition to requiring the stochastic models of both user channel condition and network congestion, the complexity of the above problem is prohibitively high because of the random arrival process and the dynamics of file sizes. Therefore, to address this issue, we propose the following index policy. We first define an index $I_i(t)$ for user i at time t:

$$I_i(t) = f(F_i(t)) \cdot g(Z_i(t)) \cdot h(D_i - t),$$

where $f(\cdot)$ indicates the importance of file size, $g(\cdot)$ represents the importance of instantaneous channel condition, and $h(\cdot)$ is the function that takes into account the urgency of the file. Examples of the above functions are $f(x) = 1$, $g(x) = x$, and $h(D_i - t) = 1/(D_i - t)$, where $D_i > t$. These function types are inspired by machine-job scheduling literature and by opportunistic scheduling literature. The intuition is to favor jobs that are close to their deadlines.

Based on the index, we allocate resource to users in a greedy manner. More specifically, we rank users according to their index from high to low. We first allocate as much resource as possible to the user with the highest index to finish its job; i.e.,

$$x_i(t) = \min\left(\frac{F_i(t)}{Z_i(t)}, \eta(t)\right)$$

If there is additional resource available, we allocate to the next user, and so on until all available resource is depleted or all jobs are scheduled. In practice, we may have extra constraint on the amount of resource allocated to users, e.g., due to device constraint such as power or upper-layer protocol constraints. We can simply include such constraints in the allocation.

4. Average Delay

In this case, we consider the average delay performance of the jobs. Recall that A_i and $F_i(A_i)$ denote the arrival time and the job size of the ith job. Let K_i be the time when job i is finished. The problem can be formally stated as:

$$\min \quad \limsup_{N \to \infty} \frac{1}{N} \sum_{i=1}^{N} (K_i - A_i) \quad (2)$$
$$\text{subject to} \quad \sum_i x_{ij} \leq \eta_j$$
$$\text{where} \quad K_i = \min\{T : \sum_{t=A_i}^{T} x_i(t) Z_i(t) \geq F_i(A_i)\}.$$

The objective of the function is to minimize average delay of all jobs. In the problem formulation, there is an implicit assumption on network stability. If the network is not stable, then the delay goes to infinity. Similar to the deadline case, in principle, the above problem can be solved using MDP, but suffers high complexity. Therefore, we propose the following index policy. We define an index $I_i(t)$ for user i at time t:

$$I_i(t) = f(F_i(t)) \cdot g(Z_i(t)),$$

We note that to minimize average delay, one would favor smaller jobs, as in machine-job scheduling literature. An example is $I_i(t) = Z_i(t)/F_i(t)$. Similar to the deadline case, a greedy allocation policy can be applied based on the index from high to low.

5. Job Completion Time

Last, we consider the static case where all jobs arrive at the beginning of time slot 1 and there are no new arrivals. In this case, we consider the problem of minimizing job completion time, which is also referred to as clearance time in machine job scheduling literature. In other words, we want to minimize the time where all jobs are finished. Define J as the clearance time, the problem can be formulated as:

$$\min \quad E(J) \quad (3)$$
$$\text{subject to} \quad \sum_i x_i(t) \leq \eta(t), \forall t, \quad (4)$$
$$\sum_{t=1}^{J} x_i(t) Z_i(t) \geq F_i(1). \quad (5)$$

In this formulation, the objective is to minimize the expected total completion time. The second constraint is the constraint that all jobs are finished by time J. Note that $F_i(1)$ represents the original file size of user i and the LHS of Eq. (5) represents the total capacity allocated to user i. In this problem formulation, because deadlines do not exist, we propose the following index $I_i(t)$:

$$I_i(t) = g(Z_i(t)). \quad (6)$$

We then rank users according to their index. The users with higher index values have higher access priority.

In the simulation, we considered $g(x) = x$. Since our goal is to finish all work earlier, favoring better channel condition improves the throughput when there is no deadline constraint. There is a little bit of subtlety in this formulation. In wireless channels, when the number of users is large, the overall (opportunistic) capacity is higher. For example, suppose that our policy is to choose the best of users to transmit. Then the larger the number of active users, the higher the capacity. It has been well studied in the opportunistic scheduling literature that this capacity gain increases as $\log(\log(n))$, where n is the number of users. Therefore, in principle, to minimize the completion time, one should also consider the number of (remaining) users in the system. However, since the capacity gain increases as $\log(\log(n))$, the difference is significant only when the number of users is small. Therefore, for simplicity, we ignore its impact in the current algorithm. It is our future work to further explore this situation.

6. Real-life Trace

For realistic evaluations on the performance of the proposed algorithm, it is crucial to collect real-life traces from the general public and conduct trace-driven simulations with diverse system parameters.

In this paper, we strive to realistically evaluate the benefit of leveraging delay tolerant traffic for the general public in real life scenarios. To achieve this goal, we collected data from the users of an Android application called PhotoSync[1]. We published PhotoSync on Google Play[11] in July 2012. There have been more than 10,000 downloads in the first eight months. Among them, with clear privacy notification, about 1700+ users participate in our data collection. After filtering out the profiles with zero-length and corrupted data, we ended up with the profiles from ~ 700 users. Among them, we have more than 100 users with 30+ days of profiles, which are used in our performance evaluations.

We collect from users, among other metrics, 3G signal strength, WiFi connectivity, and application traffic. We observed that the top 50 applications contribute $\sim 80\%$ of the traffic, and the top 10 applications contribute $\sim 60\%$ of the traffic. We also roughly classify the applications into two groups, applications that generate: (i) delay tolerant and (ii) real-time traffic. For downlink traffic, we consider Dropbox, social network content pre-fetching, and application update

---

[1] The app follows the required privacy guideline, fully discloses the information collected, and allows users to easily opt out from sending their profiles to us. We observe that a large number of users choose to opt out, an indication of user privacy-awareness and the effectiveness of the privacy disclosure. Furthermore, we only keep a hashed user ID through a one-way hash function, which is not reversible to ensure anonymity.

as delay tolerant; and Android browser and YouTube as real-time. For uplink traffic, we consider Dropbox, social network photo backup as delay tolerant; and browser, messaging, and video conferencing as real-time. Examples of the applications generating delay tolerant traffic include cloud storage applications (such as Dropbox) and social network applications (such as Facebook and Google+), which may tolerate delays in the order of minutes, and photo sharing applications (such as Instagram), which may tolerate delays in the order of hours [12]. For applications that are in the grey area, we consider them as real-time traffic, to be conservative. We then compute the per-user delay tolerant traffic fraction of the top 50 applications (in each direction). There are on average 65% (uplink) and 70% (downlink) delay tolerant traffic, which shows the potential of the proposed algorithm.

Real-life simulations use the same setup described in §7.1. The network condition $Z_i(t)$, file size $F_i(t)$, and job arrival time A_i used in the simulations are extracted from the user traces.

7. Evaluation

In this section we evaluate the proposed algorithm using simulations with both synthetic traces and real-life traces collected from the general public. We compare the performance of our algorithm under three different optimization goals, namely average delay, missed deadline, and completion time, with a naive algorithm that evenly allocates resources to all users and a baseline algorithm that allocate resources solely base on channel condition.

7.1. Simulation Setup

First, we explain the simulation setup. We assume the time is slotted. There are totally N users in the system who have exactly one file to transmit during the test. In the completion time tests, all files arrive at time slot 1.

When a job arrives, it has a deadline D_i (if necessary, not considered in the average delay case and the clearance case). At each time slot, each user experiences a channel condition $Z_i(t)$ distributed between 1 and 4. At each time slot, the available resource is between 1 and 50.

In a single test, at the beginning of each time slot j, we observe the current network condition $Z_i(t)$ for all users who have file to transmit. Each time slot, the system has $\eta(t)$ units of resources left for delay tolerant jobs. The proposed algorithm allocates $x_i(t)$ units of resources to each job i. The allocated resources for each file is guaranteed not to exceed needed resources nor the total available resource. Thus the final transmission rate per time slot for each file is $x_i(t) \cdot Z_i(t)$. The remaining size of each file is updated after transmission at the end of each time slot.

Objective	Index	Priority
Deadline	$Z_i(t)/(F_i(t)* max(D_i - j, 1))$	Jobs close to deadline
Delay	$Z_i(t)/F_i(t)$	Short jobs
Completion	$Z_i(t) * F_i(t)$	Long jobs

Table 1. Summary of different objectives and index functions.

Next, we evaluate the proposed index policies via simulation with both synthetic and real-life traces. In both simulations, we compare the index policies with different objectives against two baseline policies, a channel-only policy and an even policy. The index algorithm is noted as *index* in all figures. Channel-only policy is a special index policy with $I_i(t) = Z_i(t)$, which favors users with the best channel conditions, noted as *channel-only* in the figures. Finally the baseline policy where resource $\eta(t)$ is evenly allocated to all users, is noted as *even* in all figures. The average delay, missed deadline, and completion time of the index algorithm is compared against the previously mentioned two baseline policies. Note that different index functions are used for different objectives. We summarize the index functions used in the policies and their intuition in Table 1. In all the index functions, we favor users with good channel conditions with a linear function, which takes inspiration from the majority of optimal policies in multi-user diversity scheduling schemes. In addition, under different objective, file size and deadline are considered differently. In the simulations, different functions of file size and deadline are also considered, including log and square root functions, but not reported here individually for conciseness.

The completion time J is calculated for each test, which is the latest time all job finishes. The average delay of all jobs and the number of jobs which missed its deadline are also collected for all tests.

7.2. Synthetic Trace

First we evaluate the proposed algorithm using synthetic traces to evaluate its performance in a theoretic context where we can control system parameters and isolate specific behaviors. For example, the system load is hard to control when the file size are from real traces, and we cannot control parameters such as max file size and channel variation.

For average delay and missed deadline tests, file i has a random arrival time A_i, uniformly distributed in the test horizon (which is similar to the effect of Poisson arrival).

The deadline D_i is uniformly distributed between the current time and the horizon. Each file is of a random size uniformly distributed between the minimum file size 1 and the maximum file size 100. The difference reflects the diversity of applications. At each time slot, each user experiences a random channel condition, such that $Z_i(t)$ is uniformly distributed between 1 and 4. At each time slot, the available resource is randomly chosen between 1 and 50.

In synthetic simulations, we assume all users have homogeneous channel conditions. This is to allow us to focus on the impact of file size and load. If the users have heterogeneous channel conditions (e.g., being close to or far away from the base station), appropriate normalization may be considered; e.g., normalizing over its own average condition so that high value indicates relatively good channel condition.

Each test is repeated 500 times to be statistically meaningful. The parameters such as system load are adjust to study the impact of that particular argument. System load represents the ratio of total traffic versus available system resource available during the whole test. If we define total number of file as N and the total length of the simulation as H, the system load is defined as

$$L = \frac{N \cdot E(F_{i0})}{H \cdot E(\eta) \cdot E(Z_{ij})}. \tag{7}$$

The load is an indicator of how busy the network is.

7.3. Simulation Results

Average Delay. We first evaluate the performance of the proposed index algorithm under the objective of minimizing average delay. The index function used is

$$I_i(t) = Z_i(t)/F_i(t) \tag{8}$$

We note that the proposed index policy favors jobs with shorter remaining time and thus results in lower average delay compared to the channel only policy and even policy.

For the synthetic simulation, average delay are plotted in Figure 1 for the three policies as the system load varies from 0.7 to 1.5. For the real-life simulation, average delay are plotted in Figure 2 for the three policies as the system load varies from 0.5 to 1.5. From the figures, we observe that both index policy and channel-only policy show significant improvement over the *even* policy, which indicates the user diversity gain. The index algorithm outperform the channel only algorithm by over 20% in terms of average delay.

Figure 3 shows the throughput of all three algorithms. Throughput defined as the total size of all *completed* jobs during the simulation. While the index policy focuses on average delay in this cases, its throughput remain similar to channel only policy.

Furthermore, in Figure 4, we illustrate the traffic smoothing impact of the proposed opportunistic trough filling scheme. The solid line indicates the load without opportunistic trough filling (OTF) and the dashed line is the load after trough filling. The load before OTF

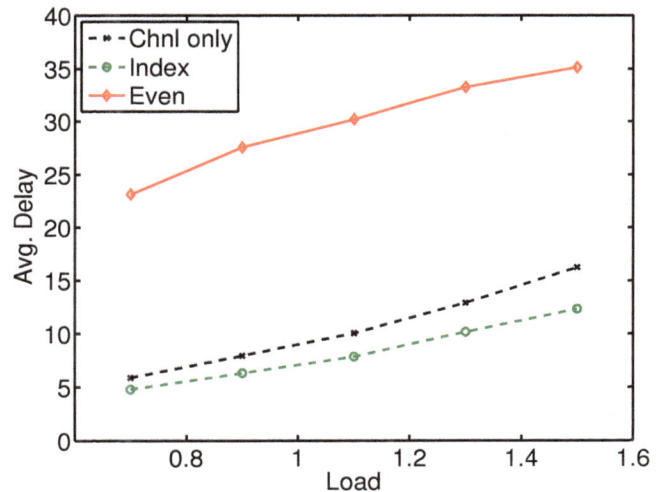

Figure 1. Average Delay when Index Policy Objective is Minimizing Average Delay, Synthetic Trace

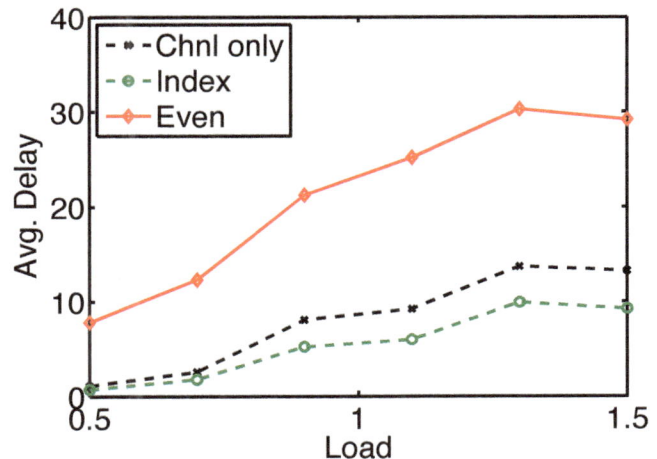

Figure 2. Average Delay when Index Policy Objective is Minimizing Average Delay, Real-life Trace

sometimes exceeds the available resource, which is capped at 50. After trough filling, the traffic is much more smooth. This figure is an intuitive illustration of "trough filling"; i.e., filling the valley and reducing the peak.

Deadline. We study the case where each job has a given deadline in this section. The index function used is

$$I_i(t) = Z_i(t)/(F_i(t) * max(D_i - j, 1)) \tag{9}$$

For synthetic simulations, missed deadlines and average delay are plotted in Figure 5 and Figure 6 respectively for the three policies as the system load varies from 0.7 to 1.5.

In synthetic simulation, we also test the impact of the file size variation. As shown in Fig. 7, the percentage of

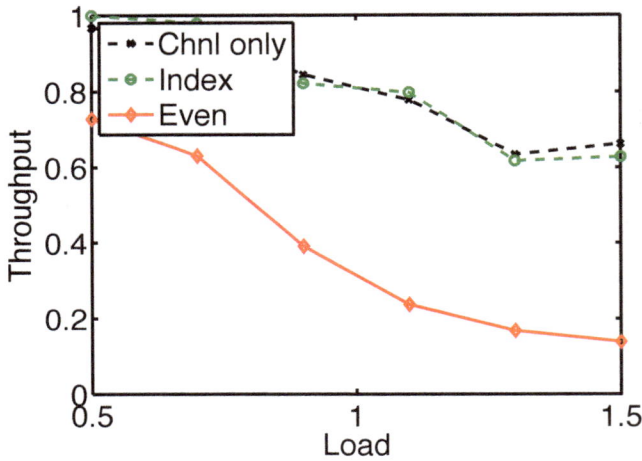

Figure 3. Throughput when Index Policy Objective is Minimizing Average Delay, Real–life Trace

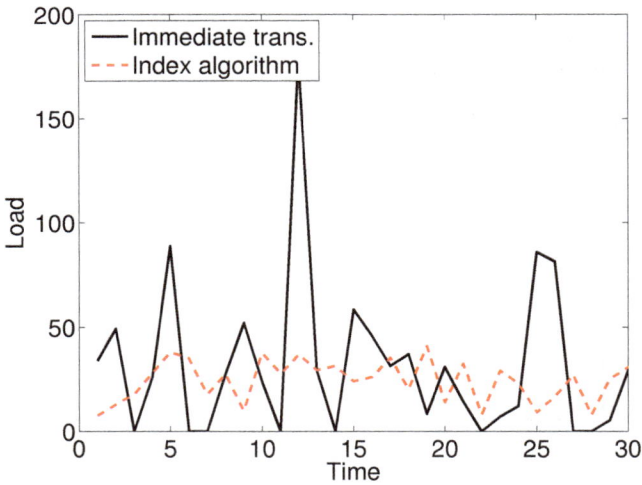

Figure 4. Traffic smoothing

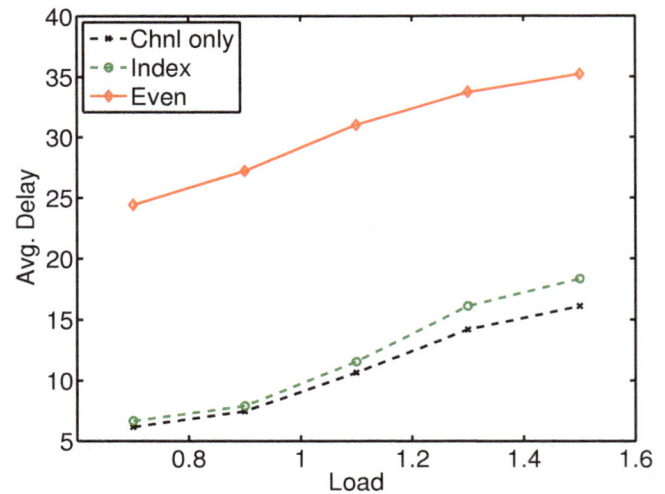

Figure 5. Percentage of Jobs Missed Deadline when Index Policy Objective is Minimizing Missed Deadlines, Synthetic Trace

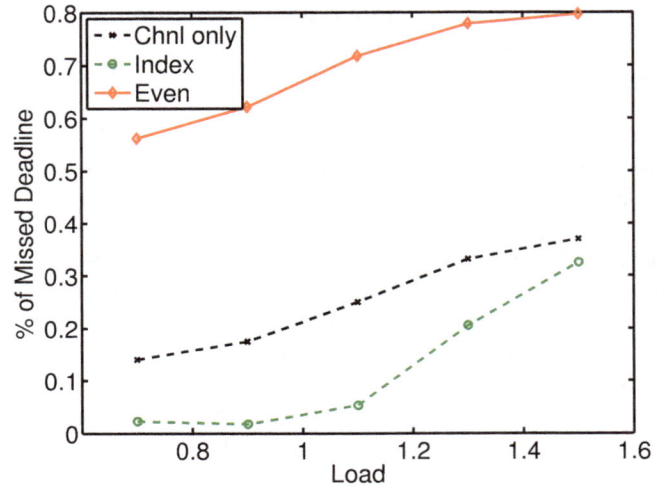

Figure 6. Average Delay when Index Policy Objective is Minimizing Missed Deadlines, Synthetic Trace

jobs missed the deadline increases as the maximum file size increases in the synthetic trace. Intuitively when the maximum and average file size increases, even the system load stay the same, the it is more likely that the larger jobs will miss the deadline.

For real-life simulations, missed deadlines and average delay are plotted in Figure 8 and Figure 9 respectively for the three policies as the system load varies from 0.7 to 1.5.

In Figure 5 and Figure 8, the x-axis is the load, the y-axis is percentage of jobs which missed their deadlines. Compared with the channel-only policy, the proposed index policy favors users close to the deadline, and thus has much lower missing rate. The cost of it is the slightly longer average delay.

In Figure 10 shows the throughput of all three algorithm in this simulation. The index policy strive

to complete jobs within their deadlines, thus having a higher throughput then the channel-only policy

Job Completion Time. As shown in Figure 11, in the job completion time simulation, our index policy is the same as channel-only policy, with $I_i(t) = Z_i(t)$. At the first sight, this may look different from the classic job scheduling literature, where the scheduling of the largest jobs often significantly impacts the completion time. However, in our case, there is one single pool of resource that is shared by all jobs similar to a single server queue with time-varying capacity. If users' channel condition does not change over time, then any work-conserving scheduling scheme will result in the same performance. Therefore, channel-only scheduling

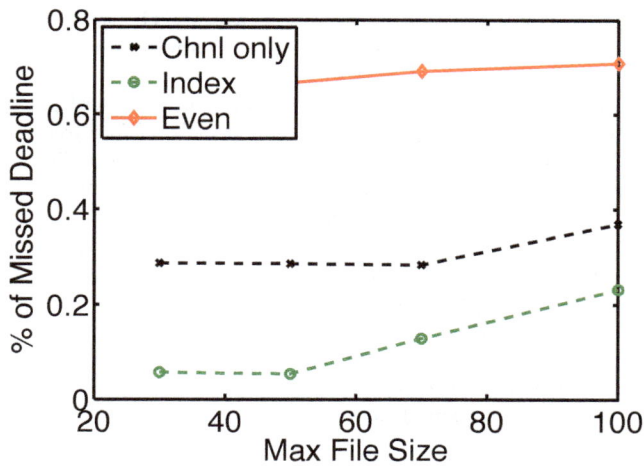

Figure 7. Percentage of Missed Deadline with Different Maximum File Size, Synthetic Trace

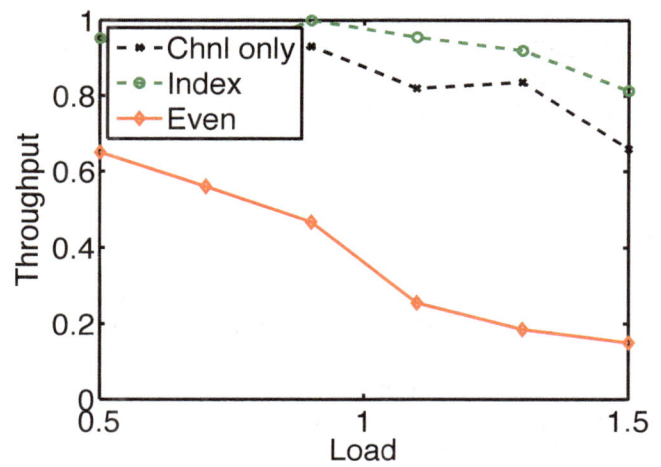

Figure 8. Percentage of Jobs Missed Deadline when Index Policy Objective is Minimizing Missed Deadlines, Real–life Trace

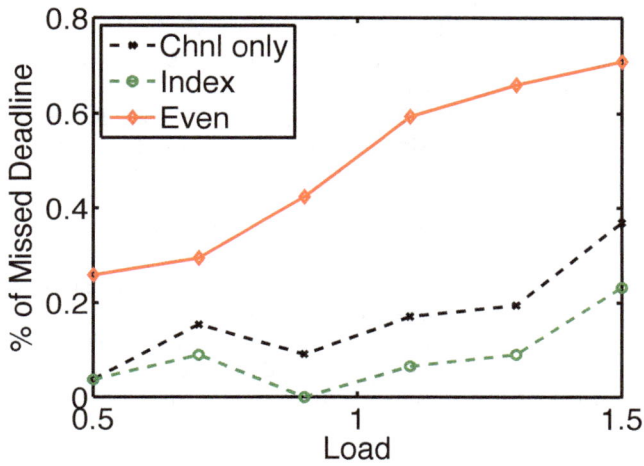

Figure 9. Average Delay when Index Policy Objective is Minimizing Missed Deadlines, Real–life Trace

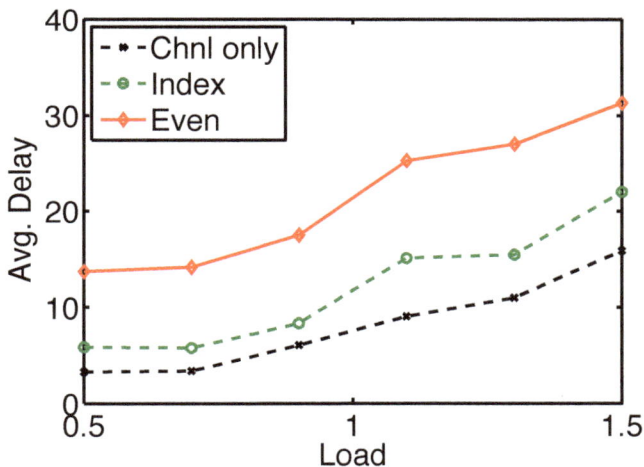

Figure 10. Throughput when Index Policy Objective is Minimizing Missed Deadlines, Real–life Trace

works well in our simulations by opportunistically favor users in good channel conditions.

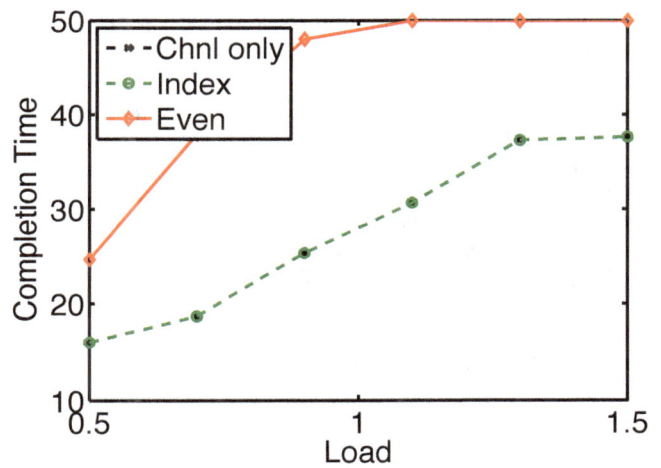

Figure 11. Job Completion Time when Index Policy Objective is Minimizing Average Delay

8. Conclusion and Future Work

This paper is a first attempt to develop large-time-scale opportunistic trough-filling schemes in cellular networks. The objective is to alleviate network congestion and improve cellular performance. Two factors are considered: first, a large amount of jobs are delay-tolerant, we can delay them to times when the network is less congested. Second, because of this delay-tolerance, users can be scheduled to transmit in relatively good channel conditions that improves spectrum efficiency; e.g., due to fading environment or user mobility. Both effects alleviate network congestion and improve user experience. We have considered three distinct objectives: minimizing delay, minimizing missed deadlines, and minimizing clearance time. Because of the practical and numerical complexity involved in their optimal solutions, we resort to heuristic index policies that take into account file size, channel condition, and deadline. Numerical results are promising, indicating large performance gain over naive algorithms that do not perform such opportunistic trough filling.

Many challenging problems are to be addressed. First, our current schemes are heuristics, mostly inspired by the existing literature on machine job scheduling and opportunistic scheduling. We hope to evaluate them in a more rigorous manner, using metrics such as throughput optimality, average delay, and average queue flow probability. We hope to further develop optimal schemes to study the impact of user channel variation in reality. In addition, we hope to consider practical issues, including signaling overhead and distributed implementations.

References

[1] (2010), World Telecommunication/ICT Indicators Database, http://www.itu.int/ITU-D/ict/statistics/.

[2] (2009), The Mobile Internet Report, http://www.morganstanley.com/institutional/techresearch/pdfs/mobile_internet_report.pdf.

[3] (2010), Cisco Visual Networking Index: Global Mobile Data Traffic Forecast Update, 2010âĂŞ2015, http://www.cisco.com/en/US/solutions/collateral/ns341/ns525/ns537/ns705/ns827/white_paper_c11-520862.html.

[4] Trestian, I., Ranjan, S., Kuzmanovic, A., and Nucci, A. (2011) Taming user-generated content in mobile networks via drop zones. In *Proc. of IEEE INFOCOM'11*.

[5] Joe-Wong, C., Ha, S. and Chiang, M. Time-dependent internet pricing.

[6] Gur and Mosheiov (2001) Scheduling problems with a learning effect. *European Journal of Operational Research* **132**(3): 687 – 693. doi:10.1016/S0377-2217(00)00175-2, URL http://www.sciencedirect.com/science/article/pii/S0377221700001752.

[7] Liu, X., Chong, E.K.P. and Shroff, N.B. (2003) A framework for opportunistic scheduling in wireless networks. *Computer Networks* **41**: 451–474.

[8] Lin, X., Shroff, N.B. and Srikant, R. A tutorial on cross-layer optimization in wireless networks. *IEEE Journal on Selected Areas in Communications, Special Issue on "Non-Linear Optimization of Communication Systems"*, vol. 24, no. 8, August 2006. .

[9] Bodas, S., Shakkottai, S., Ying, L. and Srikant, R. (2009) Scheduling in multi-channel wireless networks: rate function optimality in the small-buffer regime. In *Proceedings of the eleventh international joint conference on Measurement and modeling of computer systems*, SIGMETRICS '09 (New York, NY, USA: ACM): 121–132. doi:http://doi.acm.org/10.1145/1555349.1555364, URL http://doi.acm.org/10.1145/1555349.1555364.

[10] Shreeshankar, B., Sanjay, S., Lei, Y. and R., S. (2010) Low-complexity scheduling algorithms for multi-channel downlink wireless networks. In *Proceedings of the 29th conference on Information communications*, INFOCOM'10 (Piscataway, NJ, USA: IEEE Press): 2222–2230.

[11] Photosync. https://play.google.com/store/apps/details?id=com.metaisle.photosync.

[12] Trestian, I., Ranjan, S., Kuzmanovic, A. and Nucci, A. (2011) Taming user-generated content in mobile networks via drop zones. In *Proc. of IEEE INFOCOM'11* (Shanghai, China): 2040–2048.

9

Incentivize Spectrum Leasing in Cognitive Radio Networks by Exploiting Cooperative Retransmission

Xiaoyan Wang[1], Yusheng Ji[1,*], Hao Zhou[1], Zhi Liu[2] and Jie Li[3]

[1]Information Systems Architecture Science Research Division, National Institute of Informatics, Tokyo, Japan
[2]Global Information and Telecommunication Institute, Waseda University, Tokyo, Japan
[3]Faculty of Engineering, Information and Systems, University of Tsukuba, Tsukuba Science City, Japan

Abstract

This paper addresses the spectrum leasing issue in cognitive radio networks by exploiting the secondary user's cooperative retransmission. In contrast with the previous researches that focuses on cancellation-based or coding-based cooperative retransmissions, we propose a novel trading-based mechanism to facilitate the cooperative retransmission for cognitive radio networks. By utilizing the Stackelberg game model, we incentivize the otherwise non-cooperative users by maximizing their utilities in terms of transmission rates and economic profit. We analyze the existence of the unique Nash equilibrium of the game, and provide the optimal solutions with corresponding constraints. Numerical results demonstrate the efficiency of the proposed mechanism, under which the performance of the whole system could be substantially improved.

Keywords: spectrum leasing, cooperative retransmission, trading model, game theoretic approach

1. Introduction

Spectrum has become the scarcest resource in wireless communications due to the emerging wireless services and products in past decade. The current fixed spectrum allocation scheme is recognized to be very inefficient due to the exclusive use to licensed services (e.g., less than 20% on average in Chicago across all bands [1]). With the development of cognitive radio technologies [2], dynamic spectrum access has been suggested as a promising technique to improve the spectrum utilization efficiency. By dynamic spectrum access, an unlicensed user is capable of accessing the licensed spectrum dynamically and serving its own traffic.

The dynamic spectrum access schemes can be classified into two categories: *opportunistic spectrum access* [3–6] and *negotiated spectrum access* [7–12]. In opportunistic spectrum access, the Primary Users (PUs) are oblivious of the presence of the Secondary Users (SUs). The SUs are allowed to access the channel only in the spectrum holes (e.g., interwave schemes [3, 4])

or under a certain interference constraint (e.g., overlay schemes [5, 6]). In negotiated spectrum access, instead, the PUs are aware of the existence of the SUs. The PUs and SUs explicitly communicate with each other to reach a spectrum sharing arrangement. Specifically, PU leases part of the spectrum resources to the SU (i.e., spectrum leasing) in exchange for the improved transmission quality or appropriate economic profit.

In the context of dynamic spectrum access, most of the previous researches focused on spectrum access in the whitespace or during the PU's transmission slot [3, 4, 6–10]. However, there is a growing body of work recently that investigates the spectrum access during the PU's retransmission slot [5, 11, 12]. These work has shown that nontrivial rates for PU and SU can be achieved by careful retransmission resource management, in spite of the retransmission may be relatively infrequent. Specifically, Tannious *et al.* [5] proposed a cancellation based retransmission scheme for cognitive radio networks. This scheme allows the SU to opportunistically access the spectrum by transmitting with the PU simultaneously during the PU's retransmission period. The interference at both PU and SU sides could be canceled or mitigated by

---

*Corresponding author. Email: kei@nii.ac.jp

utilizing the knowledge of primary packet. However, the throughput of SU is extracted at the cost of reduced PU's transmission rate, which is undesired. Since PU has higher priority in cognitive radio networks, it is unrealistic to assume that PU is willing to share the spectrum if it cannot obtain any profit or at least be compensated. To incentivize the cooperation at both PU and SU sides, Stanojev et al. proposed an auction-based negotiated spectrum access scheme to reassign the retransmission slot. In this scheme, in exchange for the cooperative retransmission from the SU, PU yields part of the retransmission duration for the aided SU's own traffic. The competition of the multiple SUs are modeled by auction theory. This scheme, however, needs every SU has a minimum tolerated reliability, thus cannot be applied in a network with SUs using best-effort transmission. Our previous work [12] also investigated the negotiated spectrum access issue for cooperative retransmissions. we proposed a cooperative coding-based retransmission protocol for the cognitive radio networks. In the proposed protocol, SU relays the primary packet and serves its own traffic simultaneously. To realize that, Network Coding (NC) scheme MIMO_NC [13] is applied in the physical layer, by which the original packets can be retrieved by the corrupted and redundant packets. However, the success probability of this scheme relies on the SINR threshold of MIMO_NC, and it potentially increases the processing complexity of the wireless terminals.

In this paper, we propose a novel *trading-based* spectrum leasing mechanism for cooperative retransmission in cognitive radio networks. The proposed trading framework involves both *resource-exchange* and *money-exchange*. The resource-exchange refers to the exchange between licensed spectrum and SU's cooperating expense, and the money-exchange is executed in the form of *charge* and *reimbursement*. Motivated by the proposed trading model, a PU is willing to lease the licensed spectrum to SUs during the retransmission slot, if it can benefit from enhanced retransmission quality or considerable economic profit. And SUs may compete to access the licensed spectrum for their own transmissions in a best-effort manner, if the cost is reasonable. Under the assumption that both PU and SU are rational, we formulate the network to a Stackelberg game, which characterizes the hierarchical feature of the system. With the objective to maximizing respective utilities, we derive the optimal strategies for both PU and SU.

The main contributions of this paper are summarized as follows.

- We propose a novel trading-based spectrum leasing mechanism for cognitive radio networks by exploiting the cooperative retransmissions.

- We formulate the considered problem as a Stackelberg game, where PU and SU have their selfish objectives in terms of transmission rates and economic return.

- We analyze the unique Nash equilibrium of the game, and derive the optimal solutions with corresponding constraints. The solutions give reliable predictions of the system outcome, and are easy to implement.

- We show the efficiency of the proposed mechanism by numerical results, and analyze the impacts of different parameter settings.

The rest of the paper is organized as follows. We present the system and trading models in Section 2. In Section 3, we formulate the problem by game theory, and analyze its Nash equilibrium. Numerical results are addressed in Section 4. And finally Section 5 draws the conclusions.

2. System and Trading Models

2.1. System Model

As depicted in Fig. 1, we consider a cognitive radio network, where a PU communicates with the Base Station (BS). In the same spectrum band, a secondary network with multiple SUs tries to exploit best-effort transmissions with BS[1] by taking advantage of the opportunities that arise during PU's retransmission period (with normalized duration 1). In the case of a failed primary transmission, the SUs may hear the primary packet and be able to decode it (Fig .1(a)). Upon receiving the negative feedback message from the BS, PU involves a subset S of capable SUs[2] to perform cooperative retransmission by employing orthogonal Space-Time Codes (STCs) scheme [15] in an α fraction of the retransmission interval (Fig. 1(b)). And as a reward, the remaining $1 - \alpha$ fraction of the retransmission interval is allocated to the cooperating SUs for their own data transmissions in a Time-Division Multiplexing Access (TDMA) mode (Fig. 1(c)). The cooperation between PU and SU are incentivized by the trading model addressed in next subsection.

We assume that the channel is subject to Rayleigh fading, and is invariant within a coherence interval. The channel gain between PU and BS is denoted as h_p, and

---

[1]The considered network focuses on the primary network access, i.e., the SU accesses the primary base station through the licensed spectrum. However, it will not restrict the implementation of the proposed scheme to other scenarios, such as the secondary network access, i.e., the SU accesses the secondary base station or other SUs.

[2]Capable SU refers to the SU that can correctly decode the primary data during the transmission interval. And in implementation, this information is reported by the individual SU to PU after a random backoff time.

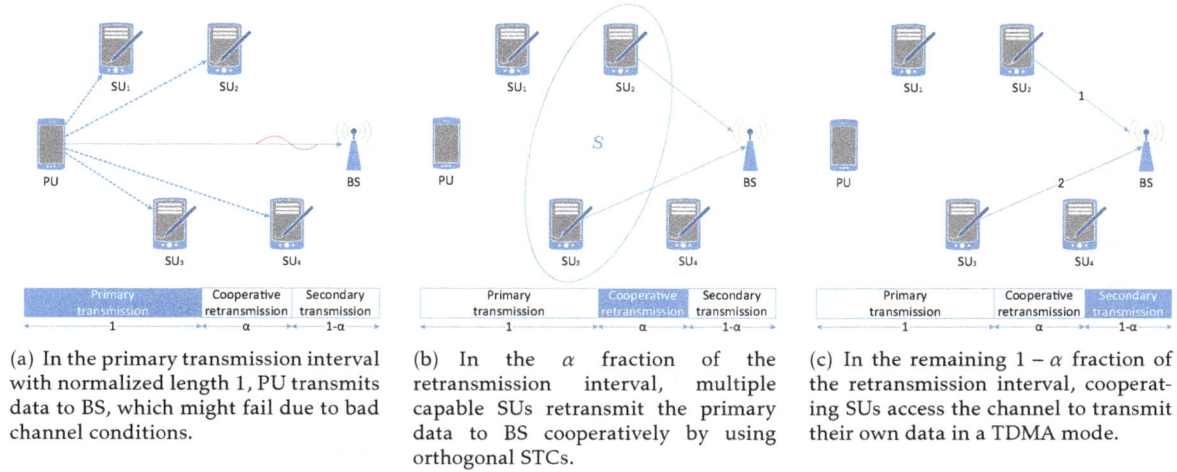

Primary transmission	Cooperative retransmission	Secondary transmission
1	α	1-α

(a) In the primary transmission interval with normalized length 1, PU transmits data to BS, which might fail due to bad channel conditions.

Primary transmission	Cooperative retransmission	Secondary transmission
1	α	1-α

(b) In the α fraction of the retransmission interval, multiple capable SUs retransmit the primary data to BS cooperatively by using orthogonal STCs.

Primary transmission	Cooperative retransmission	Secondary transmission
1	α	1-α

(c) In the remaining $1 - \alpha$ fraction of the retransmission interval, cooperating SUs access the channel to transmit their own data in a TDMA mode.

Figure 1. Transmission model for cooperative retransmission based spectrum leasing for cognitive radio networks

the channel gains between SU i ($SU_i, i \in \mathcal{S}$) and BS is denoted as h_i. It is assumed that the channel coefficient follows the Gaussian zero mean distribution, and the noises at the receivers are normally distributed as N_0.

By the cooperation of SU subset $\mathcal{S}$ using orthogonal STCs, PU's retransmission rate could be expressed as

$$R_p(\mathcal{S}) = \alpha \overline{R_p(\mathcal{S})} = \alpha \log_2 \left(1 + \frac{\sum\limits_{i \in \mathcal{S}} \left(|h_i|^2 P_i \right)}{N_0} \right), \quad (1)$$

where $\overline{R_p(\mathcal{S})}$ denotes the retransmission rate over normalized time slot, and P_i is the transmitting power at SU_i. And the transmission rate for each cooperating SU_i is expressed as

$$R_i = t_i \overline{R_i} = t_i \log_2 \left(1 + \frac{|h_i|^2 P_i}{N_0} \right), \quad (2)$$

where $\overline{R_i}$ denotes SU_i's transmission rate over normalized time slot, and t_i is the time duration granted for SU_i to access the spectrum.

2.2. Trading Model

To incentivize the user cooperation, we propose a trading model involving both resource-exchange and money-exchange. For the resource-exchange, PU leases part of the retransmission interval to the cooperating SUs in exchange for the improved retransmission quality, and SU spends its transmitting energy in cooperating behavior in exchange for the opportunity to access the licensed spectrum. However, only using resource-exchange is inadequate for the considered system. Since PU has strong position in cognitive radio network, and will always prefer to allocate the whole retransmission interval to cooperative retransmission,

which leads to a non-cooperation result. To encourage the cooperation for both PU and SU sides, market mechanisms in spectrum sharing has been exploited recently [16]. In this paper, we propose a novel money-exchange model involving both *charge* and *reimbursement*. Charge is paid by the SUs for the access right of spectrum resource, and reimbursement is paid by the PU for the cooperative retransmission. Specifically, SU is charged by the amount of time that it accesses the spectrum. Since the cooperating SUs split the accessing time $1 - \alpha$ in TDMA mode, the time duration t_i granted for SU_i to access the spectrum is defined proportional to its charge c_i as

$$t_i = \frac{c_i}{\sum\limits_{j \in \mathcal{S}} c_j} (1 - \alpha). \quad (3)$$

The reimbursement to SU is defined as the product of its cooperation time and the amount of services it provided. Specifically, the cooperation time is the allocated retransmission time α, and the amount of provided services refers to its contribution to the cooperative retransmission rate. We define the reimbursement r_i for SU_i as

$$r_i = \frac{\alpha |h_i|^2 P_i \left(R_p(\mathcal{S}) \right)}{\sum_{j \in \mathcal{S}} \left(|h_j|^2 P_j \right)} \lambda, \quad (4)$$

where $|h_i|^2 P_i \big/ \sum_{j \in \mathcal{S}} \left(|h_j|^2 P_j \right)$ is the SNR contribution of SU_i to the total received SNR at BS, and λ is the reimbursement unit price.

3. Game Theoretical Framework for Spectrum Leasing in Cooperative Retransmission

In this section, we address the spectrum leasing issue in PU's retransmission by utilizing Stackelberg game.

Based on the proposed trading model, we firstly define the utility functions for both PU and SU. Then we propose a Stackelberg game based framework, where the leader (PU) optimizes its strategy based on the knowledge of the effects of its decision on the behavior of the followers (SUs).

3.1. Utility Functions

In the considered system, PU intends to maximize its retransmission rate plus the economic profit, i.e., the total charges it earned minus the total reimbursements it paid. Therefore, the natural utility U_p for PU is defined as the cooperative retransmission rate plus the economic profit as

$$U_p = R_p(\mathcal{S}) + \sum_{j \in \mathcal{S}} c_j - \sum_{j \in \mathcal{S}} r_j. \tag{5}$$

Notice that in Eqn. (5), the equivalent economic profit per unit data rate contributes to the overall utility is set to 1 for simplicity.

Similarly, the utility U_i for SU_i is defined as its transmission rate plus the received reimbursement from PU, and minus the charge it pays. We have

$$U_i = R_i + r_i - c_i. \tag{6}$$

3.2. Stackelberg Game Based Framework

The considered system is characterized by a hierarchical structure, in which PU holds the strong position and can impose its own strategy upon the SUs. This structure could be formulated to a *Stackelberg game* [17], which consists of a leader (PU) and multiple followers (SUs). In Stackelberg game, *the leader optimizes its strategy based on the knowledge of the effects of its decision on the behavior of the followers*. Therefore, with the target to maximize its utility U_p, the leader PU declares its strategy first in terms of the time fraction α and the cooperation subset $\mathcal{S}$. Then with the objective to maximize its utility U_i, the follower SU_i reacts to the PU's declared α and $\mathcal{S}$ by deciding how much it is willing to be charged for the channel access. We solve this typical leader-follower game by backward induction as follows. Firstly, we derive the optimal solution for the SUs' non-cooperative payment decision game. Then based on this outcome, we derive the optimal strategy for the PU.

SU's Optimal Strategy. Given the time fraction α and cooperation subset $\mathcal{S}$ that decided by PU, SUs in $\mathcal{S}$ compete with each other to maximize their utilities by deciding a reasonable charge for the channel access. The payment decision game is denoted as $G = \langle \mathcal{S}, \{C_i\}_{i \in \mathcal{S}}, \{U_i\}_{i \in \mathcal{S}} \rangle$, where C_i is the strategy set, and U_i is the utility of SU_i given in Eqn. (6). For the Nash

Equilibrium (NE) existence of the payment decision game G, we have

Lemma 1. A NE exists in game G.

Proof. According to [17], A NE exists in $G = \langle \mathcal{S}, \{C_i\}_{i \in \mathcal{S}}, \{U_i\}_{i \in \mathcal{S}} \rangle$, if $\forall i \in \mathcal{S}$:

- C_i is a nonempty, convex, and compact subset of some Euclidean space R^N.

- $U_i(\mathbf{c})$ is continuous in $\mathbf{c}$ and concave in c_i.

Notice that $\mathbf{c} = (c_i, \mathbf{c}_{-i})$ is the strategy profile, where $\mathbf{c}_{-i}$ denotes the vector of strategies of all players except i. It is obvious that the first requirement in **Lemma** 1 is satisfied, and $U_i(\mathbf{c})$ is continuous in $\mathbf{c}$. We prove the concavity by taking the second order derivative of U_i with respect to c_i as

$$\frac{\partial U_i}{\partial c_i} = \frac{(1 - \alpha) \overline{R_i} \left(\sum\limits_{j \in \mathcal{S}} c_j - c_i \right)}{\left(\sum\limits_{j \in \mathcal{S}} c_j \right)^2} - 1, \tag{7}$$

$$\frac{\partial^2 U_i}{\partial^2 c_i} = \frac{2(1 - \alpha) \overline{R_i} \left(c_i - \sum\limits_{j \in \mathcal{S}} c_j \right)}{\left(\sum\limits_{j \in \mathcal{S}} c_j \right)^3}. \tag{8}$$

Obviously, $\frac{\partial^2 U_i}{\partial^2 c_i}$ is always less than 0, thus $U_i(\mathbf{c})$ is concave in c_i. Therefore, two requirements are both satisfied and, a NE exists in game G. □

Next, we analyze the uniqueness of the NE in game G and, derive the NE solution with corresponding constraint.

Theorem 1. The NE of the SUs' payment decision game G is unique, and the optimal charge is

$$c_i^* = (1 - \alpha)(|\mathcal{S}| - 1) \frac{\sum\limits_{j \in \mathcal{S}} \frac{1}{\overline{R_j}} - \frac{|\mathcal{S}| - 1}{\overline{R_i}}}{\left(\sum\limits_{j \in \mathcal{S}} \frac{1}{\overline{R_j}} \right)^2}, \tag{9}$$

under constraint

$$\sum_{j \in \mathcal{S}} \frac{1}{\overline{R_j}} > \frac{|\mathcal{S}| - 1}{\overline{R_i}}. \tag{10}$$

Proof. By definition, a strategy profile $\mathbf{c}^* = (c_i^*, \mathbf{c}_{-i}^*)$ is a NE if and only if every player's strategy is a *best response* to the other players' strategies, i.e., $c_i^* \in b(\mathbf{c}_{-i}^*)$ for every player SU_i. Due to the concavity of U_i, the best response function $b(\cdot)$ can be obtained when the first derivative of U_i with respect to c_i is zero. By solving Eqn. (7) = 0, we have

$$b(\mathbf{c}) = c_i^* = \sqrt{(1-\alpha)\overline{R_i}\left(\sum_{j\in\mathcal{S}} c_j - c_i\right)} - \left(\sum_{j\in\mathcal{S}} c_j - c_i\right), \quad (11)$$

under constraint

$$\sum_{j\in\mathcal{S}} c_j - c_i < (1-\alpha)\overline{R_i}. \quad (12)$$

Notice that for the cases that constraint Eqn. (12) does not hold, c_i^* is 0.

According to [17], if the best response functions of a non-cooperative game are *standard functions* for all players, then the game has a unique NE. The function $b(\cdot)$ is said to be standard if is has the following properties:

- *Monotonicity*: $\mathbf{c} \leq \mathbf{c}' \Rightarrow b(\mathbf{c}) \leq b(\mathbf{c}')$

- *Scalability*: $\forall \theta > 0, b(\theta\mathbf{c}) \leq \theta b(\mathbf{c})$.

Eqn. (11) can be easily proved to be monotonic and scalable. Finally, the best response function is proved to be a standard function and thus, there exists a unique NE for game G. However, Eqn. (11) cannot be directly used in real implementation, since SU_i is unaware of the charges of other SUs, i.e. c_j. To derive a practical formula for c_i^*, we solve the Eqn. (11) set with $i = (1, 2, \cdots, |\mathcal{S}|)$. By removing the term c_j, we can express c_i^* as Eqn. (9) in **Theorem** 1. Substituting Eqn. (9) into constraint Eqn. (12), we can rewrite the new constraint as Eqn. (10), which *will be used by the PU to select the optimal cooperation subset $\mathcal{S}$.* $\qquad \square$

Notice that in **Theorem** 1, to calculate c_i^*, the values of α, $|\mathcal{S}|$, $\overline{R_i}$, and $\sum_{j\in\mathcal{S}} \frac{1}{\overline{R_j}}$ are needed at SU. For implementation, besides $\overline{R_i}$, the other 3 parameters are piggybacked by PU. It is assumed that PU is aware of the individual SUs' channel conditions by periodically reporting (e.g. like BS handles the D2D link in 4G cellular systems). Thus, the value of $\sum_{j\in\mathcal{S}} \frac{1}{\overline{R_j}}$ could be easily computed by PU. Since in Eqn. (9), instead of every other $\overline{R_j}$, only the sum of all the inverse of SUs' normalized rate $\sum_{j\in\mathcal{S}} \frac{1}{\overline{R_j}}$ is required. It makes the proposed framework easy to implement and avoid large number of exchanging information. By one broadcast message with short length, every SU can calculate its optimal charge c_i^* based on Eqn. (9).

PU's Optimal Strategy. Being aware of its strategy will affect the strategy selected by SU_i (follower of the Stackelberg game), the PU (leader of the Stackelberg game) optimizes its strategy $(\alpha, \mathcal{S})$ in order to maximize its utility U_p. Substituting Eqn. (9) into Eqn. (5), the

utility of PU can be expressed as

$$U_p = (1-\alpha\lambda)\,\alpha\overline{R_p}(\mathcal{S}) + \frac{(1-\alpha)(|\mathcal{S}|-1)}{\sum_{j\in\mathcal{S}} \frac{1}{\overline{R_j}}}. \quad (13)$$

Regarding to PU's optimal strategy, we have

Theorem 2. The PU maximizes its utility when α is set to

$$\alpha^* = \frac{\overline{R_p}(\mathcal{S}) \sum_{j\in\mathcal{S}} \frac{1}{\overline{R_j}} - |\mathcal{S}| + 1}{2\lambda\overline{R_p}(\mathcal{S}) \sum_{j\in\mathcal{S}} \frac{1}{\overline{R_j}}}, \quad (14)$$

under constraint

$$\lambda > \frac{1}{2} - \frac{|\mathcal{S}|-1}{2\overline{R_p}(\mathcal{S}) \sum_{j\in\mathcal{S}} \frac{1}{\overline{R_j}}}. \quad (15)$$

Proof. By calculating the first order derivative of U_p in respective with α, we can obtain the optimal α^* given in Eqn. (14). Notice that α^* should lie in the range $[0, 1]$. Given the constraint Eqn. (10), $\alpha^* \geq 0$ is always satisfied, since

$$\overline{R_p}(\mathcal{S}) \sum_{j\in\mathcal{S}} \frac{1}{\overline{R_j}} - |\mathcal{S}| + 1 > \overline{R_p}(\mathcal{S})\frac{|\mathcal{S}|-1}{\overline{R_i}} - |\mathcal{S}| + 1 \geq 0.$$

$$(16)$$

And to ensure $\alpha^* \leq 1$, we have the constraint Eqn. (15). In implementation, Eqn. (15) is used for setting the parameter λ in the system. Specifically, we notice that Eqn. (15) is *strictly lower bounded* by 0.5, thus setting $\lambda \geq 0.5$ is appropriate regardless of the SUs' transmission rates and the size of $\mathcal{S}$. If λ is too small to satisfy Eqn. (15), PU will allocate all the retransmission interval to the cooperative retransmission (i.e., $\alpha = 1$), which results in a non-cooperation outcome. $\qquad \square$

To select the optimal cooperation subset $\mathcal{S}^*$, PU enumerates all the possible subset $\mathcal{S}$ which satisfy the constraint Eqn. (10). Based on different $\mathcal{S}$, the optimal α^* and corresponding utility U_p^* can be calculated. From all possible subsets, the $\mathcal{S}^*$ that maximizes the U_p is selected.

3.3. Implementation Details

For the network with multiple PUs transmit with one base station via distinct channels, multiple SUs could listen all the channels and randomly choose one candidate PU to cooperative, or randomly choose one channel to listen and cooperative it if possible. We give the operating details of the proposed mechanism as follows:

1. The PU transmits the primary packet to the BS, while SUs listen and store this packet. If BS

successfully decodes the packet, it sends back an ACKnowledgement (ACK) frame. Upon receiving the ACK, PU handles the next packet in the buffer if any. And SUs will discard the received primary packet.

2. If BS cannot decode the primary packet, it sends back a Negative ACK (NACK) frame. This message could be heard by SUs, who tries to decode the stored primary packet. If the decoding succeeds, it sends an Able-To-Help (ATH) frame to PU, which only consists of its identity. To avoid collision, ATH frame could be sent after a random generalized backoff time t $(0 < t < T)$, or in a particular predefined time point over time T.

3. Upon receiving the NACK frames, PU waits for $T + \delta$ time duration, where δ is the maximum propagation delay. If no transmission is overheard, PU assumes that no SU is available to provide cooperative retransmission and thus, retransmits the primary packet by itself. Otherwise, PU calculates the optimal time fraction α^* by using Eqn. (14) and determines the optimal cooperating SU set $\mathcal{S}^*$ by using Eqn. (10) over the SUs that send ATH. Then, PU could also predict the optimal charge c_i^* for SU_i by using Eqn. (9) and calculate the reimbursement r_i by Eqn. (8). Finally, SU splits the rewarding $1 - \alpha^*$ time slot to multiple durations based on Eqn. (3). PU writes α^*, $\mathcal{S}^*$, c_i^*, r_i and time splitting information (including the identity, start time, end time) into a Clear-To-Send (CTS) frame, and broadcasts it to SUs.

4. Upon receiving the CTS, selected SU_i performs the cooperative retransmission by using STCs scheme, and transmits its own message during the allocated time duration. Noted that the proposed protocol involves an extra waiting time $T + \delta$ when the secondary retransmission is unavailable. However, compared to the long payload transmission time, $T + \delta$ is extremely small and with negligible influence on the whole system.

4. Numerical Results

In this section, we show the efficiency of the proposed mechanism by numerical results. We consider the scenario that the primary transmission fails since the mutual information between PU and BS cannot support the required data rate. The network model is similar as the one in [8], where the set of SUs are all placed at approximately the same normalized distance d $(0 < d < 1)$ from the BS, and $1 - d$ from the PU. The average channel gains between SU_i and BS are assumed to be $E[|h_i|^2] = 1/d^\eta$, where $\eta = 2$ is the path loss coefficient.

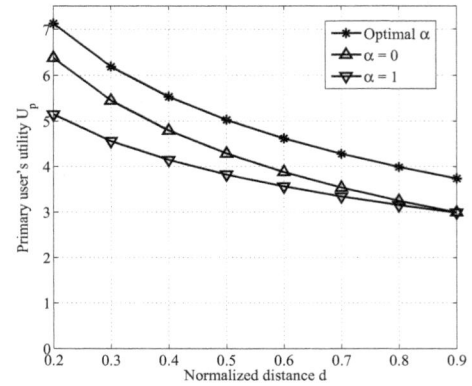

Figure 2. PU's utility by different α vs. normalized distance d, with the size of the cooperation subset $|\mathcal{S}| = 5$ and the reimbursement unit price $\lambda = 0.5$.

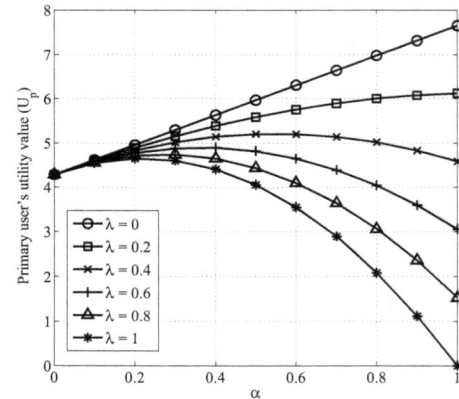

Figure 3. PU's utility at different reimbursement unit price λ vs. parameter α, with cooperation subset size $|\mathcal{S}| = 5$ and normalized distance $d = 0.5$.

Both PU and SUs transmit at a fixed transmitting power P, and the Signal to Noise Ratio (SNR) is $P/N_0 = 10$. The size of the cooperation subset $\mathcal{S}$ is 5, and reimbursement unit price λ is 0.5, unless explicitly stated otherwise.

Fig. 2 depicts the PU's utility under different schemes. Line optimal α^* denotes the proposed scheme, in which PU involves SUs to retransmit cooperatively in optimal time interval α^* and in return, leases time interval $1 - \alpha^*$ for SU's transmission. Line $\alpha = 0$ denotes the scheme that PU allocates the whole retransmission interval to the SUs' transmissions without retransmitting its own data. And line $\alpha = 1$ denotes the scheme that PU utilizes the whole retransmission interval for cooperative retransmission, without leasing any spectrum to SUs. We observe that, by optimally setting the time fraction α, the PU's utility could be maximized. Fig. 3 shows the relationship between PU's utility and parameter α at different reimbursement unit price λ. We can observe that

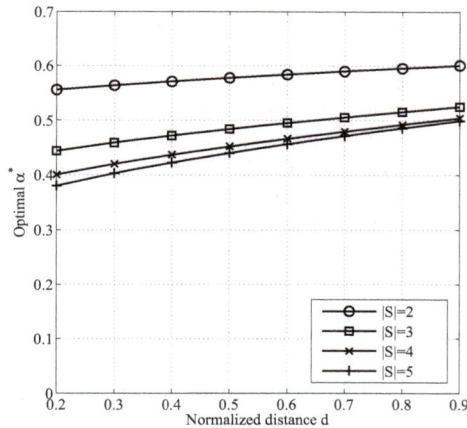

Figure 4. Optimal α^* under different cooperation subset size $|\mathcal{S}|$ vs. normalized distance d, with the reimbursement unit price $\lambda = 0.5$.

optimal α^* decreases as the λ raises. For the case $\lambda = 0$, i.e., no reimbursement is paid, PU is always willing to reserve the whole retransmission time interval to the cooperative retransmission ($\alpha^* = 1$). Actually, this is the reason why we incorporate money-exchange model to foster the cooperation. And for the high reimbursement case with $\lambda = 1$, U_p equals 0 when the whole retransmission time interval is reserved to the cooperative retransmission. It implies that in the high reimbursement case, without leasing time to SUs, PU's gain from cooperative retransmission will be totally offset by the reimbursements it pays to the SUs. Recall the Eqn. (15), setting $0.5 \leq \lambda \leq 1$ is appropriate for all the scenarios.

Fig. 4 shows the optimal α^* versus the normalized distance d at different cooperation subset size $|\mathcal{S}|$. We can observe that optimal α^* raises as the normalized distance d increases. The reason is that small d implies high channel gain and thus, leads to high cooperative retransmission rate. It is reasonable for PU to reduce α to avoid the high reimbursement and mainly rely on the charges received from SUs to maximize its utility. Moreover, the optimal α^* decreases with the size of the cooperation subset $\mathcal{S}$ raises. The reason is that with large number of cooperating SUs, the money-exchange gain from time interval $1 - \alpha$ dominates the resource-exchange gain from time interval α.

Figs. 5 and 6 depict the variation of the utilities and retransmission rates of PU with normalized distance d, respectively. It is straightforward that both U_p and $R_p(\mathcal{S})$ decrease as distance d raises. It is clear that PU tends to involve as many SUs as possible for the sake of larger utility value. Since PU is the leader of the game, it will select the maximal number of SUs to cooperate, as long as the cooperating SU satisfies constraint Eqn. (10). However, the retransmission rate does not increase as

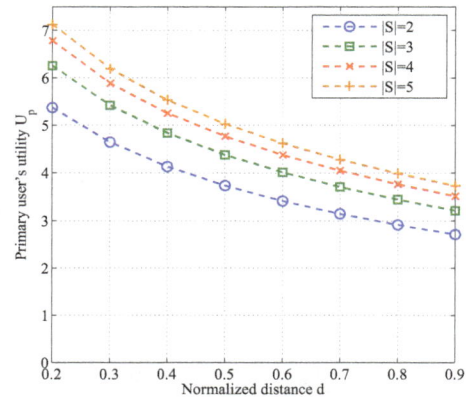

Figure 5. PU's utility at different cooperation subset size $|\mathcal{S}|$ vs. normalized distance d, with the reimbursement unit price $\lambda = 0.5$.

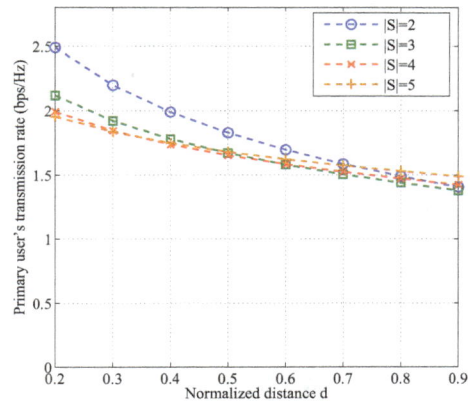

Figure 6. PU's transmission rate at different cooperation subset size $|\mathcal{S}|$ vs. normalized distance d, with the reimbursement unit price $\lambda = 0.5$.

the number of involved SUs raises. This result implies that the achieved large utility value on large $|\mathcal{S}|$ case mainly results from the economic income.

Finally, Figs. 7 and 8 depict the utilities and transmission rates of a selected SU vary with the normalized distance d, respectively. We could observe that, in contrast with the results of PU, SU prefers less contenders. This result is straightforward, since more contenders leads to smaller shared access time for each SU. However, as we mentioned, the PU is the leader in the cognitive radio networks, thus SUs have to react to the strategy of the PU.

5. Conclusions

In this paper, we have proposed a novel trading-based spectrum leasing mechanism for cooperative retransmission in cognitive radio networks. We have modeled the considered problem into a Stackelberg game. The unique Nash equilibrium of the proposed

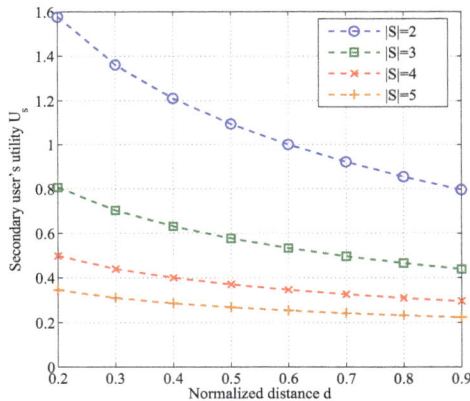

Figure 7. SU's utility at different cooperation subset size $|\mathcal{S}|$ vs. normalized distance d, with the reimbursement unit price $\lambda = 0.5$.

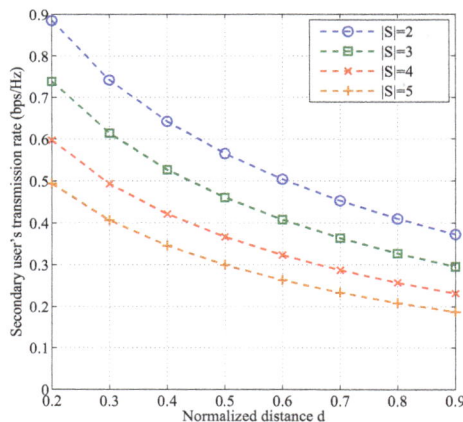

Figure 8. SU's transmission rate at different cooperation subset size $|\mathcal{S}|$ vs. normalized distance d, with the reimbursement unit price $\lambda = 0.5$.

game has been analyzed and, its optimal solutions with corresponding constraints have been derived. The proposed game theoretical framework could be implemented with low communication overhead and lead to a win-win result for both PU and SU. By numerical results, we have demonstrated that the proposed framework can substantially improve the utilities for both PU and SU in terms of data rates and economic profit.

References

[1] M. McHenry, "NSF spectrum occupancy measurements project summary," *Shared Spectrum Company*, 2005.

[2] S. Haykin, "Cognitive radio: brain-empowered wireless communications," *Selected Areas in Communications, IEEE Journal on*, vol. 23, no. 2, pp. 201–220, 2005.

[3] Q. Chen, Y.-C. Liang, M. Motani, and W.-C. Wong, "A two-level mac protocol strategy for opportunistic

spectrum access in cognitive radio networks," *Vehicular Technology, IEEE Transactions on*, vol. 60, no. 5, pp. 2164–2180, 2011.

[4] S. Liu, L. Lazos, and M. Krunz, "Cluster-based control channel allocation in opportunistic cognitive radio networks," *Mobile Computing, IEEE Transactions on*, vol. 11, no. 10, pp. 1436–1449, 2012.

[5] R. Tannious and A. Nosratinia, "Cognitive radio protocols based on exploiting hybrid arq retransmissions," *Wireless Communications, IEEE Transactions on*, vol. 9, no. 9, pp. 2833–2841, 2010.

[6] S. Sun, Y. Ju, and Y. Yamao, "Overlay cognitive radio ofdm system for 4g cellular networks," *Wireless Communications, IEEE*, vol. 20, no. 2, pp. 68–73, 2013.

[7] I. Krikidis, J. Laneman, J. Thompson, and S. McLaughlin, "Protocol design and throughput analysis for multi-user cognitive cooperative systems," *Wireless Communications, IEEE Transactions on*, vol. 8, no. 9, pp. 4740–4751, 2009.

[8] O. Simeone, I. Stanojev, S. Savazzi, Y. Bar-Ness, U. Spagnolini, and R. Pickholtz, "Spectrum leasing to cooperating secondary ad hoc networks," *Selected Areas in Communications, IEEE Journal on*, vol. 26, no. 1, pp. 203–213, 2008.

[9] J. Zhang and Q. Zhang, "Stackelberg game for utility-based cooperative cognitiveradio networks," in *Proceedings of the tenth ACM MobiHoc.* New York, NY, USA: ACM, 2009, pp. 23–32.

[10] R. Urgaonkar and M. Neely, "Opportunistic cooperation in cognitive femtocell networks," *Selected Areas in Communications, IEEE Journal on*, vol. 30, no. 3, pp. 607–616, 2012.

[11] I. Stanojev, O. Simeone, U. Spagnolini, Y. Bar-Ness, and R. Pickholtz, "Cooperative arq via auction-based spectrum leasing," *Communications, IEEE Transactions on*, vol. 58, no. 6, pp. 1843–1856, June 2010.

[12] X. Wang, Y. Ji, and J. Li, "Cooperative coding based retransmission protocol for cognitive radio networks by exploiting hybrid arq," in *Wireless Communications and Mobile Computing Conference (IWCMC), 2014 International*, Aug 2014, pp. 399–404.

[13] E. Fasolo, F. Rossetto, and M. Zorzi, "Network coding meets mimo," in *Network Coding, Theory and Applications, 2008. Fourth Workshop on*, 2008, pp. 1–6.

[14] B. Zhao and M. Valenti, "Practical relay networks: a generalization of hybrid-arq," *Selected Areas in Communications, IEEE Journal on*, vol. 23, no. 1, pp. 7–18, Jan 2005.

[15] J. Laneman and G. W. Wornell, "Distributed space-time-coded protocols for exploiting cooperative diversity in wireless networks," *Information Theory, IEEE Transactions on*, vol. 49, no. 10, pp. 2415–2425, 2003.

[16] J. Huang, "Market mechanisms for cooperative spectrum trading with incomplete network information," *Communications Magazine, IEEE*, vol. 51, no. 10, pp. 201–207, October 2013.

[17] Z. Han, *Game Theory in Wireless and Communication Networks: Theory, Models, and Applications*, ser. Cambridge books online. Cambridge University Press, 2011.

10

Spectrum Trading for Efficient Spectrum Utilization

Cong Xiong[1,*], Geoffrey Ye Li[1], Lu Lu[1], Daquan Feng[1,2], Zhi Ding[3], and Helena Mitchell[4]

[1]School of Electrical and Computer Engineering, Georgia Institute of Technology, Atlanta, GA, USA
[2]National Key Lab on Commun., University of Electronic Science and Technology of China, Chengdu, China
[3]School of Electrical and Computer Engineering, University of California, Davis, CA, USA
[4]Center for Advanced Communications Policy, Georgia Institute of Technology, Atlanta, GA, USA

Abstract

The conventional command and control based spectrum management has led to substantial underutilization of some spectrum bands while severely crowding others due to the uneven and dynamic needs that vary over time and at different locations. Spectrum trading has emerged as a promising management approach to substantially improve spectrum utilization and user experience in wireless communications by taking advantage of market-based mechanisms. This article presents an overview of spectrum trading, including the fundamental characteristics of spectrum trading markets, the state-of-the-art techniques for modeling and resolving various spectrum trading issues, and trading based dynamic spectrum sharing and access. Moreover, some open issues in spectrum trading are identified for future research in this area.

Keywords: spectrum market, spectrum trading, dynamic spectrum sharing and access

1. Introduction

Traditionally, radio spectrum is a highly regulated commodity whose management and coordination involve a central government entity, such as the *Federal Communications Commission* (FCC) in the United States. Such centralized command and control based spectrum assignment mechanism typically predicts and predetermines static bands for their respective specific usage without accounting for the dynamic nature of *radio frequency* (RF) spectrum quality and the service requirements. As a result, most preassigned spectrum bands can be substantially underutilized while many highly popular civilian bands are increasingly overcrowded, particularly due to the recent boom of many wireless applications [1–4]. For example, TV broadcast bands are routinely underutilized while cellular mobile bands are becoming overly crowded. In brief, such a centralized spectrum planning is simplistic, often leading to poor spectral efficiency, and is difficult to attain the full social and economic values for the spectrum.

Auction is a traditional market mechanism used by regulatory agencies to release the unallocated spectrum to bidders. Once the initial auction is over, the winners are licensed to use the auctioned spectrum bands. Generally, they cannot further trade their acquired spectrum on the market [5, 6]. To fully involve economic incentives in spectrum trading to achieve high spectrum utilization efficiency, subsequent markets that permit spectrum licensees to flexibly choose among capital investment, spectrum utilization, or profit trading on a dynamic basis are desired beyond the initial auction and assignment [5–8].

Unlike many popular approaches to dynamic spectrum assignment, with focus either on finding vacant spectrum for unlicensed systems or on noncooperative sharing between licensed and unlicensed users, spectrum trading is a more proactive and open form of cooperative spectrum sharing that involves temporary or long-term spectrum license transfers for economic reasons [5–8]. In essence, spectrum trading presents various spectrum seekers an opportunity to take over spectrum bands and deliver better values, thereby facilitating the establishment of dynamic incentive-driven and competitive wireless communication markets.

*Corresponding author. Email: xiongcong@gatech.edu

Liberalized trading of spectrum licenses can potentially take better advantages of market-based economic mechanisms, leading to maximum social and economic utilities of spectrum [5–10]. Specifically, in establishing a fully competitive and cooperative wireless market, wireless service providers are expected to increase their profits by thoroughly exploiting the benefits and characteristics of different spectrum bands and appropriately allocating resources across a wide swath of multi-spectrum bands. Users will have more choices for better services at lower prices. Society can better fulfill practical RF spectrum needs, including public safety, telemedicine, and social services. Thus, spectrum trading based on market supply and demand is of growing interest to regulatory agencies, social and economic studies, as well as academic and industrial wireless research groups.

This article focuses on spectrum trading in wireless communications. We present an overview on recent development of spectrum trading market including known techniques and also point out some open challenges. The rest of this article is organized as follows. Section 2 introduces several models for spectrum trading markets. Sections 3, 4, and 5 investigate several important issues of spectrum trading, models and methods for spectrum trading, and spectrum trading based dynamic spectrum sharing and access, respectively. We also highlight some open issues in Section 6. Finally, we conclude this paper in Section 7.

2. Spectrum Trading Markets

The desire to use market-based mechanisms instead of traditional command and control methods for spectrum management has motivated studies on the forms and viability of future spectrum trading markets. In general, the spectrum trading market may consist of entities, such as spectrum sellers, spectrum buyers, spectrum brokers, and a regulator, as illustrated in Figure 1. Their functions and the interaction among them depend on the spectrum trading market models. In this section, we introduce three existing spectrum trading market models, including the roles of the involved entities, which can be used to analyze the potential of spectrum trading markets. These spectrum trading market models are briefly summarized in Table 1.

2.1. Secondary Spectrum Trading Markets

In the traditional centrally controlled spectrum market, the assignment of newly released spectrum is usually static, lasting for a long term (e.g. tens of years) and covering a large geographic area (e.g., national-wide). However, such an approach has caused substantial underutilization of some licensed spectrum bands while badly crowding others [1].

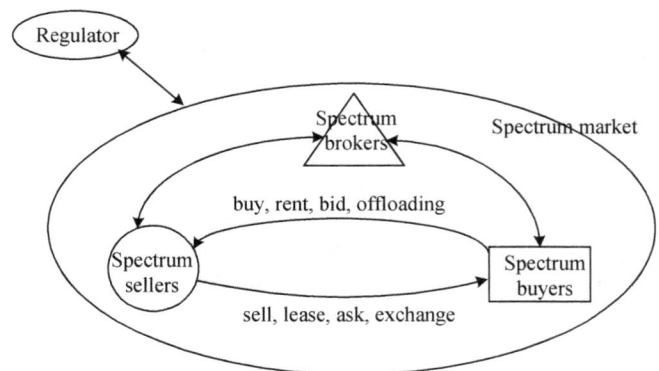

Figure 1. Structure of spectrum trading markets.

Realizing the potential power of market forces in improving the overall utilization efficiency of spectrum, governmental regulatory agencies have acknowledged the possibility for existing spectrum licensees to trade their unused spare spectrum through secondary spectrum markets under certain rules [2–4]. A secondary spectrum market allows and encourages existing spectrum licensees to trade away their rights to unused spectrum, either leasing it temporarily, or on a longer-term basis, or even outright selling their rights, to other organizations that badly need this spectrum. We believe that an effectively functioning secondary spectrum market can increase the amount of spectrum capacity available to prospective wireless service demands and new wireless technologies by substantially improving the utilization of spectrum currently assigned to less active license holders.

To facilitate successful development of secondary spectrum markets, two categories of spectrum leasing options have been proposed [2, 3]. One leasing option, namely "spectrum manager" leasing, permits entities to engage in leasing arrangements without prior governmental approval as long as the original licensee retains both *de jure control* of the license and *de facto control* over the leased spectrum. The other option, known as "de facto transfer" leasing, creates a streamlined approval process for leases that involve a transfer of the de facto control of the license.

The spectrum market based on the concept of secondary spectrum market adopts some market-driven mechanisms for spectrum management and is a smooth evolution of the command and control based spectrum market. However, it still lacks the high-level liberalization of commodity market and involves a certain level of governmental regulation and restriction, which may hinder the efficient reallocation of spectrum.

Table. 1. Comparison of different models for spectrum trading market

Model	Features
Secondary spectrum trading market	spectrum licensees trade with buyers directly spectrum leasing without transfer of de jure control of license low-level of market-based mechanisms, high-level of regulation
Exchanged-based spectrum trading market	market maker facilitates trading as a dealer spectrum leasing with and without transfer of de jure control of license medium-level of market-based mechanisms, medium-level of regulation
Two-tire spectrum trading market	separation of spectrum ownership from provision of wireless service spectrum broker manages spectrum assets high-level of market-based mechanisms, low-level of regulation

2.2. Exchange-based Spectrum Trading Markets

To introduce more market-forces into spectrum trading, an exchange-based spectrum trading market is proposed in [5]. The entities participating in the exchange-based spectrum market include spectrum exchange, spectrum license holders, spectrum license requestors, spectrum regulators, and market makers. Spectrum exchange maintains facilities for spectrum trading; spectrum license holders own spectrum license and want to sell it; spectrum license requestors want to buy spectrum licenses; spectrum regulators oversee the spectrum trading market; and market makers facilitate trading and acts as a dealer that holds an inventory of spectrum. In a typical trade of the exchange-based spectrum market, the spectrum exchange collects offers to sell spectrum from spectrum license holders and offers to buy spectrum from spectrum license requestors, determines the winning bid, and transfers a spectrum usage right from its holder to its winning requestor.

Compared to the secondary spectrum trading markets, exchange-based spectrum trading markets have more liberalized market structure and may exploit more benefits from general market forces.

2.3. Two-Tier Spectrum Trading Markets

To fully exploit the power of market-based mechanisms, a two-tier spectrum trading market [6] is suggested, which advocates the separation of spectrum ownership from the provision of wireless services as a reasonable outcome of trading spectrum property right. The upper-tier market involves spectrum owners trading spectrum property rights similar to other commodity rights, whereas the lower-tier market consists of spot markets for service providers to acquire limited-duration rentals of spectrum assets from owners at particular locations. Spectrum brokers managing spectrum assets at particular locations play the role of middlemen for spectrum owners of the upper tier and service providers of the lower tier. Note that the two market tiers could run at different time-scales: spectrum assets at the upper tier might be traded on a long-term scale, whereas rented spectrum assets at the lower tier could be negotiated over short-term time scales.

The two-tier spectrum market potentially enables relatively low trading/transcation and maintenance costs for wireless service providers, especially for small and short-term acquisition. The entry barriers for wireless service market may also be lowered, leading to more competitive and diversified services and more efficient utilization of the entire spectrum. However, whether such high-level liberalization of the spectrum will damage some social value aspects, such as fairness, of spectrum utilization and how to avoid the possible damage need to be carefully addressed.

3. Important Issues for Spectrum Trading

In this section, we discuss some important issues in various spectrum trading activities, including pricing mechanism and utility functions.

3.1. Pricing Mechanism

In spectrum trading, price reflects the value of the spectrum bands to their sellers and buyers. Therefore, price mechanism plays an important role. Buyers want to choose spectrum that can accommodate required service with the lowest price while sellers wish to maximize their profits through selling (or leasing) the spectrum. Hence, the pricing mechanism in spectrum trading should be designed to offer profitable business to the sellers and create favorable services for the buyers. On the one hand, the spectrum price also depends on buyers' demands of spectrum and sellers' inventory of spectrum. When there are multiple buyers trying to bid for the same spectrum band, the price would rise. If multiple sellers can provide spectrum bands of similar service capacity, the price would fall.

On the other hand, competition and cooperation among spectrum sellers and buyers in spectrum trading also significantly affect the pricing mechanism. A more liberalized spectrum market will spontaneously encourage competition and cooperation among different sellers and buyers. Competition allows spectrum sellers to survive and to make more profits by providing spectrum at a more reasonable price. Meanwhile, cooperation in both pricing and spectrum assignment among the sellers is helpful for their own benefits, considering the dynamics in spectrum spatio-temporal availability, user need, channel quality, and the potential multi-user diversity gain. For example, if the sellers cooperatively match high quality spectrum bands to the buyers requiring high quality bands, the overall spectrum efficiency is high, which makes it possible to accommodate more buyers and higher profits.

In [11], the monopolist-dominated quality-price contract is introduced as the pricing mechanism. It is offered by the spectrum seller and contains a set of quality-price combinations, each intended for a type of consumers. In [12], knapsack-based auction is employed to develop a dynamic pricing strategy for competitive spectrum sellers, which stimulates sellers to upgrade their network resources and offers better services for more profits. In [13], pricing dynamics in a competitive spectrum market consisting of selfish spectrum sellers are investigated. As a result of non-cooperative nature of the sellers, price war occurs when the buyers are price-sensitive. In [14], an equilibrium pricing scheme is developed through a game-theoretic model for an environment in which multiple spectrum sellers compete mutually to offer buyers spectrum. Pricing mechanisms for spectrum trading with bandwidth uncertainty and spatial reuse are analyzed via game models in [15, 16]. In [17], the optimal investment and pricing decisions of a spectrum seller under spectrum supply uncertainty is investigated through a Stackelberg game.

If commodities in addition to spectrum bands, such as *base stations* (BSs) and power supply, are also involved in the spectrum trading, the adopted pricing mechanisms need to consider all such factors and generally become more complicated. A recent survey of pricing schemes in wireless environments can be found in [18], where pricing schemes are discussed and classified according to the factors involved in the price calculation, e.g., the available bandwidth and frequency of the spectrum, the negotiation capabilities between spectrum sellers and buyers, and the network facilities.

3.2. Utility Functions

Another critical issue in spectrum trading involves finding proper utility functions for optimizing resource allocation under practical constraints, such as bandwidth, power, data rate, and channel quality. Although price as discussed earlier is an important user consideration, user experience and satisfaction for a specific service need play an equally important role in resource allocation decision by spectrum trading.

Basically, utility functions map resource use (bandwidth, power, etc.) or performance criteria (data rate, delay, etc.) to their corresponding utility. The introduction of utility function allows its optimization as a metric for resource allocation. There are usually two approaches to obtaining or defining utility functions [19]. For a specific type of applications, the utility function can be obtained by sophisticated subjective surveys. For example, a utility function to indicate users' satisfaction toward throughput for best-effort services has been obtained in [20] through surveys. Another simpler and more popular method is to design utility functions based on engineers' quantified expectation of *quality-of-service* (QoS) with respect to objective indexes for delivered data streams, such as speed, accuracy, and latency, plus appropriate fairness in the network. The most commonly used utility functions of this category include throughput [21, 22], delay [23, 24], fairness [25–28], net profit [29–31], energy efficiency [32–34], as well as sigmoid utility functions [35], etc.

Mathematical properties and tractability of utility functions are also crucial and need to be carefully considered when designing or choosing utility functions. For example, utility functions exhibiting convexity are generally easier to deal with than nonconvex ones. Moreover, in some game scenarios, utility functions satisfying certain conditions can guarantee network convergence and stability [36].

Another important issue on utility functions arises when optimizing the total utility of multiple entities. A common approach is to use the (weighted) sum or product of all individual utility functions as the overall utility function, where each individual utility function purely represents user's self-interest [37, 38]. This approach is reasonable when each individual is rational and committed to achieving the best outcome for itself regardless of impact on others. However, the rationality assumption about individuals is somewhat asocial and may not be adequate for truly cooperative scenarios. To well characterize the interplay among multiple entities in cooperative systems, *satisficing theory* [39, 40] and *social utility functions* [41, 42] are introduced as alternatives. Satisficing theory is used to explain the behavior of decision makers under circumstances in which an optimal solution cannot be determined or is not worth the efforts for pursuing it, considering that human beings can hardly evaluate all outcomes with sufficient precision or know the relevant probabilities of outcomes. It emphasizes that in many social scenarios the only information we can draw from for decision-making are the preferences of individuals. Social utility functions are constructed around multiple decision-maker preferences rather than actions, thereby letting each individual expand one's sphere of interest beyond the self. Thus it can characterize complex interrelationship among individuals, such as cooperation, compromise, negotiation, and altruism.

4. Models and Methods for Spectrum Trading

In this section, we review some classical and state-of-the-art models and methods that can be applied to various spectrum trading scenarios, which are first briefly summarizes in Table 2, and then discussed in details.

4.1. Conventional Optimization for Spectrum Trading

A spectrum trading process can be simply regarded as an optimization problem, where spectrum sellers and/or buyers aim to maximize their respective utilities under certain constraints by finding out the optimal operating parameters [22, 34, 43–48]. Depending on spectrum trading scenario, various conventional optimization techniques, including *linear programming*, *mixed integer linear programming*, *convex programming*, *mixed-integer convex programming*, etc., may be applied to model and solve the problem.

There are two useful tricks for solving established optimization problems by classical optimization techniques. One is through transformation to simplify the problem. The other is through relaxation to provide better mathematical tractability. For example, when spectrum buyers maximize their throughput under interference constraints, the optimization problem can be established as a mixed-integer convex programming problem and then be solved by *Lagrange dual decomposition* [49] as in [22]. In [34], when the spectrum buyers optimize their energy efficiency, the problem can be set up as a mixed-integer nonconvex programming problem before being solved through a branch-and-bound algorithm after relaxation and transformation.

4.2. Game for Spectrum Trading

Game is a mutually developed mathematical theory for understanding the strategic interplay among rational entities [50–52]. Because of underlying economic incentive structures in spectrum trading, it is quite natural to apply game theory there to model and analyze individual and group behaviors of entities involved, e.g., spectrum sellers and buyers, as widely discussed in some surveys [53–57] and more original contributions [15, 16, 25, 27, 58–73]. Using appropriate games to model various spectrum trading scenarios, the involved entities' strategic behaviors can be

analyzed in formalized game structures. Moreover, game models can provide well defined equilibria and Pareto optimality/efficiency to characterize spectrum trading and may offer distributed solutions for non-centralized scenarios.

Mathematically, a game problem formulation consists of *players*, their *strategies*, and their *payoff functions*. A spectrum trading process can be modeled as a game, where the involved entities, e.g., spectrum sellers and buyers, are treated as the players, the feasible actions related to spectrum trading, e.g., spectrum to trade, transmission parameters, and pricing mechanisms, are treated as the strategies. The chosen utility functions, discussed in Section 3.2, are treated as payoff functions. When predicting outcomes of games, equilibria are an important family of solutions in which a unilateral deviation from the equilibrium strategy by one player would result in a lower payoff for the deviating player. Pareto optimality/efficiency is another important family of solutions in which it is impossible to make any player strictly better off without making at least one other player strictly worse off.

Depending on the availability of coordinators, spectrum trading can be modeled and analyzed as either non-cooperative spectrum trading game or cooperative spectrum trading game.

Non-Cooperative Spectrum Trading Game. In a *non-cooperative game* [52], self-interest players have no enforceable cooperation mechanisms and make decisions independently. While players could cooperate, any cooperation must be *self-enforcing* [52]. If the strategies and payoff functions of the players are commonly known in a game[1], it is a game of *complete information*; otherwise, it is called a game of *incomplete information* [52]. For a non-cooperative game of complete information, *Nash equilibrium* [51], if exists, is an important stable solution (to evaluate game outcome) in which no player can attain a higher payoff by unilaterally deviating his own strategy. Similarly, for a non-cooperative game of incomplete information, *Bayesian Nash equilibrium* [74], if exists, is an important stable solution. Because of the non-cooperative nature, non-cooperative games are suitable to model and solve the non-cooperative spectrum trading problems, which primarily focus on distributed design and cooperation stimulation. For example, the competitive behaviors of multiple spectrum sellers with complete information [16, 25] and with incomplete information [15, 62] as well as the competition among spectrum buyers [63] are

---

[1]This corresponds to the spectrum trading scenarios in which all network parameters including constraints, bandwidth, powers, channel gains, etc. are common knowledge to all those entities involved.

Table. 2. Models and methods for spectrum trading

Model	Method
Conventional optimization	linear programming [45] convex programming [43, 44, 46, 47] nonconvex programming [22, 34]
Game [53–57]	noncooperative game: static [15, 16, 25, 62, 63], dynamic (repeated [65], multi-stage [64]), Bayesian [66], sequential, Stackelberg [67, 68] cooperative game: coalition formation [69–71], Nash bargaining [27, 72, 73, 77]
Auction [79–81]	one-sided [29, 82, 83] vs. two-sided [84, 85] single-unit [86] vs. multi-unit [87] auction game: noncooperative [54, 58, 89] vs. cooperative [30]
Graph	conflict graph [90]: multi-point, dynamic [91] factor graph [93] bipartite graph [94–96] layer-graph [97]

modeled as non-cooperative games and then analyzed through Nash equilibria.

In practice, some spectrum trading processes may last over such long periods that the spectrum spatio-temporal availability varies. An appropriate game model should enable its players to make timely decisions in respond to such network dynamics. In a *static (non-cooperative) game* (also namely *strategic (non-cooperative) game* [52]), however, players make one-time decisions at the beginning of the game without change over time. On the other hand, a *dynamic (non-cooperative) game* (also namely *extensive (non-cooperative) game*) allows players to make decisions not only at the beginning of the game but also at any time when they have to. If each player is perfectly informed of previous actions by all other players at each stage in the dynamic game, it is a game of *perfect information*; otherwise, it is a game of *imperfect information*. To properly model and analyze players' interplay evolving over time in dynamic spectrum trading, dynamic game models, such as *repeated game, multi-stage game, Bayesian game,* and *sequential game* [52], should be applied instead of static models. Depending on the availability of complete and perfect information, dynamic spectrum trading can be further divided into dynamic spectrum trading of complete information[2] and dynamic spectrum trading of incomplete information.

- For dynamic spectrum trading scenarios of complete information, repeated game and multi-stage game

can be used as models. For these games, *subgame perfect equilibrium* [52] is widely used to predict the outcome. For example, the competition between two spectrum sellers [64] and the competitive behaviors of multiple selfish spectrum buyers [65] are modeled by a three-stage dynamic game and a repeated dynamic game, respectively, before analysis based on the subgame perfect equilibria.

- For dynamic spectrum trading scenarios of incomplete information, Bayesian game and sequential game, where players generate, and rely on, their self-renewing beliefs (i.e., probability distribution) over observation uncertainties regarding the actions of others, can be applied as models. For these games, *perfect Bayesian equilibrium* and *sequential equilibrium* [52] are commonly used to measure the outcome. For example, the competition for shared bandwidth among multiple spectrum buyers who are unable to completely observe other buyer behaviors (e.g., speed of movement, bandwidth demand) is modeled by a Bayesian game and subsequently measured by Bayesian Nash equilibrium in [66].

Another practical consideration in spectrum trading is that different entities may have different levels of priorities, e.g., some entities may be primary while others are subsidiary. Such spectrum trading scenarios can be well modeled by Stackelberg games. *Stackelberg game* [52] is a two-player extensive game with perfect information where a leader chooses an action from his strategy set and a follower, informed of the leader's choice, selects an action from its strategy set. The usual solution to such games is that of subgame perfect equilibrium. For example, a spectrum trading scenario with a single spectrum seller (aiming to maximize

---

[2]In a slight abuse of terminology, spectrum trading of complete information refers to spectrum trading environments having *complete* and *perfect* information in the sense of game throughout this manuscript.

Table. 3. Game modeling and equilibria for non-cooperative spectrum trading.

	Spectrum trading of complete information	Spectrum trading of incomplete information
Static spectrum trading	Static game of complete information *Nash equilibrium*	Static game of incomplete information *Bayesian Nash equilibrium*
Dynamic spectrum trading	Dynamic game of complete information *Subgame perfect equilibrium*	Dynamic game of incomplete information *Perfect Bayesian equilibrium, sequential equilibrium*

its revenue by properly adjusting its load across the spectrum) and multiple selfish spectrum buyers (aiming to maximize their own payoffs) is modeled as a Stackelberg game, in which the seller is treated as the leader while the buyers are viewed as the followers [67]. In [68], multiple legitimate parties (i.e., spectrum sellers), facing malicious eavesdroppers who also hold spectrum, allow multiple other terminals, (i.e., spectrum buyers), to rent spectrum if they help combat the malicious parties by acting as cooperative jammers. The spectrum trading process is modeled and analyzed as a Stackelberg game with legitimate parties as the leaders and friendly jammers as the followers.

In Table 3, we summarize the game modeling and equilibria for non-cooperative spectrum trading. Note that non-cooperative spectrum trading modeled by non-cooperative games may suffer low efficiency in spectrum utilization, a problem inherent to the equilibria in those games. As an alternative with possibly high spectrum utilization efficiency, cooperative spectrum trading also has attracted substantial attention.

Cooperative Spectrum Trading Game. A *cooperative game* [52] is a game, in which groups of players (also namely *coalitions* in game terminology) may enforce cooperative behaviors to maximize a common payoff of a coalition. Because of the cooperative nature, cooperative games can well model and solve the cooperative spectrum trading problems, in which cooperation among entities is enforced through centralized control. Coalition formation game and Nash bargaining game are two commonly used cooperative game models [52].

Coalition formation game [52] is a cooperative game admitting a set of players seeking to form coalitions to improve their payoffs. Classical coalitional problems are typically modeled in the *characteristic form* [75], in which the utility of a coalition is not affected by the formation of other distinct coalitions. In contrast, for coalitional games in *partition form* [76], the value of any coalition strongly depends on how other players outside the coalition are organized. The solution for coalition formation game is called the *core* [52], which is defined as the set of all *undominated imputations* [52]. Coalition formation game is adequate to model and analyze the spectrum trading where some involved entities want to cooperative to improve their transmission opportunities [69–71].

Nash bargaining game [52] can appropriately model and analyze the negotiation/bargaining processes of involved entities in the spectrum trading [27, 72, 73, 77]. And a *Nash bargaining solution* [52] is Pareto optimal for a Nash bargaining game.

4.3. Auction for Spectrum Trading

Auction theory [78], an applied branch of economics, deals with how entities act in auction markets and studies the properties of auction markets. In general, auction suits a situation where the commodity value to the buyers is uncertain and may even vary. For example, auction has been used historically by government authorities to sell licenses for some spectrum bands, such as bands for mobile communications. An auction is an event to trade/exchange commodities and may involve *bidder(s)*, *seller(s)*, and *auctioneer(s)* [78]. A bidder submits *bids* (i.e., bidding prices) to buy the commodities; a seller owns the commodities and offers *asks* (i.e., asking prices) to sell them; an auctioneer is an intermediate agent betweens bidders and sellers, and is in charge of the auction process.

Recently, auction theory has been recognized as a promising tool to efficiently solve various emerging spectrum trading problems, which has been extensively covered in some surveys and tutorials [79–81] as well as related literature [29, 30, 58, 82–89]. Spectrum trading can be modeled, analyzed, and implemented as auctions, where the spectrum buyers are bidders, the spectrum sellers remain as sellers, and the spectrum exchange or spectrum brokers or even the spectrum sellers act as auctioneers. Using auction models, we can analyze the outcome, efficiency, fairness, and optimal and equilibrium bidding and asking strategies of the spectrum trading.

Spectrum trading auctions can be categorized based on different criteria, including one-sided and two-sided spectrum trading auctions, single-unit and multi-unit spectrum trading auctions, and non-cooperative and cooperative spectrum trading auction games.

One-Sided and Two-Sided Spectrum Trading Auctions. Depending on the number of spectrum sellers and the number of spectrum buyers in the market, spectrum trading auctions can be categorized as one-sided

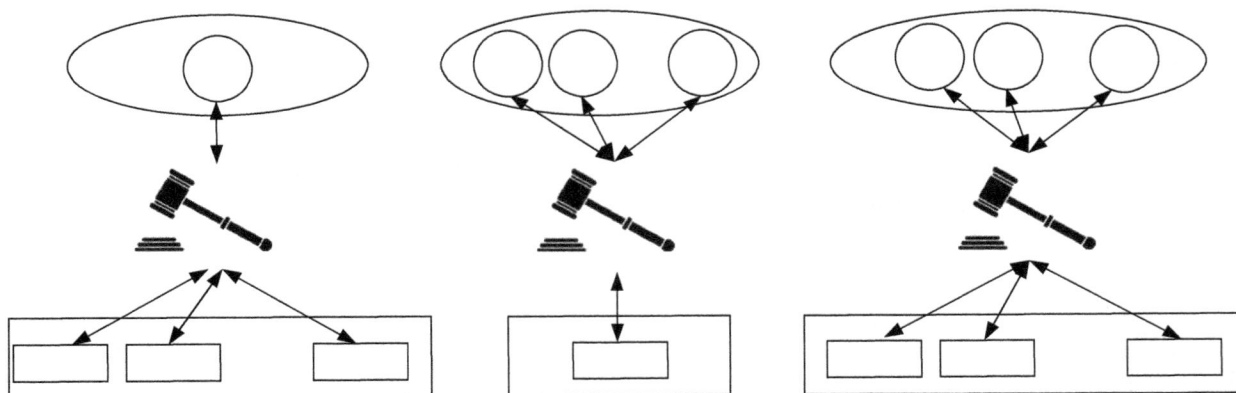

Figure 2. Spectrum trading auctions: (a) forward one-sided spectrum trading auction; (b) reverse one-sided spectrum trading; (c) two-sided spectrum trading auction.

spectrum trading auctions and two-sided spectrum trading auctions, as shown in Figure 2.

Usually, in a forward one-sided auction, the bidders submit their bids to a single seller; in a reverse one-sided auction, the sellers offer their asks to a single bidder. Spectrum trading with a single spectrum seller and multiple spectrum buyers can be well modeled and analyzed as forward a one-sided auction while spectrum trading with a single spectrum buyer and multiple spectrum sellers can be properly modeled and analyzed as a reverse one-sided auction. For example, spectrum trading with a single spectrum seller and multiple spectrum buyers is modeled and analyzed as a forward one-sided auction in [82]. In [83], the spectrum trading process with multiple competing spectrum sellers is investigated as a reverse auction. *Vickrey-Clarke-Groves (VCG) auction* [78] is used as a forward one-sided spectrum trading auction for competitive spectrum sellers to offer spectrum bands to buyers with and without contracts simultaneously in [29].

In a two-sided auction, bidders and sellers submit their bids and asks, respectively. The auctioneer then matches the asks and the bids before exchanging payment and commodity for each matching pair of bidder and seller. Spectrum trading with multiple spectrum sellers and multiple spectrum buyers can appropriately modeled and investigated as a two-sided auction. For example, a general framework for *truthful* two-sided spectrum trading auctions for homogeneous spectrum bands is established for spectrum trading with multiple sellers and buyers in [84]. In [85], a truthful two-sided auction considering the heterogeneity of spectrum bands is developed for spectrum trading with multiple sellers and buyers.

Single-Unit and Multi-Unit Spectrum Trading Auctions. As a result of the dynamic nature of spectrum spatio-temporal availability, there could be multiple isolated unused spectrum bands available for spectrum trading. According to the number of spectrum bands that are auctioned at a time, spectrum trading auctions can be divided into single-unit spectrum trading auctions and multi-unit spectrum trading auctions.

In a *single-unit auction* [78], only one commodity is auctioned at a time. Spectrum trading with multiple spectrum bands for sale sequentially can be well modeled and analyzed as a single-unit spectrum trading auction as in [86].

In a *multi-unit auction* [78], multiple commodities are auctioned at a time. Spectrum trading with multiple spectrum bands for sale simultaneously can be appropriately modeled and investigated as a multi-unit spectrum trading auction shown in [87].

Non-Cooperative and Cooperative Spectrum Trading Auction Games. To model and analyze the strategic asking and bidding behaviors of spectrum sellers and buyers, and to enable decentralized design in spectrum trading auction, game theory can be naturally incorporated to form an auction game. In a spectrum trading auction game, participants' strategies are limited by the adopted auction machinery (i.e., the bidding/asking languages) while the auction outcome can be evaluated by equilibrium bidding and asking strategies. Depending on the level of cooperation among sellers and among bidders, spectrum trading auction games can be classified as either non-cooperative spectrum trading auction games or cooperative spectrum trading auction games.

For a one-sided spectrum trading auction game with non-cooperative bidders or non-cooperative sellers, Nash equilibrium is commonly viewed as the optimal outcome of the auction. For a two-sided spectrum trading auction game involving non-cooperative sellers and non-cooperative buyers, *competitive equilibrium* [54], a price at which the number of buyers willing to pay is equal to the number of sellers willing to accept, is a well defined theoretical prediction of the outcome. Moreover, considering spectrum trading auction over long periods, dynamic game can be employed to incorporate the spectrum dynamics and participants' bidding/asking history into the auction. The equilibria, such as perfect Bayesian equilibrium and sequential equilibrium or subgame perfect equilibrium, can be utilized to anticipate the outcome of dynamic spectrum trading auction games. For example, a double auction is established among multiple TV broadcasters (i.e., spectrum sellers) and WRAN service providers (i.e., spectrum buyers) for TV radio spectrum bands in [58]. The WRAN service providers compete with each other by adjusting the service price charged to WRAN users. A non-cooperative game is formulated to obtain the solution in terms of the number of TV bands and the service price of a service provider for maximum profit. In [89], a repeated auction game of incomplete information is established to characterize the strategic competitions among the multiple spectrum buyers for spectrum bands from a spectrum seller, where knowledge regarding other buyers is limited due to the distributed nature of the network.

In a cooperative spectrum trading auction environment, coalition formation game can be applied to maximize the sum expected payoff of the spectrum sellers or buyers in a coalition. The solution for the cooperative auction game is the core. For example, in [30] the interaction among the sellers in a secondary spectrum trading market is modeled as a cooperative auction game. It is further demonstrated that, for any fixed strategies of the buyers, the core of the cooperative auction game is nonempty.

4.4. Graph for Spectrum Trading

Graph theory can also be used for spectrum trading. To capture different features of different spectrum trading scenarios, various of graph structures can be utilized.

One of the most common approaches to spectrum trading is based on *conflict graphs* [90]. Conflict graphs can capture the interference relationship among the entities involved in spectrum trading. For example, each spectrum buyer is represented by one node. If two nodes interfere with each other, one edge is linked between them. Weights on edges can measure the interference levels, where different utility functions (e.g., those in Section 3.2) can be used to capture various requirements. When one node can be assigned to more than one spectrum bands, a *multi-point conflict graph* can be used. To capture the dynamics of interference,

the spectrum trading system can be modeled by a *dynamic conflict graph* [91]. After the conflict graph is constructed, spectrum will be assigned accordingly. To avoid interference, connected nodes are normally assigned with different spectrum bands. To fulfill this goal, *graph-coloring* can be utilized [92], such that one color is used to represent one spectrum band and connected nodes may not have the same color. Besides using one node and one color to represent one spectrum buyer and one spectrum band, respectively, the conflict graph can be generalized to capture the spectrum trading system where a group of buyers share the same set of spectrum bands, with one node and one color representing a group of buyers and a set of spectrum bands, respectively. In spectrum trading, the interference information may be unknown or not accurately known to spectrum buyers (nodes). *Factor graph* [93] can model and tackle such spectrum trading scenarios using probabilistic inference approaches, such as belief propagation.

Another type of graph structures often used is the *bipartite graph*, where nodes representing buyers are put on one side whereas nodes representing spectrum bands are put on another side. This structure captures the actions between buyers and spectrum bands. An edge is connected between a buyer and a spectrum band if the buyer is a potential candidate for the spectrum. Similar to the conflict graph, weights can be put on edge to characterize the system requirements, such as throughput and energy efficiency. There are two known spectrum trading methods based on the bipartite graph, *maximal matching* and *stable matching*. Maximal matching methods, such as the *Hungarian algorithm* [94], lead to a maximal sum-weight spectrum trading result [95]. Unlike maximal matching methods, stable matching algorithms, such as the *Gale-Shapley algorithm* [94], take the preferences of nodes on both sides into account [96]. Nodes can build their preference lists according to their own requirements, not necessarily according to the same utility function.

Besides conflict graph, factor graph, and bipartite graph, other graph structures, such as *layered graph* [97] can also be used to analyze spectrum trading. A layered graph can capture the connection between spectrum assignment and routing path computation and may adequately model multi-hop spectrum trading environments.

5. Spectrum Trading Based Dynamic Spectrum Sharing and Access

In this section, we focus on spectrum trading based dynamic spectrum sharing and access in various wireless environments, including the highly popular cases of *cognitive radio* (CR), *heterogeneous networks* (HetNets), and *device-to-device* (D2D) transmissions.

5.1. Spectrum Trading for CR

The scarcity of spectrum and the underutilization of most licensed spectrum bands have motivated the popularity of CR over the past decade [98–100]. CR may be one promising technology for spectrum efficient wireless access. A typical CR network often consists of three main entities: spectrum authorities, licensed users, and unlicensed users. The spectrum authorities normally refer to government agencies regulating spectral bands. Licensed users, also called *primary users* (PUs), hold licenses of spectrum bands issued by the spectrum authorities. PUs can be television stations, wireless telecom operators, and other related parties. Unlicensed users, also called *secondary users* (SUs), have lower priorities than the PUs but eagerly seek transmission opportunities on the licensed bands whenever possible. The concept of spectrum trading is apparently inherent to CR, considering traffic dynamics and spectrum underutilization of PUs and transmission requirements of SUs. In other words, CR is simply one particular form of spectrum trading. Moreover, the secondary spectrum trading market model originates from the concept of CR. Next, we primarily focus on the spectrum trading activities among PUs and SUs in the secondary spectrum trading market [2, 3].

In the secondary spectrum trading market, PUs can sell or lease parts of their own bands either exclusively to the SUs or coexisting with them. Their objective is to maximize the utility of the licensed spectrum. In general, utility function of one PU will be an increasing function of its income from selling the bands while a decreasing function of its performance loss due to allowing SUs to use its bands. SUs want to buy or rent spectrum bands from PUs. Their goal is to minimize the cost while maximizing their payoff, such as throughput. The utility function of one SU will be an increasing function of its own performance while a decreasing function of its cost. Based on these fundamental observations, there exist a number of works on spectrum trading in CR, including several survey [53–55, 57], pricing mechanisms [11, 14–17], utility functions [36], spectrum trading games [15, 16, 25, 27, 58, 59, 62, 67], spectrum trading auctions [58, 82, 89], graph-based spectrum trading [90, 91, 93, 96, 97], among others.

5.2. Spectrum Trading for HetNets

HetNets may consist of macro BSs and micro BSs, which typically transmit at high power levels, overlaid with several pico BSs, femto BSs, and relay BSs, that transmit at substantially lower power levels [101] for local coverage. The high-power BSs are in charge of the general coverage area while the low-power BSs can be deployed to eliminate coverage holes in the macro-only systems and to improve quality and throughput

in high activity (hot) spots. However, to suppress interference among different BSs, the current spectrum sharing policy in HetNets is static or inflexible, which may sacrifice the full capacity of spectrum. To tackle this issue, spectrum trading has also been introduced into HetNets for more flexible and dynamic spectrum sharing [71, 77, 102–105].

In [71], a framework for macrocell-femtocell cooperation under a closed access policy is proposed, where a femtocell user may act as a relay for macrocell users and is rewarded with a fraction of the frame duration used by macrocell users for transmission. A coalitional formation game is used to model and analyze the spectrum trading process and is solved by the recursive core of the game. The data rates for both macrocell and femtocell users can be greatly improved.

In a two-tier HetNet, spectrum trading, in which small cells may dynamically open portion of the access opportunities to macrocells for profits, is investigated as a *hierarchical dynamic game* [102]. At the lower level, an *evolutionary game* is formulated to model and analyze the adaptive service selection of users. At the upper level, a Stackelberg game, in which macrocells are leaders while small cells are followers, is formulated to characterize the pricing strategy of macrocells and the open access ratio strategy of small cells. The resulting dynamic control outperforms its static counterpart.

In [103], a femtocell expects to rent spectrum from a coexisting macrocell to serve its end users and also allows hybrid access of the macrocell users to improve utilities of both femtocell and macrocell. The whole spectrum trading procedure is modeled and analyzed as a three-stage Stackelberg game, in which the macrocell and the femtocell determine the spectrum leasing ratio, spectrum leasing price, and open access ratio sequentially. It is shown that both tiers can benefit from the spectrum leasing. In fact, the hybrid access of femtocell can further improve their utilities.

5.3. Spectrum Trading for D2D

In cellular networks, nearby mobile terminals may communicate directly without going through the base station (as a relay). Such direct terminal-to-terminal link is known as D2D communications [95]. D2D communication as an underlaying network to cellular networks can share the use of cellular resources for better spectral utilization so long as its interference to co-channel cellular users is appropriately contained. Because of the underlaying nature of D2D transmission, a similar spectrum trading scenario as in a secondary spectrum trading in CR arises from cellular networks to support underlying D2D mechanisms.

In [43], a spectrum trading problem of D2D users and prioritized cellular users is investigated. It is demonstrated that optimized spectrum trading improves the total throughput without generating harmful interference to cellular networks.

A spectrum trading environment where D2D users seek transmission opportunities when sharing spectrum with uplink cellular users while guaranteeing the QoS of normal cellular users is studied in [44]. Despite the nonconvexity of the optimization problem, an analytical characterization of the globally optimal non-orthogonal spectrum sharing strategy is given.

In [60], a spectrum trading scenario focusing on mode selection for energy-efficient D2D communications is modeled as a coalitional game. Stable coalitions, where no D2D link can change its communication mode to lower transmission cost without making others worse off, are found as optimal solutions.

6. Open Issues

In the previous sections, we have provided a comprehensive overview of spectrum trading that can significantly improve spectrum efficiency. However, there are still some open issues on the development of more flexible and more efficient spectrum trading.

6.1. Novel Modeling of Spectrum Trading Market

The development of spectrum markets for wireless service provision requires an evolving spectrum trading market model. This model should consider the latest advancement in wireless technology and be adequately supported by the latest wireless technological infrastructure. For example, in HetNets terminals from different cells (i.e., different spectrum buyers) may operate on the same band. The current spectrum trading market models seldom consider the trading of a spectrum to multiple spectrum buyers simultaneously or say the coexistence of multiple spectrum buyers within the same band. How to manage such coexistence needs to be well incorporated in new spectrum trading market models.

6.2. Spectrum Trading for Emergency Response

Under severe conditions or extreme cases, e.g., major disasters (e.g. earthquake and hurricane), first responders need to locate and rescue trapped survivors and maintain reliable communications between responders and public safety agencies at a time when regular civilian communications may be jammed or be out of service. The infrastructure of currently existing wireless communication systems is inadequate to fully meet the future demands of such emergency responses from the perspectives of security, reliability, and robustness. Contrarily, trading based spectrum paradigm potentially can realize efficient and reliable emergency network transmission, by offloading and redistributing regular non-urgent traffics to spare or low quality spectrum bands while establishing at the same time a coordinated inter-agency rapid response network. However, how to use spectrum trading to develop interoperable communications system that would integrate existing radio frequency infrastructures for public services, including fire, police and emergency medical services, hospitals, emergency rooms and trauma centers, with an *Internet Protocol* (IP) backbone while maintaining connectivity for mobile terminals remains to be another challenge.

6.3. Spectrum Trading for Social Benefit Network

Social benefit networks, which cost users less than regular commercial one, can be set up by opportunistically exploiting the spectrum vacancies, without the necessity of strictly ensuring a certain level of QoS. For example, when the high-rate spectrum bands are lightly-loaded with regular users, low-income users may enjoy high-rate services for skill training and commerce at a deep discount; when the high-rate spectrum bands become crowded with regular user, low-income users can be handed over to other low-rate spectrum bands without considering its QoS. How to schedule the overall traffic and adjust the resources allocated to socially or economically disadvantaged users to balance the social and economic values of spectrum is an interesting challenge.

6.4. Transmission across Broad Multi-Spectrum Bands

Given highly dynamic user requirements and spectrum availability, resource allocation in flexible spectrum trading based systems is less likely to assign contiguous spectrum to a single user whose service request often consists of multiple data flows of varying QoS needs. First, acquisition of contiguous broad spectrum is more difficult and costly. In fact, costs at different wavelengths often differ as some bands are more suitable for mobile users while others are more suitable for high-rate low mobility users. Second, different user data flows may demand different QoS levels that are not monolithic. The cost can be substantially reduced when users are assigned fragmented, multi-spectrum resources to match their data QoS needs. Another important reason for fragmented multi-spectrum resource allocation lies in the broad bandwidth ranging from the lower *very high frequency* (VHF) (54-88MHz, 174-216MHz) once vacated by terrestrial TV stations up to millimeter wave V-band (60GHz). These bands are inherently fragmented with multiple special purpose licensees.

6.5. Channel Acquisition across Broad Multi-Spectrum Bands

In flexible spectrum trading based systems, broadband channel acquisition/estimation is essential for achieving multi-spectrum diversity. Spectrum sellers that dynamically allocate their users across broad multi-spectrum bands plus mobile users flexibly switching among multi-spectrum bands in a proactive manner will result in a higher spectral efficiency and/or better QoS. Normally, pilots and channel estimation for a single-band are specifically designed according to prior historic information on channel characteristics. For example, more pilots are required for a high-frequency band to account for faster fading and longer delay-spread channel states whereas fewer pilots are deployed for a low-frequency band that tends to exhibit slower fading and shorter delay spread. However, to acquire/estimate (instantaneous) *channel state information* (CSI) of different channels across broad multi-spectrum bands simultaneously is much more challenging.

6.6. Broadband Sensing

Till now, there has only been limited works on wideband sensing for CR networks [106–108]. To adapt to various applications in free spectrum trading, system feature should be taken into account. In order to reduce wideband processing time, service searching procedure must be shortened. Without negotiating with many spectrum sellers at the same time, it is more efficient for each spectrum seller to pronounce a range about their service quality and the corresponding charge. With such information, users can choose among a small number of spectrum sellers based on the coarse information and perform additional price and service negotiating processes with only a small number of remaining spectrum sellers. Only when the coarse service searching is done across a broadband spectrum, will the ensuing procedure proceed on a set of given spectrum bands. This process can reduce complexity. In brief, different features need to be taken into account to make the system practical.

7. Conclusions

Because of the non-uniform need and heterogeneity of radio wireless communications temporally and spatially, traditional rigid spectrum management relying on centralized command and control tends to be inefficient. In fact, such centralized approach has resulted in massive underutilization of some spectrum bands while overcrowding others. Spectrum trading can improve social and economic values of the spectrum by taking advantage of market-based mechanisms and thereby becomes a promising spectrum management approach

in wireless communications. This article presents an overview of spectrum trading, ranging from the fundamental structure of spectrum trading markets to the state-of-the-art techniques for modeling and solving various spectrum trading problems, including spectrum trading based dynamic spectrum sharing and access. Furthermore, we identify some open challenges in spectrum trading for future research in this area.

References

[1] FCC (2002) *Spectrum Policy Task Force Report*. Tech. rep., FCC ET Docket 02-155.

[2] FCC (2003) *The development of secondary markets-Report and Order and further notice of proposed rule marketing*. Tech. rep., FCC 03-113.

[3] FCC (2004) *Second report and order: Promoting efficient use of spectrum through elimination of barrier to the development of secondary markets*. Tech. rep., FCC 04-167.

[4] FCC (2010) *Unlicensed Operations in the TV Broadcast Bands, Second Memorandum Opinion and Order*. Tech. rep., FCC 10-174.

[5] Caicedo, C.E. and Weiss, M.B.H. (2011) The viability of spectrum trading markets. *IEEE Commun. Mag.* **49**(3): 46–62.

[6] Berry, R., Honig, M. and Vohra, R. (2010) Spectrum markets: Motivation, challenges and implications. *IEEE Commun. Mag.* **48**(11): 146–155.

[7] Caicedo, C.E. and Weiss, M. (2008) An analysis of market structures and implementation architectures for spectrum trading markets. In *Telecommun. Policy Research Conf.* (Fairfax, VA).

[8] Berry, R. (2012) Network market design part II: Spectrum markets. *IEEE Commun. Mag.* **50**(11): 84–90.

[9] Tonmukayakul, A. and Weiss, M. (2008) A study of secondary spectrum use using agent-based computational economics. *Netnomics* **9**(2): 125–151.

[10] Noam, E. (2012) The economists' contribution to radio spectrum access: The past, the present, and the future. *IEEE Proc.* **100**(Special Centennial Issue): 1692–1697.

[11] Gao, L., Wang, X.B., Xu, Y.Y. and Zhang, Q. (2011) Spectrum trading in cognitive radio networks: A contract-theoretic modeling approach. *IEEE. J. Sel. Areas Commun.* **29**(4): 843–855.

[12] Sengupta, S. and Chatterjee, M. (2009) An economic framework for dynamic spectrum access and service pricing. *IEEE/ACM Trans. Netw.* **17**(4): 1200–1213.

[13] Xing, Y., Chandramouli, R. and Cordeiro, C.M. (2007) Price dynamics in competitive agile spectrum access markets. *IEEE J. Sel. Areas Commun.* **25**(3): 613–621.

[14] Niyato, D. and Hossain, E. (2008) Competitive pricing for spectrum sharing in cognitive radio networks: Dynamic game, inefficiency of Nash equilibrium, and collusion. *IEEE J. Sel. Areas Commun.* **26**(1): 192–202.

[15] Kasbekar, G. and Sarkar, S. (2010) Spectrum pricing games with bandwidth uncertainty and spatial reuse in cognitive radio networks. *Proc. ACM Int. Symp. Mobile Ad Hoc Netw & Comput. (MobiHoc'10)* .

[16] Kasbekar, G.S. and Sarkar, S. (2012) Spectrum pricing games with spatial reuse in cognitive radio networks. *IEEE J. Sel. Areas Commun.* **30**(1): 153–164.

[17] Duan, L., Huang, J. and Shou, B. (2011) Investment and pricing with spectrum uncertainty: A cognitive operators perspective. *IEEE Trans. Mobile Comput.* **10**(11): 1590–1604.

[18] Gizelis, C.A. and Vergados, D.D. (2011) A survey of pricing schemes in wireless networks. *IEEE Commun. Surveys Tutorials* **13**(1): 126–145.

[19] Song, G.C. and Li, G.Y. (2005) Utility-based resource allocation and scheudling in OFDM-based wireless broadband networks. *IEEE Commun. Mag.* : 127–134.

[20] Jiang, Z., Ge, Y. and Li, G.Y. (2005) Max-utility wireless resource management for best-effort traffic. *IEEE Trans. Wireless Commun.* **04**(1): 100–111.

[21] Eraslan, B., Gozupek, D. and Alagoz, F. (2011) An acution theory based algorithm for throughput maximizing scheduling in centralized cognitive raido networks. *IEEE Commun. Lett.* **15**(7): 734–736.

[22] Zhou, X., Li, G., Li, D., Wang, D. and Soong, A. (2010) Probabilistic resource allocation for opportunistic spectrum access. *IEEE Trans. Wireless Commun.* **9**(9): 2870–2879.

[23] Cui, Y., Lau, V.K.N. and R. Wang, H.H. (2012) A survey on delay-aware resource control for wireless systems-Large deviation theory, stochastic lyapunov drift, and distributed stochastic learning. *IEEE Trans. Info.* **58**(3): 1677–1701.

[24] Shi, X., Medard, M., and Lucani, D. E. (2013) Spectrum assignment and sharing for delay minimization in multi-hop multi-flow CRNs. *IEEE J. Sel. Areas Commun.* **31**(11): 2483–2493.

[25] Niyato, D., Hossain, E. and Han, Z. (2009) Dynamics of multiple-seller and multiple-buyer spectrum trading in cognitive radio networks: A game-theoretic modeling approach. *IEEE Trans. Mobile Comput.* **8**(8): 1009–1022.

[26] Zheng, L. and Tan, C.W. (2013) Cognitve raido network duality and algorithms for utlity maximization. *IEEE J. Sel. Areas Commun.* **31**(3): 500–513.

[27] Ni, Q. and Zarakovitis, C.C. (2012) Nash bargaining game theoretic scheduling for joint channel and power allocation in cognitive radio systems. *IEEE J. Sel. Areas Commun.* **30**(1): 70–71.

[28] Bian, K., Park, J.M., Du, X. and Li, X. (2013) Enabling fair spectrum sharing: Mitigating selfish misbehaviors in spectrum contention. *IEEE Netw.* **27**(3): 16–21.

[29] Gao, L., Huang, J., Chen, Y.J. and Shou, B. (2013) An integrated contract and auction design for secondary spectrum trading. *IEEE J. Sel. Areas Commun.* **31**(3): 581–592.

[30] Chun, S.H. and La, R.J. (2013) Secondary spectrum trading-auction-based framework for spectrum allocation and profit sharing. *IEEE/ACM Trans. Netw.* **21**(1): 176–189.

[31] Xu, D., Liu, X. and Han, Z. (2013) Decentralized bargain: A two-tier market for efficient and flexible dynamic spectrum access. *IEEE Trans. Mobile Comput.* **12**(9): 1697–1711.

[32] Li, G., Xu, Z., Xiong, C., Yang, C., Zhang, S., Chen, Y. and Xu, S. (2011) Energy-efficient wireless communications: Tutorial, survey, and open issues. *IEEE Wireless Commun.* **18**(6): 28–35.

[33] Li, L., Zhou, X., Xu, H., Li, Y.G., Wang, D. and Soong, A. (2010) Energy-efficient transmission in cognitive radio networks. In *Proc. IEEE Consumer Commun. Netw. Conf. (CCNC'10)*.

[34] Xiong, C., Lu, L. and Li, G.Y. (2014) Energy-efficient spectrum access in cognitive radios. *IEEE J. Sel. Areas. Commun.* **32**(3).

[35] Kuo, W.H. and Liao, W. (2008) Utility-based radio resource allocation for QoS traffic in wireless networks. *IEEE Trans. Wireless Commun.* **7**(7): 2714–2722.

[36] Zhao, Y., Mao, S., Neel, J.O. and Reed, J.H. (2009) Performance evaluation of cognitive radios: Metrics, utility functions, and methodology. *IEEE Proc.* **97**(4): 642–659.

[37] Song, G.C. and Li, G.Y. (2005) Cross-layer optimization for OFDM wireless networks-part I: Theoretical framework. *IEEE Trans. Wireless Commun.* **4**(2): 614–624.

[38] Song, G.C. and Li, G.Y. (2005) Cross-layer optimization for OFDM wireless networks-part II: Algorithm development. *IEEE Trans. Wireless Commun.* **4**(2): 625–634.

[39] Simon, H. (1957) A behavioral model of rational choices. In *Models of Man, social and rational: mathematical essays on rational human behavior in a social setting* (New York: Wiley).

[40] Stirling, W.C. (2003) *Satisficing Games and Decision Making: With Applications to Engineering and Computer Science* (Cambridge University Press).

[41] Stirling, W.C. (2005) Social utility functions-part I: Theory. *IEEE Trans. Sys. Man Cybernetics-Part C: Applications & Reviews* **35**(4): 522–532.

[42] Stirling, W.C. and Frost, R.L. (2005) Social utility functions-part I: Applications. *IEEE Trans. Sys. Man Cybernetics-Part C: Applications & Reviews* **35**(4): 533–543.

[43] Yu, C.H., Doppler, K., Ribeiro, C. and Tirkkonen, O. (2011) Resource sharing optimization for device-to-device communication underlaying cellular networks. *IEEE Trans. Wireless Commun.* **10**(8): 2752–2763.

[44] Wang, J., Zhu, D., Zhao, C., Li, J. and Lei, M. (2013) Resource sharing of underlaying device-to-device and uplink cellular communications. *IEEE Commun. Lett.* **17**(6): 1148–1151.

[45] Hoang, A.T. and Liang, Y.C. (2008) Downlink channel assignment and power control for cognitive radio networks. *IEEE Trans. Wireless Commun.* **7**(8): 3106–3117.

[46] Pei, Y. and Liang, Y.C. (2013) Resource allocation for device-to-device communications overlaying two-way cellular networks. *IEEE Trans. Wireless Commun.* **12**(7): 3611–3621.

[47] Zhang, R. and Liang, Y.C. (2008) Exploiting multi-antennas for opportunistic spectrum sharing in cognitive radio networks. *IEEE J. Sel. Areas Commun.* **2**(1): 88–102.

[48] Zhai, C., Zhang, W. and Ching, P. (2013) Cooperative spectrum sharing based on two-path successive relaying. *IEEE Trans. Commun.* **61**(6): 2260–2270.

[49] BOYD, S. and VANDENBERGHE, L. (2004) *Convex Optimization* (Cambridge University Press).

[50] NEUMANN, J.V. and MORGENSTERN, O. (1944) *Theory of Games and Economic Behavior* (Princeton University Press).

[51] NASH, J. (1950) Non-cooperative games. *Ann. Math.* **54**(2): 286–295.

[52] OSBORNE, M.J. and RUBINSTEIN, A. (1994) *A Course in Game Theory* (The MIT Press).

[53] ZHAO, Q. and SADLER, B.M. (2007) A survey of dynamic spectrum access. *IEEE Signal Process. Mag.* **24**(3): 79–89.

[54] JI, Z. and LIU, K.J.R. (2007) Cognitive radios for dynamic spectrum access - Dynamic spectrum sharing: A game theoretical overview. *IEEE Commun. Mag.* **45**(5): 88–94.

[55] NIYATO, D. and HOSSAIN, E. (2008) Spectrum trading in cognitive radio networks: A market-equilibrium-based approach. *IEEE Wireless Commun.* **15**(12): 71–80.

[56] AKKARAJITSAKUL, K., HOSSAIN, E., NIYATO, D. and KIM, D.I. (2011) Game theoretic approaches for multiple access in wireless networks: A survey. *IEEE Commun. Surveys & Tutorials* **13**(3): 372–395.

[57] WANG, B., WU, Y. and LIU, K. J. (2010) Game theory for cognitive radio networks: An overview. *Computer Networks,* **54**(14): 2537–2561.

[58] NIYATO, D., HOSSAIN, E. and HAN, Z. (2009) Dynamic spectrum access in IEEE 802.22-based cognitive wireless networks: A game theoretic model for competitive spectrum bidding and pricing. *IEEE Wireless Comm.* **16**(2): 16–23.

[59] KASBEKAR, G. and SARKAR, S. (2011) Spectrum pricing games with arbitrary bandwidth availability probabilities. In *Proc. IEEE Int. Symp. Inf. Theory (ISIT'11).*

[60] AKKARAJITSAKUL, K., PHUNCHONGHARN, P. and HOSSAIN, E. (2012) Mode selection for energy-efficient D2D communications in LTE-advanced networks: A coalitional game approach. In *Proc. IEEE Int. Conf. Commun. Sys. (ICCS'12)* (Singapore).

[61] WANG, F., SONG, L., HAN, Z., ZHAO, Q. and WANG, X. (2013) Joint scheduling and resource allocation for device-to-device underlay communications. In *IEEE Wireless Commun. & Netw. Conf. (WCNC'13)* (Shanghai, China).

[62] KASBEKAR, G.S. and SARKAR, S. (2012) Spectrum pricing games with random valuations of secondary users. *IEEE J. Sel. Areas Commun.* **30**(11): 2262–2273.

[63] HAN, Z., JI, Z. and LIU, K.J.R. (2007) Non-cooperative resource competition game by virtual referee in multi-cell OFDMA networks. *IEEE J. Sel. Areas Commun.* **25**(6): 1079–1090.

[64] DUAN, L., HUANG, J. and SHOU, B. (2012) Duopoly competition in dynamic spectrum leasing and pricing. *IEEE Trans. Mobile Comput.* **11**(11): 1706–1719.

[65] ETKIN, R., PAREKH, A. and TSE, D. (2007) Spectrum sharing for unlicensed bands. *IEEE J. Sel. Areas Commun.* **25**(3): 517–528.

[66] AKKARAJITSAKUL, K., HOSSAIN, E. and NIYATO, D. (2011) Distributed resource allocation in wireless networks under uncertainty and application of Bayesian game. *IEEE Commmun. Mag.* **49**(8): 120–127.

[67] HADDAD, M., ELAYOUBI, S.E., ALTMAN, E. and ALTMAN, Z. (2011) A hybrid approach for radio resource management in heterogeneous cognitive networks. *IEEE J. Sel. Areas Commun.* **29**(4): 831–842.

[68] STANOJEV, I. and YENER, A. (2013) Improving secrecy rate via spectrum leasing for friendly jamming. *IEEE Trans. Wireless Commun.* **12**(1): 134–145.

[69] SAAD, W., HAN, Z., DEBBAH, M. and HJORUNGNES, A. (2009) A distributed coalition formation framework for fair user cooperation in wireless networks. *IEEE Trans. Wireless Commun.* **8**(9): 4580–4593.

[70] SAAD, W., HAN, Z., DEBBAH, M., BAŞAR, T. and HJØRUNGNES, A. (2009) Coalitional game theory for communication networks. *IEEE Signal Process. Mag.* **26**(6): 77–97.

[71] PANTISANO, F., BENNIS, M., SAAD, W. and DEBBAH, M. (2012) Spectrum leasing as an incentive towards uplink macrocell and femtocell cooperation. *IEEE J. Sel. Areas Commun.* **30**(3): 617–630.

[72] HAN, Z., JI, Z. and LIU, K. (2005) Fair multiuser channel allocation for OFDMA networks using Nash bargaining solutions and coalitions. *IEEE Trans. Commun.* **53**(8): 1366–1376.

[73] ZHANG, Z., SHI, J., CHEN, H.H., GUIZANI, M. and QIU, P. (2008) A cooperation strategy based on Nash bargaining solution in cooperative relay networks. *IEEE Trans. Veh. Techonol.* **57**(4): 2570–2577.

[74] HARSANYI, J.C. (1967) Games with incomplete information played by "Bayesian" players, I-III: Part I. The basic model. *Management Science* **14**(159-182).

[75] RAY, D. (2007) *A Game-Theoretic Perspective on Coalition Formation* (New York: USA: Oxford University Press).

[76] HAN, Z., NIYATO, D., SAAD, W., BAŞAR, T. and HJØRUNGNES, A. (2011) *Game theory in wireless and communications networks: Theory, models and applications* (Cambridge, UK: Cambridge Univeristy Press).

[77] LIN, P., ZHANG, J., ZHANG, Q. and HAMDI, M. (2013) Enabling the femtocells: A cooperation framework for mobile and fixed-line operators. *IEEE Trans. Wireless Commun.* **12**(1): 158–167.

[78] KLEMPERER, P. (2004) *Auctions: Theory and Practice* (Princeton Univeristy Press).

[79] ZHANG, Y., LEE, C., NIYATO, D. and WANG, P. (2013) Auction approaches for resource allocation in wireless systems: A survey. *IEEE Commun. Surveys & Tutorials* **15**(3): 1020–1041.

[80] ZHANG, Y., NIYATO, D., WANG, P. and HOSSAIN, E. (2012) Auction-based resource allocation in cognitive radio systems. *IEEE Commun. Mag.* **50**(11): 108–120.

[81] IOSIFIDIS, G. and KOUTSOPOULOS, I. (2011) Challenges in auction theory driven spectrum management. *IEEE Commun. Mag.* **49**(8): 128–135.

[82] WANG, X., LI, Z., XU, P., XU, Y., GAO, X. and CHEN, H.H. (2010) Spectrum sharing in cognitive radio networks-An auction-based approach. *IEEE Trans. Sys. Man Cybernetics, Part B: Cybernetics* **40**(3): 587–596.

[83] GAO, L., XU, Y. and WANG, X. (2011) MAP: Multi-auctioneer progressive auction in dynamic spectrum access. *IEEE Trans. Mobile Comput.* **(99)**.

[84] ZHOU, X. and ZHENG, H. (2009) TRUST: A general framework for truthful double spectrum auctions.

In *Proc. IEEE Int. Conf. Computer Commun. (INFO-COM'09)*.

[85] FENG, X., CHEN, Y., ZHANG, J., ZHANG, Q. and LI, B. (2012) TAHES: A truthful double auction mechanism for heterogeneous spectrums. *Wireless Communications, IEEE Transactions on* 11(11): 4038–4047.

[86] BAE, J., BEIGMAN, E., BERRY, R.A., HONIG, M.L. and VOHRA, R. (2008) Sequential bandwidth and power auctions for distributed spectrum sharing. *IEEE J. Sel. Areas Commun.* 26(7): 1193–1203.

[87] CHEN, Z., HUANG, H., SUN, Y.E. and HUANG, L. (2013) True-MCSA: A framework for truthful double multi-channel spectrum auctions. *IEEE Trans. Wireless Commun.* 12(8): 3838–3850.

[88] SUN, Y., JOVER, R. and WANG, X. (2012) Uplink interference mitigation for OFDMA femtocell networks. *IEEE Trans. Wireless Commun.* 11(2): 614–625.

[89] HAN, Z., ZHENG, R. and POOR, H.V. (2011) Repeated auctions with Bayesian nonparametric learning for spectrum access in cognitive radio networks. *IEEE Trans. Wireless Commun.* 10(3): 890–900.

[90] PLUMMER, A. and BISWAS, S. (2011) Distributed spectrum assignment for cognitive networks with heterogeneous spectrum opportunities. *Wireless Commun. and Mobile Computing* 11(9): 1239–1253.

[91] HOANG, A. and LIANG, Y. (2008) Downlink channel assignment and power control for cognitive radio networks. *IEEE Trans. Wireless Commun.* 7(8): 3106–3117.

[92] PENG, C., ZHENG, H. and ZHAO, B.Y. (2006) Utilization and fairness in spectrum assignment for opportunistic spectrum access. *Mobile Networks and Applications* 11(4): 555–576.

[93] CHEN, S., HUANG, Y. and NAMUDURI, K. (2011) A factor graph based dynamic spectrum allocation approach for cognitive network. In *Proc. IEEE Wireless Commun. Netw. Conf.*: 850–855.

[94] WEST, D.B. (2001) *Introduction to Graph Theory* (Prentice Hall), 2nd ed.

[95] FENG, D., LU, L., YUAN-WU, Y., LI, G.Y., FENG, G. and LI, S. (2013) Device-to-device communications underlaying cellular networks. *IEEE Trans. Commun.* 61(8): 3541–3551.

[96] LU, L., HE, D., YU, X. and LI, G.Y. (2013) Energy-efficient resource allocation for cognitive radio networks. In *Proc. IEEE Global Commun. Conf.*

[97] XIN, C., MA, L. and SHEN, C.C. (2008) A path-centric channel assignment framework for cognitive radio wireless networks. *Mobile Networks and Applications* 13(5): 463–476.

[98] MITOLA, J. and MAGUIRE, G.Q. (1999) Cognitive radio: Making software radios more personal. *IEEE Pers. Commun.* 6(4): 13–18.

[99] MA, J., LI, G.Y. and JUANG, B.H. (2009) Signal processing in cognitive radio. *Proc. IEEE* 97(5): 805–823.

[100] LU, L., ZHOU, X., ONUNKWO, U. and LI, G.Y. (2012) Ten years of research in spectrum sensing and sharing in cognitive radio. *EURASIP J. Wireless Commun. and Netw. 2012*.

[101] DAMNJANOVIC, A., MONTOJO, J., WEI, Y., JI, T., LUO, T., VAJAPEYAM, M., YOO, T. *et al.* (2011) A survey on 3GPP heterogeneous networks. *IEEE Wireless Commun.* 18(3): 10–21.

[102] ZHU, K., HOSSAIN, E. and NIYATO, D. Pricing, spectrum sharing, and service selection in two-tier small cell networks: A hierarchical dynamic game approach. *to appear in IEEE Trans. Mobile Comput.*.

[103] YI, Y., ZHANG, J., ZHANG, Q. and JIANG, T. (2012) Spectrum leasing to femto service provider with hybrid access. In *Proc. IEEE Int. Conf. Computer Commun. (INFOCOM'12)*.

[104] CHENG, S.M., LIEN, S.Y., CHU, F.S. and CHEN, K.C. (2011) On exploiting cognitive radio to mitigate interference in macro/femto heterogeneous networks. *IEEE Wireless Commun.* 18(3): 40–47.

[105] GHAREHSHIRAN, O., ATTAR, A. and KRISHNAMURTHY, V. (2013) Collaborative sub-channel allocation in cognitive LTE femto-cells: A cooperative game-theoretic approach. *IEEE Trans. Commun.* 61(1): 325–334.

[106] QUAN, Z., CUI, S., SAYED, A.H. and POOR, H.V. (2009) Optimal multiband joint detection for spectrum sensing in cognitive radio networks. *IEEE Trans. Signal Process.* 57: 1128–1140.

[107] PAYSARVI-HOSEINI, P. and BEAULIEU, N.C. (2011) Optimal wideband spectrum sensing framework for cognitive radio systems. *IEEE Trans. Signal Process.* 59: 1170–1182.

[108] SUN, H., NALLANATHAN, A., WANG, C.X. and CHEN, Y. (2013) Wideband spectrum sensing for cognitive radio networks: A survey. *IEEE Wireless Commun.* 20(2): 74–81.

11

Service Co-evolution in the Internet of Things

Huu Tam Tran*, Harun Baraki and Kurt Geihs

University of Kassel, Distributed Systems Group, Wilhelmshöher Allee 73, Germany

Abstract

The envisioned Internet of Things (IoT) foresees a future Internet incorporating smart physical objects that offer hosted functionality as IoT services. This service-based integration of IoT will be smarter, easier to communicate with and more valuable for enriching our environment. However, the interfaces and services can be modified due to updates and amendments. Such modifications require adaptations in all participating parties. Therefore, the aim of this research is to present a vision of service co-evolution in IoT. Moreover, we propose a novel agent architecture which supports the evolution by controlling service versions, updating local service instances and enabling the collaboration of agents. In this way, the service co-evolution can make systems more adaptive, efficient and reduce costs to manage maintenance.

Keywords: Service co-evolution, Coordinating agents, IoT services, Web services, Adaptive services

1. Introduction

In this paper, we address the challenge of coordinated services in the scope of IoT by employing an agent-based approach. Service providers may depend on third party services to deliver quality products to customers and to other service providers as well. To prevent outages and failures by individual service modifications and updates coordinated evolution (hereafter co-evolution) is required in such complex systems, i.e. they need a co-evolution for services in order to ensure that no interruptions occur. A centralized solution would not be realizable due to administrative and technical reasons. It would not be scalable, in particular, in the area of IoT, and security issues would complicate the whole approach. Consequently, service providers have to be responsible for the evolution of their own services. The required actions have to be coordinated with other providers in the IoT environment. The objective is to automate the coordinated evolution as much as possible.

With the emergence of Internet Protocol-based IoT devices [9] and the concept of embedded Web services [16, 19], new and highly interconnected IoT-aware applications can be created. Nonetheless, changes can happen at any stage in the service life cycle and have

unpredictable impact on the service stake-holders [23]. Though Web services bring more flexibility, they also create new challenges for change management in the Web service lifecycle. How to handle those changes for each Web service consumer as well as facilitate the client application updates on the consumer side? This question has become an emerging concern for Web service providers and Web service consumers. Being therefore able to control how changes manifest in the service life cycle is essential for both service providers and service consumers [23]. In fact, changes of Web service can occur in three aspects of services: change in the functional behavior of the service; change in the non-functional behavior of the service; and syntactical changes in the service interface.

Recently, agent-based models have been suggested for IoT as they can capture autonomy, and proactive and reactive features. Besides that, they can include ontologies for cooperation and different contexts [2, 3]. Within the scope of IoT, agent approaches address application levels and can use services provided by smart objects in order to achieve co-evolution.

Service co-evolution in IoT has received barely attention so far. Thus, there are some needs for detailing the vision of service co-evolution and solutions to provide benefits for IoT users. However, there are many challenges and requirements to tackle to meet an overall trade-off between aspects like the satisfaction

---

*Corresponding author. Email: tran@vs.uni-kassel.de

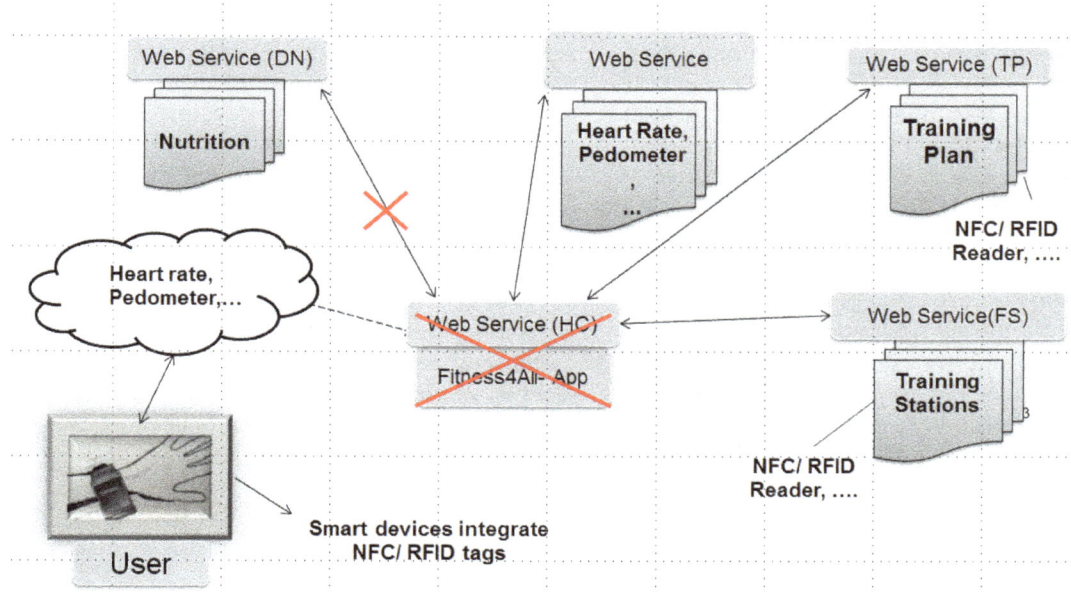

Figure 1. Fitness4All disrupts due to service updates.

of clients, the resource consumption of provided interface versions and the efforts to update them. Consequently, this paper will analyse the roles of this evolution regarding potential results, challenges and its requirements as well as the solution. It is not the intention of this paper to present details of Web service evolution as that has been done elsewhere [11–13, 15]. This paper aims at promoting the idea of co-evolution of web services in IoT by (i) illustrating how a service co-evolution is carried out, what should be involved, why it is essential, and what should be prepared in order to meet the co-evolution requirements, (ii) highlighting a novel agent architecture for service providers in the IoT environment and explaining how these agents can be used in service co-evolution environments, (iii) discussing some potential research challenges of service co-evolution. Thus, the main contribution of this paper is to make software engineers aware of the power of service co-evolution and make systems more adaptive, efficient and reliable.

The rest of the paper is structured as follows. Section 2 provides a motivating example in the healthcare area. Section 3 illustrates an overview of our solution and its key components. Section 4 analyses the coordination of services. Section 5 introduces a number of existing researches and compares them with our approach, and finally section 6 draws conclusions on our current results and provides an outlook for future work.

2. A motivating example

To clarify and illustrate the idea of co-evolution, we first present a scenario set on Health Care applications, and some important concerns for application development in the IoT environment. HealthCareCo (HCC) is a company specializing in providing healthcare services for customers. Through its Web service, the company provides a service named Fitness4All to customers (users). Users can employ Fitness4All to obtain information about health indicators such as heart rate, number of daily steps, insulin level, arterial blood pressure as well as daily calorie consumption. Fitness4All can also provide diagnosis of the illness as well as consulting users by offering a user guide on nutrition and exercises to promote health. Users will be advised to practise in health centers like Fitness Stations Co (FSC) or Training Plan Co (TPC).

The company's customers are provided with a smart device which can access remotely Fitness4All services. This smart device has an integrated RFID tag. When customers go into FSC or TPC, local RFID readers will identify the user. At the same time, the smart device will connect to the center's Web service. After that, FSC or TPC will provide the appropriate training for all users through personal health information based on the Fitness4All. During training, the device will signal an alert warning to the users if there are any health issues. In addition, Fitness4All can advise on nutrition information to the customers. This service is supported from another company named Daily Nutrition Co (DNC) which provides various nutritious foods to people. Suddenly one day, the Web services from service providers like FSC, TPC and DNC updated to new version without notifying HCC. This leads to the disruption of Fitness4All. Users can not access the information from the web service. It created many claims and the company had to return money back to users. HCC had to invest many financial resources and human efforts to continue providing Fitness4All. HCC Board of Directors understood that the emerging social Internet sector was still immature because of its lack of efficacy in dealing with changes while ensuring the quality of the composite. After some time, customers no longer trust the Fitness4All service as several interruptions occurred 1. As a consequence, the company had to stop the service and its activities.

After HCC Board's initial problems they decided to contact a consultancy firm called IoT-SU(an IoT-support Company), specialized in supporting companies that use third party internet services to implement robust systems. IoT-SU knows that when they publish a certain service through an interface, they have to be ready to maintain different versions of that interface over time, in order to not deny service to hundreds of thousands of subscribers.

However, service providers would like to publish new interfaces and forget about maintaining old ones. This would allow them to evolve the services faster, and continuously offer improved services for their subscribers. The solution to face this problem is called service co-evolution. Service Co-evolution artifacts are also installed on the consumer's side so that changes in the interfaces can be managed mostly automatically. If it cannot be adjusted automatically, then the development team of HCC will be informed. In this case, the service provider may offer the old version in parallel to the new version for a specified period of time to avoid an abrupt interruption of HCC.

IoT-SU helped the HCC service to adopt the service co-evolution approach. The sustainability of the selected solution allowed them to continue the provision of their services and to expand their user base.

Figure 2 illustrates a possible service co-evolution scenario with support by Evolution Agents (EVAs).

Third party service providers usually update interfaces in order to make their services more reliable and faster or to provide further functionality. In the introductory example the developers forgot about maintaining the previous versions for the Fitness4All service. In that case, for a short-term solution, the technical group of HCC should contact soon the third party service providers to get support in updating the interfaces. In the longer term, the technical group should deploy EVA for their Web services. When an update appears from the Web service providers, the Fitness4All Web service will be update almost simultaneously if appropriate update mechanisms are available or it will inform developers immediately.

It is imperative that an application has to cope with such evolutions, so that its business continuity is not compromised. Unfortunately, we cannot rely on a centralized evolution manager, since it would represent a traffic bottleneck and a central point of failure. Having a centralized manager might not even be feasible, for technical or administrative reasons.

Evolution also cannot be fully automated. In general, it is a multi-step process that a service must go through to transition from a problematic configuration, to a more acceptable new one. This transition may involve adaptation mechanisms that are already in-place, as well as offline activities, such as requirements gathering and software development. Although software evolution mechanisms have been deeply studied in the last decades, service co-evolution offers many new research challenges:

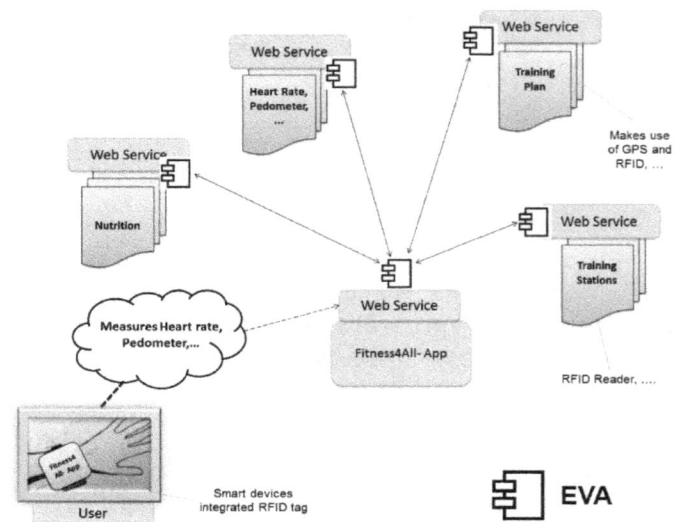

Figure 2. The service co-evolution with EVAs in the scenario.

(i) Heterogeneous services in IoT have de-coupled lifecycles, meaning that single services may be updated, or newly developed, while others are still in operation. Any evolution that we perform on a service requires that this action be coordinated with other actions paramount if we want to preserve the applications overall functionality and quality of service.

(ii) The evolution of such complex systems will require that we harness and understand the horizontal and vertical relationships that exist between services, so that we can have them evolve in a coordinated fashion. This can be achieved through modelling and analytics, and through detailed runtime analysis, e.g., runtime testing and formal verification. Given the decentralized nature of the application environment, all these

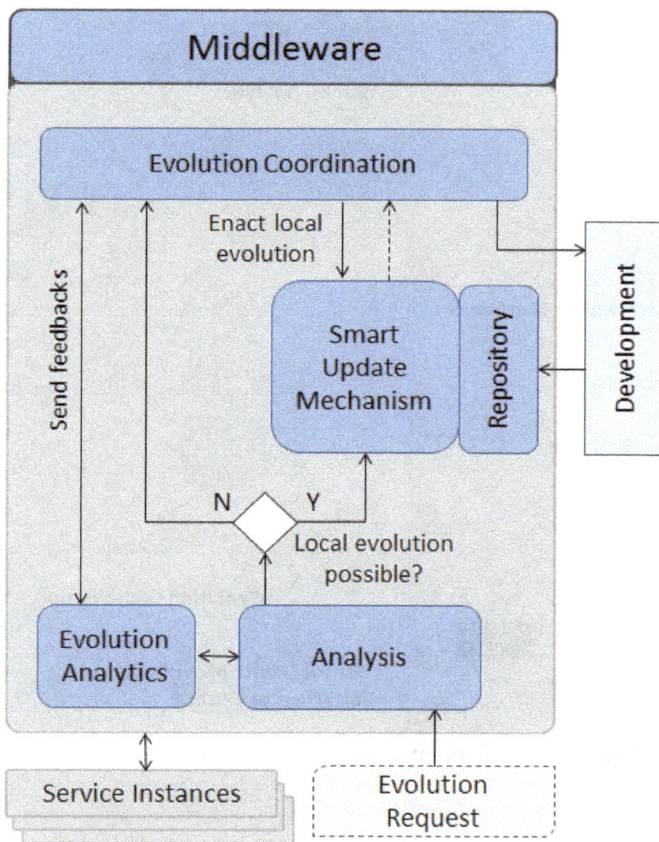

Figure 3. Architecture of the EVA

tools need to rely on local knowledge of the service itself and of its surroundings.

3. Solution Overview

Services running on heterogeneous systems and offered by different providers have de-coupled lifecycles, in particular, in IoT. Single services will be updated due to amendments or refinements or to provide further functionalities. Other providers may cut back the functionalities without taking notice of remaining clients that try to apply the removed functions. Business processes and applications that depend on services require appropriate coordination and adaptation by the participating parties. The solution we worked out equips every service with an agent, called EVA (Evolution Agent) that is capable to undertake these tasks. The internal structure and the rough composition of an EVA are depicted in figure 3. The next sections introduce the main components of an EVA and their interactions.

3.1. Analysis

The information interaction flow within our model is as follows. When an EVA receives first an Evolution Request, it is analysed by the Analysis module. An Evolution Request demands for adaptation to be able to take part in future interactions. In case a service provider wants to update his service, the Evolution Request can be sent by its EVA to the EVAs of the clients. This scenario is discussed in section 4. A further scenario is that the EVA of a service that is composed of other services and depends on them, demands one of his service providers to evolve to be able to update his own service.

In the latter case, the analysis module of the receiving EVA has to decide whether an evolution should take place and, if so, whether a local evolution is possible or whether the evolution has to be coordinated with other EVAs. For this reason, it assesses firstly the significance of the Evolution Request by evaluating the importance, the reputation and the number of partners who sent the request. The importance of a partner will increase, the more clients are affected by him. The significance will rise too, if the local service strongly depends on the other service and if there are no alternative services available. If either resources are becoming scarce or if it takes high efforts to satisfy the request, then lowly rated Evolution Requests may be rejected. Service instances not requested for a long time can be switched off to free resources for crucial service instances.

To estimate the efforts required for adaptation, the Analysis module considers initially local knowledge that includes information about locally available update mechanisms, the different service instances realizing different versions of the service, and the dependencies that the service versions might have towards other services. In case the Analysis module accepts the Evolution Request and a local update would satisfy the request, it will instruct the Smart Update Mechanism module, as presented below, to execute the local update and to provide eventually a new service instance. If a pure local update is not available or not sufficient due to interplay between several services, the Evolution Coordination module has to deal with a coordinated evolution and possibly ask software developers for further configurations.

3.2. Evolution Analytics

As time passes, the Analysis and the Evolution Coordination module can take more sophisticated decisions. The Evolution Analytics module collects runtime data about successful and unsuccessful evolution procedures. These data include information about local and coordinated evolutions since both modules feed the Evolution Analytics module. The goal is to discover promising evolution patterns by fostering successful and proven evolution procedures and preventing unsuccessful ones. Success does not only depend on smooth running in a technical sense, but has to consider the cost-performance ratio, the revenue, the reputation and QoS (Quality of Service) parameters too. Costs comprise, for instance, hardware and human resources which can be estimated hardly in the very beginning. If a new configuration has been implemented, the developer specifies the total man-hours spent. By means of Evolution Analytics EVA will learn to predict worthwhile evolutions while minimizing costs and time and maximizing the own revenue and reputation. The reputation of an EVA may decrease if it denies regularly Evolution Requests. Here, Evolution Analytics has to weigh the reputation against other factors like the costs for updates and the future revenue. To estimate reputation, costs and QoS, we will make use of our two prediction algorithms presented in [4].

For reasons of bootstrapping, EVAs are allowed to share parts of their knowledge with other EVAs. Special know how that affects only the service supervised by the EVA, has to be left out.

3.3. Evolution Coordination

In the event that a pure local evolution is not applicable, the Evolution Coordination module will co-operate with other EVAs and possibly interact with software developers. For example, the service is providing a method that depends on data delivered by a third party. To customize the interface for the client sending the Evolution Request, the Evolution Coordination will determine first the involved third parties and send them an Evolution Request. A continuous feedback between the EVAs is required to keep all parties up-to-date and to recognize future developments early. If a third party rejects the Evolution Request or if it is not available anymore, the Evolution Coordination can start a search for suitable services. To this end, we will adopt our service selection algorithms proposed in [7].

If the latter fails due to a lack of matching services, the Evolution Coordination will instruct the service provider or a responsible software developer to adapt the service. For this purpose, the developer may implement a configuration that is subsequently executed by the Smart Update Mechanism.

3.4. Smart Update Mechanism

The Smart Update Mechanism encompasses mainly two types of evolution capabilities. Firstly, it is aware of the different versions of the services running as service instances on the local machine and the versions used in the past. If one of them is fulfilling the conditions required, then it will be assigned to the requesting party. The second approach is a specification of the evolution rules and constraints that represent the possible service re-configurations and adaptations. In MUSIC [10] application developers specified the possible variants of an application and

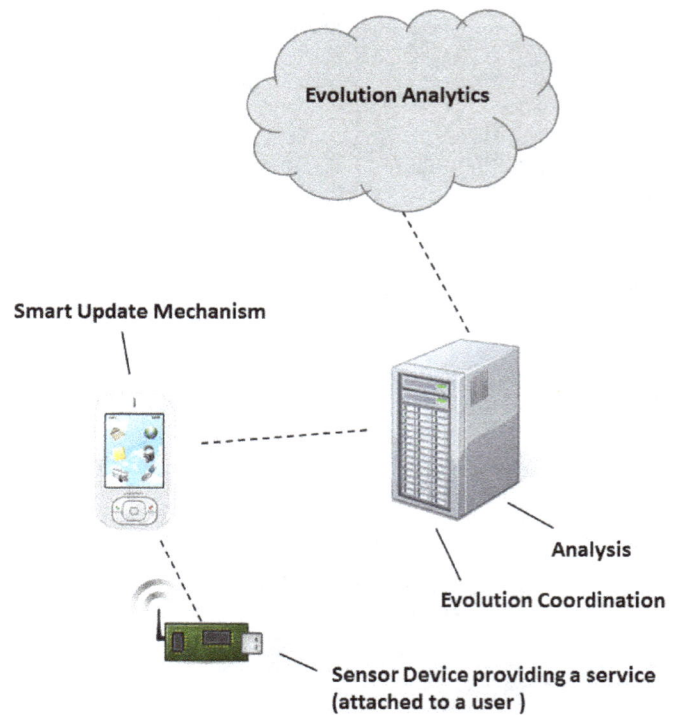

Figure 4. Deployment of an EVA in an IoT scenario

their dependencies on the runtime context; this was exploited by the adaptation manager in the middleware to achieve optimized application adaptation in different situations.

An EVA maintains up-to-date evolution models of its services. The models expose the possible configuration and adaptation paths. The EVA may govern multiple instances and versions of the same service at the same time, in order to accommodate different applications that may have different needs with respect to the service. Eventually, out-dated alternatives will be slowly retired.

The Analysis and Evolution Coordination modules introduced in the previous sections decide which configuration or version will be used for a specific client. In this connection, they do not only consider the possibilities offered by the Smart Update Mechanism, but take also into account the Evolution Analytics to optimize criteria like revenue, reputation, response time and own operability.

3.5. Repository

The Smart Update Mechanism makes use of a repository where several configurations were made available by developers. Developers can add new configurations to the repository during the lifetime of a service, for instance, if the Smart Update Mechanism did not find appropriate ones to update the service.

3.6. Middleware

Since objects or mobile devices are free to enter or leave the system, the middleware enables EVAs to communicate with each other in an asynchronous and loosely coupled manner. Besides that, the EVA itself can be divided into its modules such that each module may run on another device. This allows to make use of powerful runtime environments while energy constrained IoT devices that deliver the data offered by the service are spared. Small services, such as an object in IoT, will not have the processing power or storage needed to implement a full EVA. Figure 4 shows a low-level object that has outsourced its EVA components. This is a conservative deployment scenario since we assume that the on-site server and the cloud are always available.

However, more opportunistic approaches are also conceivable. For example, an object might rely on the availability of mobile devices that can enter or leave the system freely, to provide the resources it needs to communicate with the cloud.

4. Coordination of EVAs

Coordinating the evolution of services is a major challenge since it is a complex process that requires multiple interactions, as well as continuous feedback to understand whether the distributed evolution is proceeding as desired. To prevent never-ending negotiations between service providers about which service has to adapt first or to change at all, we introduce an algorithm that gives a clear path for the evolution. Therefore, we include the number of clients of each concerned service and their overall reputation. Figure 5 shows the process of taking into account the feedback received in response to an evolution request, particular from EVA x which sends the requests to its clients.

To tackle the aforementioned challenge, we define the following terms to coordinate and adapt EVAs.

WEIGHT OF A SERVICE (w):

The weight of a service is defined as the product of the reputation r of the service (range [0,1]) and its number of clients n_c (scaled into the range [0,1] by incorporating the max and min values of all services considered). The reputation r of the service and the weight w of a service can be defined respectively as follows:

$$r = \frac{\sum_1^n rating_i}{n} \qquad (1)$$

$$w = r \times n_c \qquad (2)$$

VOTE OF A SERVICE (v): The EVA that is managing an affected service is either interested in an adaptation or rejects it. For this reason, an EVA can vote for or against the evolution of a used service. Hence, we adopt the values of votes:

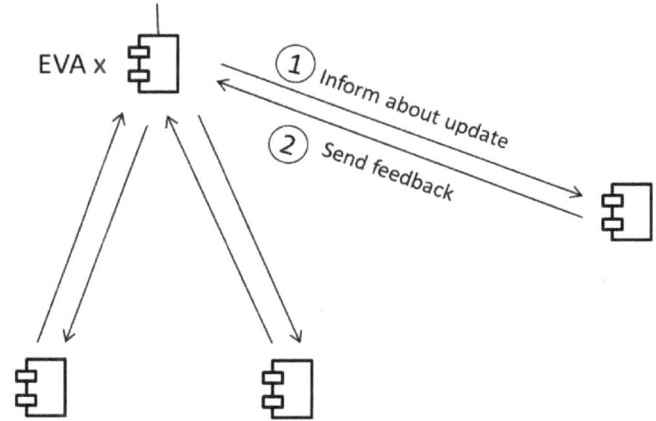

Service wants to update interface

Figure 5. Coordination of EVAs based on client's feedback

- Vote $(v_i) = +1$ votes for accepting the new service version.

- Vote $(v_i) = -1$ votes for not updating the interface or do update but keep the old version.

FEEDBACK (F): The higher the reputation of a service and the higher its number of clients, the higher the vote of the EVA that is managing the service is weighted. Thus, the overall feedback is comprised of the multiplication of the vote and the weight that consists of the reputation and the number of clients. This means that services that satisfy and affect more clients have a higher impact.

Feedback of one client (f_i):

$$f_i(w_i, v_i) = w_i \times v_i \qquad (3)$$

Feedback of all n clients (f_{agg}):

$$f_{agg} = \sum_i^n w_i \times v_i \qquad (4)$$

4.1. Coordination algorithm

The co-evolution will be executed if

$$f_{agg} \geq \epsilon(thresholdvalue) \qquad (5)$$

A step-wise structure of the proposed algorithm that encompasses the equations from (1) to (5), is given in the following:

Input: Evolution request of EVA x to EVAs $c \in C$; number of clients and the reputation of the EVAs $c \in C$.

Step 1: The service managed by EVA x will be updated by the provider or EVA x received an evolution request from another EVA y.

Step 2: x is asking the EVAs $c \in C$ of its clients whether they would accept or reject the required adaptation.

Step 3: x is summing up the feedbacks of $c \in C$ by considering their vote and their reputation and number of clients that are both scaled into the range [0,1].

Step 4: x is dividing the summed up feedbacks by the number of clients to obtain f_{agg} and compares f_{agg} with a predefined threshold value ϵ.

Step 5: The co-evolution will be executed if $f_{agg} \geq \epsilon$. In this case, the update mechanisms will be executed.

Step 6: Otherwise, the evolution requests will be rejected.

Output: Accept or reject the evolution requests

4.2. Optimization Problem

Web services may be composed of several other Web services and, hence, stay in contact to different EVAs. This scenario is similar to that of business processes in SOA (Service-oriented Architectures), where a central process orchestrates Web services to realize a certain functionality. The orchestration is formulated as a process that may start with the arrival of a message like a receive or a pick activity. It may send an answer back to the requester with a reply activity. Activities and service invocations may be grounded inside a sequence or a flow structure.

In service co-evolution, we are searching for the optimal interface. A service may be a composition of multiple EVAs where each EVA represents a Web service. The orchestration itself is monitored and managed by an EVA. It tries to optimize the own revenue and the rate of satisfaction of clients by providing a suitable interface for a given functionality. This interface is built by a composition of the interfaces of other EVAs (services). The goal of an selection algorithm is to find first an optimal choice of interfaces that realizes the most preferred interface for the orchestration. This is done by the aforementioned coordination algorithm. With EVA x, for instance, we define the set I that contains the set of requested service interfaces. The set I depends on the clients who vote whether they would accept or reject the interface of a composition of certain interfaces.

After finding the set I of matching interfaces, the challenge is to find those services that optimize the overall quality of the orchestration for the clients and the revenue for the provider. Finding the optimal solution means now to maximize the overall satisfaction (S) of the own clients and to select those EVAs that will provide the required interfaces and that maximizes the own revenue (R).

We consider the following objective function for realizing the EVA orchestration:

$$MaximizeF_{obj}(I) = F_{obj1}(R) \times F_{obj2}(S) \qquad (6)$$

$F_{obj1}(R)$ and $F_{obj2}(S)$ are explained below. $F_{obj1}(R)$ is an objective function for maximizing the provider's

revenue, $F_{obj2}(S)$ has the goal to maximize the clients' satisfaction.

We incorporate the following quality dimensions to compute the overall satisfaction for the clients:

- $throughput(t)$: number of service invocations per time unit.

- $reliability(l)$: the probability that the service executes successfully.

- $executiontime(e)$: the time it takes to execute the service.

- $availability(a)$: the percentage of time during which the service is available.

The vector Q (t, l, e, a) contains the quality of service (QoS) dimensions.

The value of functions F_{obj1} and F_{obj2} are weighted and combined to make the final decision. To maximize $F_{obj}(I)$ in (6) both objective functions below has to be maximized:

$$F_{obj1}(R) = \sum_{i=1}^{n} r_i, r_i \in R \qquad (7)$$

$$F_{obj2}(S) = k_1 \times t + k_2 \times l + k_3 \times \frac{1}{e} + k_4 \times a \qquad (8)$$

Therefore:

$$F_{obj}(I) = (\sum_{i=1}^{n} r_i) \times (k_1 \times t + k_2 \times l + k_3 \times \frac{1}{e} + k_4 \times a) \qquad (9)$$

To keep the description as simple and clear as possible, the normalization steps to the range of [0,1] are not included in the formula. The factors $k_i, i = 1...4$ represent the weights for the quality dimensions depending on the preferences of the EVA. Let us assume that the current interface of EVA x can provide a revenue of R_0 and an average satisfaction of (t_0, l_0, e_0, a_0). The goal of interface selection would be to find a pair (R_i, S_i) better than the pair (R_0, S_0).

In the event of pure Web service selection for business processes, this issue can be solved by our approach in the paper [7] or by other popular approaches like integer linear programming [24] and genetic algorithms [6].

5. Related Work

Over the last decades, the service evolution raised to a more and more important topic that brings many new challenges to software engineering. This section will present some frequently cited works related to our research.

One of first works handling the problem of service evolution is developed by Forkaefs et al. [11]. Their tool VTracker is designed to analyse the evolution of WSDL interfaces. The idea of Vtracker is based on the Zhang-Shashas tree-edit distance [25] which calculates the minimum edit distance between to trees. In this study the WSDL interfaces are considered as usual XML files. Specifically the authors created an intermediate XML representation to reduce the verbosity of the WSDL specification. In this simplified XML representation, among other transformations, the authors trace the references between message parameters and data types and replace the references with the data types themselves. However, VTracker does not take into account the syntax of WSDL interfaces. As consequence, their approach outputs only the percentage of differences between XML elements. In addition, this approach of transforming a WSDL interface into a simplified representation can lead to the detection of multiple changes while there has been only one change.

Similarly, Romano and M. Pinzger [18] presented an outstanding work called WSDLiff that compares subsequent versions of WSDL interfaces to automatically extract the changes. This approach takes into account the syntax of the WSDL file and the schema file XSD [18] that is used to define the data types of the WSDL interface. In particular, WSDLDiff extracts the types of the elements affected by changes (e.g., Operation, Message) and the types of changes (e.g., removal, addition, move, attribute value update). Romano at al. refer to these changes as fine-grained changes. The fine-grained changes extraction process of WSDLDiff is based on the UMLDiff algorithm [21]. This proposed tool is a useful means to understand how a particular Web service evolves over time. This approach is relevant for Web service subscribers who want to compare the evolution of different Web services with similar features or to analyze the most frequent changes affecting a WSDL interface. By applying this approach, Web service subscribers can estimate the risk associated to the usage of a certain Web service. Nonetheless, the authors did not investigate the co-evolution of different Web service which differs from our approach.

Other well-known research results come from M.P. Papazoglou and V. Andrikopoulos [1, 15] with analyzing shallow changes and deep changes. In their papers, they developed a set of theories and modes that unify different aspects of services (description, versioning, and compatibility) to assist service developers in controlling and managing service changes. They distinguished between shallow changes (small-scale, localized) and deep changes (large-scale, cascading) for service compatibility and reasoning mechanisms for delimiting the effect of changes which can keep local to and consistent with a service description. They discussed when a change in a service is triggered, how to analyse its impact, and the possible implications of the

implementation of the change for the service providers and consumers. However, the authors only focused to deal with shallow changes. Additionally, their approach did not mention about the IoT environment and its services. But some lessons can be learned from their formal principles and theories for the description of the coordination of EVAs, for instance.

Design patterns have been widely used for software development for structuring solutions [8, 20]. S. Wang et al [20] focuses on a common evolution scenario where a single service, provided by a single provider, is used by many different and possibly unknown consumers, as is the case of most current Web services, such as Google Maps, eBay Trading, and Amazon E-Commerce. In the scenario of [20], the services usually face large and frequent changes as a result of an increasing need to conform to changing business and technological requirements [20]. In particular, the paper proposed four patterns involving compatibility, transition, split-map, and merge-map. These patterns provide generic and reusable strategies for service evolution. These patterns can be involved to deploy and support our agents as they can be used to derive the set of changes.

Another important service evolution approach is the analysis of service dependencies. Basu et al. [5] introduced a tool that can extract dependencies from log files. Their technique could be adapted in order to infer a set of dependent service consumers. One the dependencies are understood, it is also important to infer the impact of service changes on the dependent applications. The Chain of Adapters technique [12] is an alternative approach for deploying multiple versions of a Web service in the face of independently developed unsupervised clients. The basic idea is to resolve the mismatches between the expectations of the consumers and the supported versions and configurations of the services. This can prove useful in self-configurations.

It is worth mentioning the work on service compatibility with the WS2JADE tool [14] which is based on an agent approach. Xuan Thang Nguyen and Ryszard Kowalczyk proposed the WS2JSADE toolkit for integrating Web services and the Jade agent platform. This tool provides facilities to deploy and control Web services as agent services at run time for deployment flexibility and active service discovery. The authors also discuss different ways how Web services can be visible to agents and how they can be accessed and used by agents. Although WS2JADE offers many advantages over other existing tool, it still lacks substantial theoretical work with respect to agent to Web service integration like, for instance, service co-evolution.

In fact, there are many agent-based approaches available to support interoperable IoT devices and their services nowadays [2, 3, 17, 22]. Nonetheless, the adaptation mechanisms and the collaboration characteristics in these agents are not sufficient in order to achieve coordinated service evolution. Furthermore, it needs a global vision which can predict potential effects, challenges and requirements for participating service providers.

6. Conclusion and Future Work

This paper introduces a new vision of service co-evolution in IoT. It provides developers a common evolution management model and reference architecture and represents a focused effort to provide a foundation for realizing the full potential of service-based architectures. For this reason, the challenges in the co-evolution of services that cover the wide spectrum from IoT to Cloud Computing are analyzed as well.

This paper also adopts a novel conceptual agent as a solution for service co-evolution. Evolution tasks like the assessment and coordination of evolution requests, updating and versioning the interfaces and selecting matching services can be performed automatically or semi-automatically by EVAs.

Furthermore, the paper also proposes an approach for coordinating EVAs in service co-evolution. Besides that, it proposes a first approach for the selection of interfaces in orchestrations that is required to satisfy revenue and satisfaction requirements. In this way, systems can be made more adaptive, efficient and reduce costs to manage maintenance.

In summary, the paper is a step forward to service co-evolution in IoT. The results of this research will provide support to professional service providers and business process engineers. In future, first research prototypes for the coordination of EVAs shall be delivered to evaluate the prospect of this approach.

Acknowledgment

The authors would like to acknowledge the generous support of DAAD (Deutscher Akademischer Austauschdienst).

References

[1] Andrikopoulos, V., Benbernou, S., Papazoglou, M. P., 2012. On the evolution of services. Software Engineering, IEEE Transactions on 38 (3), 609–628.

[2] Atzori, L., Iera, A., Morabito, G., 2010. The internet of things: A survey. Computer Networks 54 (15), 2787–2805.

[3] Ayala, I., Amor, M., Fuentes, L., 2012. An agent platform for self-configuring agents in the internet of things. INFRASTRUCTURES AND TOOLS FOR MULTIAGENT SYSTEMS, 65–78.

[4] Baraki, H., Comes, D., Geihs, K., 2013. Context-aware prediction of qos and qoe properties for web services. In: Networked Systems (NetSys), 2013 Conference on. IEEE, pp. 102–109.

[5] Basu, S., Casati, F., Daniel, F., 2008. Toward web service dependency discovery for soa management. In: Services Computing, 2008. SCC'08. IEEE International Conference on. Vol. 2. IEEE, pp. 422–429.

[6] Canfora, G., Di Penta, M., Esposito, R., Villani, M. L., 2005. An approach for qos-aware service composition based on genetic algorithms. In: Proceedings of the 7th annual conference on Genetic and evolutionary computation. ACM, pp. 1069–1075.

[7] Comes, D., Baraki, H., Reichle, R., Zapf, M., Geihs, K., 2010. Heuristic approaches for qos-based service selection. In: Service-Oriented Computing. Springer, pp. 441–455.

[8] Daigneau, R., 2012. Service Design Patterns: fundamental design solutions for SOAP/WSDL and restful Web Services. Addison-Wesley.

[9] Dunkels, A., et al., 2009. Efficient application integration in ip-based sensor networks. In: Proceedings of the First ACM Workshop on Embedded Sensing Systems for Energy-Efficiency in Buildings. ACM, pp. 43–48.

[10] Floch, J., Frà, C., Fricke, R., Geihs, K., Wagner, M., Gallardo, J. L., Cantero, E. S., Mehlhase, S., Paspallis, N., Rahnama, H., Ruiz, P. A., Scholz, U., 2013. Playing music - building context-aware and self-adaptive mobile applications. Softw., Pract. Exper. 43 (3), 359–388.

[11] Fokaefs, M., Mikhaiel, R., Tsantalis, N., Stroulia, E., Lau, A., 2011. An empirical study on web service evolution. In: Web Services (ICWS), 2011 IEEE International Conference on. IEEE, pp. 49–56.

[12] Kaminski, P., Müller, H., Litoiu, M., 2006. A design for adaptive web service evolution. In: Proceedings of the 2006 international workshop on Self-adaptation and self-managing systems. ACM, pp. 86–92.

[13] Leitner, P., Michlmayr, A., Rosenberg, F., Dustdar, S., 2008. End-to-end versioning support for web services. In: Services Computing, 2008. SCC'08. IEEE International Conference on. Vol. 1. IEEE, pp. 59–66.

[14] Nguyen, X. T., Kowalczyk, R., 2007. Ws2jade: Integrating web service with jade agents. Springer.

[15] Papazoglou, M. P., Andrikopoulos, V., Benbernou, S., 2011. Managing evolving services. Software, IEEE 28 (3), 49–55.

[16] Priyantha, N. B., Kansal, A., Goraczko, M., Zhao, F., 2008. Tiny web services: design and implementation of interoperable and evolvable sensor networks. In: Proceedings of the 6th ACM conference on Embedded network sensor systems. ACM, pp. 253–266.

[17] Roalter, L., Kranz, M., Möller, A., 2010. A middleware for intelligent environments and the internet of things. In: Ubiquitous Intelligence and Computing. Springer, pp. 267–281.

[18] Romano, D., Pinzger, M., 2012. Analyzing the evolution of web services using fine-grained changes. In: Web Services (ICWS), 2012 IEEE 19th International Conference on. IEEE, pp. 392–399.

[19] Shelby, Z., 2010. Embedded web services. Wireless Communications, IEEE 17 (6), 52–57.

[20] Wang, S., Higashino, W., Hayes, M., Capretz, M. A., 2014. Service evolution patterns. Proceedings of the 21st IEEE International Conference on Web Services.

[21] Xing, Z., Stroulia, E., 2005. Umldiff: an algorithm for object-oriented design differencing. In: Proceedings of the 20th IEEE/ACM international Conference on Automated software engineering. ACM, pp. 54–65.

[22] Yu, H., Shen, Z., Leung, C., 2013. From internet of things to internet of agents. In: Green Computing and Communications (GreenCom), 2013 IEEE and Internet of Things (iThings/CPSCom), IEEE International Conference on and IEEE Cyber, Physical and Social Computing. IEEE, pp. 1054–1057.

[23] Yu, Q., Liu, X., Bouguettaya, A., Medjahed, B., 2008. Deploying and managing web services: issues, solutions,

and directions. The VLDB JournalâĂŢThe International Journal on Very Large Data Bases 17 (3), 537–572.

[24] Zeng, L., Benatallah, B., Ngu, A. H., Dumas, M., Kalagnanam, J., Chang, H., 2004. Qos-aware middleware for web services composition. Software Engineering, IEEE Transactions on 30 (5), 311–327.

[25] Zhang, K., Shasha, D., 1989. Simple fast algorithms for the editing distance between trees and related problems. SIAM journal on computing 18 (6), 1245–1262.

12

Portfolio Optimization in Secondary Spectrum Markets

Praveen K. Muthuswamy[1,*], Koushik Kar[1], Aparna Gupta[2], Saswati Sarkar[3], Gaurav Kasbekar[4]

[1]Department of Electrical, Computer, and Systems Engineering, Rensselaer Polytechnic Institute
[2]Lally School of Management and Technology, Rensselaer Polytechnic Institute,
[3]Deptartment of Electrical and Systems Engg, University of Pennsylvania
[4]Deptartment of Electrical Engg, Indian Institute of Technology, Bombay, India.

Abstract

In this paper, we address the spectrum portfolio optimization (SPO) question in the context of secondary spectrum markets, where bandwidth (spectrum access rights) can be bought in the form of *primary* and *secondary* contracts. While a primary contract on a channel provides guaranteed access to the channel bandwidth (possibly at a higher per-unit price), the bandwidth available to use from a secondary contract (possibly at a discounted price) is typically uncertain/stochastic. The key problem for the buyer (service provider) in this market is to determine the amount of primary and secondary contract units needed to satisfy its uncertain user demand. We formulate single and multi-region spectrum portfolio optimization problems as one of minimizing the cost of the spectrum portfolio subject to constraints on bandwidth shortage. Two different forms of bandwidth shortage constraints are considered, namely, the demand satisfaction rate constraint, and the demand satisfaction probability constraint. While the SPO problem under demand satisfaction rate constraint is shown to be convex for all density functions, the SPO problem under demand satisfaction probability constraint is not convex in general. We derive some sufficient conditions for convexity in this case. We also discuss application of the Bernstein approximation technique to approximate a non-convex demand satisfaction probability constraint by a convex constraint. The SPO problems can therefore be solved efficiently using standard convex optimization techniques. We then consider a discrete version of the SPO problem, in which the primary and secondary contracts can bought/sold in discrete units. We study the NP-hardness submodularity property of the discrete SPO problem and discuss a branch-and-bound algorithm to obtain the optimal solution for this problem. Finally, we perform a thorough simulation-based study of the single-region and the multiple-region problems for different choices of the problem parameters, and provide key insights regarding the portfolio composition, the efficiency of the Bernstein convex approximation technique, and the closeness of the optimal discrete spectrum portfolio solutions to their continuous approximations. We provide several insights about the scaling behavior of the unit prices of the secondary contracts, as the stochastic characterization of the bandwidth available from secondary contracts change.

Keywords: Spectrum Markets, Portfolio Optimization, Bandwidth contracting.

*Corresponding Author. Email: muthusamy.pk@gmail.com, kark@rpi.edu

1. Introduction

The number of users of the wireless spectrum, as well as the demand for bandwidth per user, has been growing at an enormous pace in recent years. Since spectrum is limited, its effective management is vitally important to meet this growing demand. The spectrum available for public use can be broadly categorized into the unlicensed and licensed zones. In the unlicensed part of the spectrum, any wireless device is allowed to transmit. To use the licensed part, however, license must be obtained from appropriate government authority – the Federal Communications Commission (FCC) in the United States, for example – for the exclusive right to transmit in a certain block of the spectrum over the license time period, typically for a fee. While spectrum management in licensed bands has mostly been controlled by responsible government bodies, the need for bringing market based reform in spectrum trading is being increasingly recognized [1], [2], [3]. In order to achieve spectrum-usage efficiency, spectrum markets should allow dynamic trading of spectral resources and derived contracts of different risk-return characteristics. Providers can then choose to buy/sell one or more of these spectrum contracts depending on the level of service they wish to provide to their customers.

We consider a spectrum market in which a wireless service provider (buyer) can purchase spectrum access rights from another provider (seller) in the form of two types of spectrum contracts: *primary contract* and *secondary contract*. Typically, the buyer will be a smaller local or regional provider, buying access rights over its operational area from a larger regional or national provider which acts as the seller, although the framework and results that we present in this paper does not make any such assumption. Primary contract offers unrestricted access rights on a channel – a specific channel or one of a set of channels "owned" by the seller. On the other hand, secondary contract offers restricted access rights on a channel or a set of channels – it provides access to the "leftover" bandwidth on the channel(s) that the primary users of the channel(s) do not need at that specific time. At their core, primary and secondary contracts differ in the risk-return tradeoff that they provide. A primary contract represents a risk-free contract in terms of its bandwidth return characteristics, while the secondary contract is inherently risky in terms of the bandwidth it can provide. Primary contracts would generally be more expensive (in terms of cost per unit contract), since they provide full access rights. Secondary contracts would typically be cheaper due to their riskiness. These two contracts represent two fundamental forms of spectrum access contracts – analogous to bonds and stocks in terms of the risk characteristics. In financial markets, it is well known that bonds and stocks help investors achieve their desired risk-return tradeoff on investment. Similarly, we envisage that the wireless service providers can efficiently tradeoff the level of

service they wish to provide against their cost by using these two types of contracts.

A key challenge for a provider in this market is to determine an appropriate mix (i.e. a portfolio) of primary and secondary contracts that can provide the desired level of service to its users at a low cost. We formulate and study this *Spectrum Portfolio Optimization (SPO)* problem from the perspective of a buyer. In standard financial portfolio optimization, the objective is to maximize the expected portfolio return while satisfying a constraint on the variance of return. In the spectrum market context, minimizing the cost of the portfolio is a more reasonable objective. Furthermore, the constraint in the SPO problem can be specified meaningfully in two ways – either in terms of the expected bandwidth shortage, or in terms of the probability of bandwidth shortage. We refer to these constraints as the *demand satisfaction rate constraint* and the *demand satisfaction probability constraint*, respectively. We study the SPO problem under the two constraints separately.

The technical contributions of this paper are as follows. Firstly, we show that the SPO problem under demand satisfaction rate constraint is convex under any assumptions on the user demand and the bandwidth return distributions. Secondly, we show that the SPO problem under demand satisfaction probability constraint is not convex in general, and also derive sufficient conditions on the demand density functions for convexity to hold. The motivation behind showing convexity of the optimization problems is that convex problems can be solved efficiently using standard techniques such as gradient descent and Newton's methods, whereas there are no general techniques for solving non-convex problems efficiently. We also discuss application of the Bernstein convex approximation technique in cases where the demand satisfaction probability constraint is non-convex; this technique approximates a non-convex probability constraint by a convex expectation constraint. In the next step, we extend the SPO problem and the convexity results to a multiple-region scenario, where the buyer's portfolio is intended to serve a set of disjoint geographical locations, each having its own user demand, using available primary and secondary contracts that provide access rights only over subsets of all locations of interest. We then consider an integer programming formulation of the SPO problem, since the primary and secondary contracts can be bought/sold only in discrete units in an operational spectrum market. We show that the discrete SPO problem is not only NP-hard, but the constraint function in both versions is not submodular either. We then discuss a branch-and-bound algorithm algorithm to solve it efficiently (although the procedure is naturally not polynomial time, given the NP-hardness of the problem). Finally, we perform a detailed simulation-based study of the single and the multiple-region SPO problems and provide insights about the portfolio composition and the price characteristics of the secondary contracts.

Broadly speaking, one of the main contributions of the work is deriving sufficient conditions for convexity of the SPO problem, which would ensure that optimal solutions of the problem could be computed in an efficient manner. We do not propose specific algorithms/approaches for solving the SPO problems for this case - any standard convex optimization algorithm/toolbox could be used. For the case when the SPO problem is non-convex, we apply and evaluate the application of the Bernstein approximation (convexification) method to the problem, which can be used in solving the problem efficiently (albeit approximately). The NP-hardness and non submodularity of the SPO problem under discrete portfolio constraints make the existence of polynomial or pseudo-polynomial solutions to the discrete SPO problem extremely unlikely. Yet, the branch-and-bound method outlined for that case allows us to compute the optimal solution in a manner that seems reasonably efficient in practice (although not polynomial complexity).

Economics of spectrum allocation and auction mechanisms have been discussed widely in the literature [4], [5], [6], [7]. Spectrum sharing games and/or pricing issues have been considered in [8], [9], [10], [11]. Discussions and recommendations for transition to spectrum markets and secondary markets for spectrum trading have emerged [12], [13], [14]. In [14], the authors consider a spectrum secondary market analogous to the stock market for dynamically trading their channel holdings. The proposed auction-based market mechanism is shown to improve user performance and spectrum utilization. However, a clear design of the contract types and tradeoff analysis using portfolio theory have not been considered before. In [15], the authors propose a wireless spectrum market with two types of contracts, namely, the long-term and the short-term contract, and study the structural properties of the optimal dynamic trading strategy. Unlike the short-term contract defined in [15], the amount of bandwidth available for access from a secondary contract is a random variable. Moreover, the problem addressed in this paper is the spectrum portfolio optimization question over a single period and is different from the multi-period trading question considered in [15]. The spectrum trading question has also been addressed from game theoretic perspectives. In particular, [16] analyzes the spectrum trading problem between one primary and multiple secondary users as a mechanism design question, and discusses the feasibility of contract formation, and the properties of the optimal contract. The authors in [17], on the other hand, models and analyzes the spectrum trading question as a non-cooperative evolutionary game between multiple primary users (sellers) and multiple secondary users (buyers).

Portfolio optimization problem has been studied extensively in finance since the development of the mean-variance optimization framework in [18]. Several attempts have been made to improve the model and the risk measure [19], [20], [21], [22]. In [22], the authors propose a new measure of risk, namely, the expected shortfall and show that the problem of minimizing expected shortfall subject to a linear equality constraint is convex. The expected shortfall function considered in [22] measures the shortfall of return with respect to the α-quantile of the return distribution. But the demand satisfaction rate constraint that we consider measures the shortfall of the bandwidth return relative to a stochastic quantity, and is therefore different from the shortfall function in [22]. However, we are still able to make use of some of their analysis techniques to our problem. Probabilistic constraints have not been studied much, until recently in [23] and [24]. In [23], the authors study probabilistically constrained linear programs and present conditions for convexity of the constraint. While we apply some of their results in our context, we also provide additional conditions for convexity on the SPO problem with the demand satisfaction probability constraint.

The novelty of our contribution stems from the following aspects. Though the notion of primary and secondary users and their spectrum access rights have been extensively discussed recently, our modeling of these access rights as bond-like riskless and stock-like risky contracts, and the rigorous formulation of the spectrum portfolio optimization problem, are novel. Convexity of various versions of the portfolio optimization question have been studied in the finance and optimization literature; however, very limited results exist on the specific demand satisfaction constraints that appear meaningful in the spectrum access context. We provide several interesting results for the SPO problem with such constraints in this paper. The formulation and analysis of the multi-region SPO problem, and the insights obtained from our numerical studies, also constitute novel contributions of this work.

In recent years, spectrum regulatory agencies in various countries across the globe have taken bold steps towards fostering spectrum sharing/trading in secondary spectrum markets. However, dynamic spectrum markets, and active trading in such markets, still remain in their infancy. As dynamic spectrum trading across service providers picks up, the solutions to the SPO problem that we provide here would be useful to the spectrum service providers to optimally balance their overall revenue with risk of customer dissatisfaction. The solutions and algorithms we provide could be used by spectrum service providers to determine how many primary and secondary licenses to purchase to attain its desired risk-return tradeoff point. Other than providing convexity conditions, we also provide convexification based approximation algorithms for non-convex cases, and a method for finding optimal portfolios under discreteness constraints. Which of these results or algorithms are useful for a spectrum service provider will depend on the nature of the bandwidth demand and return processes (which is expected to evolve over time), and the constraints on the specific provider's portfolio. In this paper, we provide a broad set of results covering different possibilities; a subset of these are likely to find use in any specific spectrum portfolio planning context.

The rest of the paper is organized as follows. In Section 2, we formally define the SPO problems under demand satisfaction rate and probability constraints. In Sections 3 and 4, we study the convexity of the two SPO problems under the two types of constraints. In Section 6, we consider a discrete version of the SPO problem and study its submodularity properties. In Section 5, we study the multiple-region SPO problem. In Section 6, we study the SPO problem with discrete portfolio constraints. Finally, in Section 7, we present the simulation results.

2. Spectrum Portfolio Optimization Problem Formulation

In this section, we formally define the spectrum portfolio optimization (SPO) problem for a single region. The formulation and discussion of the multi-region SPO problem is deferred to Section 5. Although not necessary for the mathematical formulation or subsequent analytical treatment of the SPO problem, it is easy to motivate the development of the framework by considering a (secondary) spectrum market in which N "higher level" spectrum providers are selling access contracts in the form of primary and secondary contracts to other "lower level" providers. These seller spectrum providers will typically be large providers (like VerizonWireless, AT&T, and Sprint in the US for example) who have directly licensed spectrum from the governing body (like FCC), and might want to offer their excess bandwidth in the form of primary and secondary contracts. The buyers of the contracts can be smaller, possibly local or smaller regional wireless spectrum service providers who are trying to obtain bandwidth at the cheapest price to serve their user (customer) demand. We assume that primary and secondary contracts can be obtained in multiple units. Without loss of generality, we can assume that each unit of primary contract provides exclusive access to 1 unit of bandwidth in some channel that the seller provider operates on. On the other hand, each unit of secondary

contract provides exclusive access to bandwidth that is a random variable varying between 0 and 1 unit. While this assumption is for the ease of exposition, it can be easily generalized. A simple way to view this setting would be to consider a seller provider having C units of bandwidth, offering C units of primary and C units of secondary contracts. If in any time slot, the primary contract holders in totality use $\alpha < C$ units of bandwidth, each unit of secondary contract has access to $0 < (C - \alpha)/C < 1$ units of bandwidth. A buyer holding x units of secondary contracts with this seller provider will then have access to $x(C - \alpha)/C$ units of bandwidth in that time slot.

Note that we are associating contracts – primary or secondary – with the seller providers, not specific channels. *All primary contracts (no matter which seller provider provides it) can be considered equivalent, since they offer the same bandwidth return (one unit, guaranteed).* This also argues for the fact that they must be priced the same; without loss of generality, we assume that the cost of one unit of any primary contract is unity. Secondary contracts offered by different seller providers will differ from one another, depending on the access pattern of the primary members of the seller provider, and their price per unit will also differ. However, since each unit of secondary contract offers an average return of less than one unit bandwidth, and have some risk associated with the return, the price per unit for each secondary contract should be less than unity (the price of a unit of primary contract).

With this abstraction, the SPO problem can be viewed in the context of a market where a single type of primary contract, and N different types of secondary contracts, are being offered.[1] Each unit of primary contract sold in the secondary spectrum market offers guaranteed access to 1 unit of bandwidth at a cost of 1. The secondary contract offered by the provider i can be described by the pair (p_i, B_i), where, p_i is the unit price of the secondary contracts offered by the ith seller provider and B_i is the random variable (varying between 0 and 1) characterizing the bandwidth return from one unit of secondary contract of the ith provider. From the above discussion, $p_i < 1, \forall i$.

In the following, we assume that each seller provider has a large pool of available bandwidth, and so any amount of primary or secondary contract units can be bought from the providers. This is for ease of exposition, and can be easily generalized by incorporating into the SPO problem additional upper bounds on the number of primary and secondary contract units available from a seller provider.

---

[1]Note that the basic portfolio optimization question in financial markets, while considering multiple risky (stock) assets, assumes only a single risk-free (bond) asset, for similar reasons.

Now we are ready to formally define the SPO problem from the perspective of a single buyer provider. The buyer's objective is to create a spectrum portfolio consisting of primary and secondary contract units from the N seller providers in order to provide service to its customer base. Let $x_i, 1 \leq i \leq N$ denote the amount of secondary contract units purchased from the ith seller provider. Since the primary contracts offered by all the N providers are identical, we only need to keep track of the total amount of primary contract units bought, which we denote by x_0. We assume a relaxation that $x_0, x_1, .., x_N$ are non-negative real numbers, not necessarily integers. Let the vector $\overline{x} = (x_0, x_1, .., x_N)$, denote the buyer's spectrum portfolio. The buyer wishes to satisfy its customers' demand for bandwidth using the spectrum portfolio, $\overline{x}$. The customer demand is modeled as a random variable Q, as it is often unknown in advance. The bandwidth return or the actual units of bandwidth available from a spectrum portfolio $\overline{x}$, is uncertain, due to presence of the secondary contracts. The bandwidth return of the portfolio $\overline{x}$, $B(\overline{x})$, is defined as $B(\overline{x}) = x_0 + \sum_{i=1}^{N} x_i \times B_i$.

Since the bandwidth return and the demand are stochastic, it is impossible or highly expensive to construct a portfolio that always offers enough bandwidth to satisfy the customer demand. However, it is desirable to construct portfolios with low levels of bandwidth shortage. Let us define $S(\overline{x}) = Q - B(\overline{x})$. Then the bandwidth shortage of a portfolio, denoted by $S(\overline{x})^+$, is given as $S(\overline{x})^+ = \max(S(\overline{x}), 0) = \max(Q - B(\overline{x}), 0)$. Note that the shortage, $S(\overline{x})^+$, is also a stochastic quantity as both Q and $B(\overline{x})$ are random variables.

The spectrum portfolio optimization (SPO) problem for the buyer is to find the least costly portfolio with low levels of bandwidth shortage. The SPO objective is

$$\text{minimize } C(\overline{x}) = x_0 + \sum_{i=1}^{N} x_i \times p_i. \tag{1}$$

The constraint on bandwidth shortage can be specified either in terms of expected shortage or probability of shortage. Therefore, we consider two versions of constraints for the SPO problem – the *Demand Satisfaction Rate (DSR)* constraint, and the *Demand Satisfaction Probability (DSP)* constraint, as expressed below:

DSR Constraint: $E[S(\overline{x})^+]$ $< \delta;$ (2)

DSP Constraint: $Pr(S(\overline{x}) > 0)$ $< \epsilon.$ (3)

Here $C(\overline{x}) = x_0 + \sum_{i=1}^{N} x_i \times p_i$ is the cost of the spectrum portfolio $\overline{x}$. The DSR constraint ensures that the expected amount of bandwidth shortage is below a certain acceptable level δ. On the other hand, the DSP constraint bounds the probability of shortage to a low value ϵ. Note here that $Pr(S(\overline{x}) > 0)$ is the same as

$Pr(S(\overline{x})^+ > 0)$. We devote the following sections to the study of the SPO problem under these two types of constraints.

An alternative formulation (version) of the problem would have been to reverse the constraint and objective functions. In particular, we could minimize the bandwidth shortage probability (in expectation or probability) subject to maximum limit on the cost of the spectrum portfolio. The fundamental complexity/computability of the optimal solution, does not change with this reversal however, as the optimal solution of one version of the problem could be translated to the optimal solution of the other in polynomial time. Furthermore, the main complexity of the SPO problem comes from that of the non-linear functions in (2) and (3), the function in (1) being is a simple linear function. The main theoretical results that we show in this paper - on the convexity of the SPO problem under the DSR and DSP constraints - are equally applicable to the alternative (reversed) version of the problem as well.

Finally, note that our cost function in (1) assumes that the price per unit bandwidth remains the same irrespective of the quantity bought. In practice, however, the per-unit price would vary with the quantity bought. Since bandwidth is a limited quantity, it is reasonable to assume that the marginal price increases with the quantity purchased [25]. In other words, if $P_i(x_i)$ denotes the total price to be paid for buying x_i units of commodity, we can assume that P_i is an increasing convex function in its argument. Then the cost function in (1) can be written as $C(\overline{x}) = x_0 + \sum_{i=1}^{N} P_i(x_i)$, which is a convex function. Therefore, in that case too, the complexity of the problem is dictated by that of the functions in the constraints (2) and (3), which is what we address in this paper. For the sake of simplicity, however, in the rest if the paper we assume that $P_i(x_i)$ is linear in x_i, as given in (1).

3. SPO under Demand Satisfaction Rate (DSR) Constraint

In this section, we study the properties of the SPO problem under demand satisfaction rate constraint, and provide the expressions for certain useful quantities that can be utilized to compute the optimal portfolio solution efficiently. The objective function of the SPO problem (Equation 1) is linear and therefore convex. The demand satisfaction rate function (i.e. $E[S(\overline{x})^+]$), however, is non-linear in $\overline{x}$. Borrowing from the analysis techniques in [22], we show below that $E[S(\overline{x})^+]$ is also convex in $\overline{x}$. This implies that the feasibility set represented by the DSR constraint (Equation 2) is also convex, and therefore the SPO problem under DSR constraint is a convex problem.

Theorem 1. $E[S(\overline{x})^+]$ is convex in $\overline{x}$.

Proof: We show that the Hessian of the function $E[S(\overline{x})^+]$ is positive semi-definite. We obtain the gradient and Hessian of $E[S(\overline{x})^+]$ as follows.

Let $g(\overline{x}) = E[S(\overline{x})^+] = E[S(\overline{x}) \times I(S(\overline{x}) > 0)]$, where $I(\cdot)$ is an indicator function and $S(\overline{x}) = Q - (x_0 + \sum_{i=1}^{N} x_i \times B_i)$. Also let random vector $\overline{B} = [B_1 \ B_2 \ ... \ B_N]$. We first obtain $\frac{\partial g(\overline{x})}{\partial x_i}$, for $i = 1$ to N. Given i, define $u = Q - x_0 - \sum_{j \neq i} x_j \times B_j$ and $v = B_i$.[2] Note that $S(\overline{x}) = u - x_i v$. Now,

$$g(\overline{x}) = \int_0^\infty \int_{x_i v}^\infty (u - x_i v) f_{U,V}(u,v) du dv,$$

where $f_{U,V}$ denotes the joint density function of the random variables U and V.

$$\begin{aligned}
\frac{\partial g(\overline{x})}{\partial x_i} &= \frac{\partial}{\partial x_i} \int_0^\infty \int_{x_i v}^\infty (u - x_i v) f_{U,V}(u,v) du dv \\
&= \int_0^\infty \int_{x_i v}^\infty (-v) f_{U,V}(u,v) du dv \\
&= -E[B_i \times I(S(\overline{x}) > 0)].
\end{aligned} \qquad (4)$$

$\frac{\partial g(\overline{x})}{\partial x_0}$ can be obtained similarly by defining $u = Q - \sum_j x_j \times B_j$.

$$\begin{aligned}
\frac{\partial g(\overline{x})}{\partial x_0} &= \frac{\partial}{\partial x_0} \int_{u=x_0}^\infty (u - x_0) f_U(u) du \\
&= \int_{u=x_0}^\infty (-1) f_U(u) du \\
&= -E[I(S(\overline{x}) > 0)].
\end{aligned} \qquad (5)$$

In the above, f_U denotes the density function of the random variable U. We next obtain the Hessian of the shortfall constraint, i.e. $\nabla^2 g(\overline{x})$, using a similar approach. First, we find $\frac{\partial^2 g(\overline{x})}{\partial x_k \partial x_i}$, where $k \neq i$ and $k, i \geq 1$. Define $u = Q - x_0 - \sum_{j \neq i,k} x_j \times B_j$, $v = B_k$, $w = B_i$, which have the joint density $f_{U,V,W}(.,.,.)$. Now, $S(\overline{x}) = u - x_k v - x_i w$ and

$$\begin{aligned}
\frac{\partial g(\overline{x})}{\partial x_i} &= \int_0^\infty \int_0^\infty \int_{-\infty}^\infty w I(S(\overline{x}) > 0) f_{U,V,W}() du dv dw. \\
\frac{\partial^2 g(\overline{x})}{\partial x_k \partial x_i} &= \frac{\partial}{\partial x_k} \int_0^\infty w \int_0^\infty \int_{x_k v + x_i w}^\infty f_{U,V,W} du dv dw \\
&= f_{S(\overline{x})}(0) E[B_i B_k | S(\overline{x}) = 0]
\end{aligned}$$

$\frac{\partial^2 g(\overline{x})}{\partial x_0^2}$ can be obtained by defining $u = Q - \sum_{j \neq k} x_j \times B_j$, $w = B_k$, for some k.

$$\begin{aligned}
\frac{\partial^2 g(\overline{x})}{\partial x_0^2} &= -\frac{\partial}{\partial x_0} \int_{-\infty}^\infty \int_{x_0 + x_k w}^\infty f_{U,W}(u,w) du dw \\
&= -\int_{-\infty}^\infty (-1) f_{U,W}(x_0 + x_k w, w) dw \\
&= f_{S(\overline{x})}(0) \int f_{\overline{B}|S(\overline{x})}(\overline{b}|0) d\overline{b} = f_{S(\overline{x})}(0),
\end{aligned}$$

where $f_{\overline{B}}$ is the joint density function of the bandwidth return vector $\overline{B}$. Similarly, we can show that: $\frac{\partial^2 g(\overline{x})}{\partial x_0 \partial x_k} = f_{S(\overline{x})}(0) E[B_k | S(\overline{x}) = 0]$ and $\frac{\partial^2 g(\overline{x})}{\partial x_k^2} = f_{S(\overline{x})}(0) E[B_k^2 | S(\overline{x}) = 0]$. Thus, the Hessian of the constraint can be written as,

$$\nabla^2 g(\overline{x}) = f_{S(\overline{x})}(0) \times E[\overline{A A}^T | S(\overline{x}) = 0], \qquad (6)$$

where $\overline{A} = [1 \ B_1 \ B_2 \ ... \ B_N]^T$.

Since $f_{S(\overline{x})}(0) \geq 0$ and $E[\overline{A A}^T | S(\overline{x}) = 0]$ is positive semi-definite, $\nabla^2 E[S(\overline{x})^+]$ is also positive semi-definite. Therefore, $E[S(\overline{x})^+]$ is convex.

4. SPO under Demand Satisfaction Probability (DSP) Constraint

Next, we study the convexity properties of the SPO problem under the DSP constraint. We first show that the DSP constraint is non-convex, without any assumptions on the distribution of the demand Q and the bandwidth return variables B_i. Later, we present the conditions under which the constraint and therefore the SPO problem becomes convex.

4.1. Non-convexity of SPO

We present an example where the feasible set of the SPO problem under the DSP constraint (Equation 3) is non-convex. Consider a simple case, when there are two secondary contracts, i.e $N = 2$. Let the B_1 and B_2 be uniformly distributed between 0 and 1. Let Q have a triangular density function given by, $f_Q(q) = 2 \times q, 0 \leq q \leq 1$. Note that $Pr(S(\overline{x}) > 0) = Pr(B(\overline{x}) < Q)$, where $S(\overline{x}) = Q - B(\overline{x})$ and $B(\overline{x}) = x_0 + \sum_{i=1}^{N} x_i \times B_i$. Consider the portfolio vectors $\overline{x}_1 = (0, 1, 0)$, $\overline{x}_2 = (0, 0, 1)$. We have $Pr(S(\overline{x}_1) > 0) = Pr(B_1 < Q) = \frac{2}{3} = Pr(S(\overline{x}_2) > 0)$. Choose $\epsilon = 0.67$, and denote the feasibility set by $\mathcal{X}_{0.67} = \{\overline{x} : Pr(S(\overline{x}) > 0) < 0.67\}$. We see that $\overline{x}_1, \overline{x}_2 \in \mathcal{X}_{0.67}$. However, for the convex combination, $\overline{x}_3 = \frac{1}{2} \times \overline{x}_1 + \frac{1}{2} \times \overline{x}_2$, $Pr(S(\overline{x}_3) > 0) = Pr(\frac{1}{2} \times B_1 + \frac{1}{2} \times B_2 < Q) = \frac{17}{24} > 0.67$. That is, $\overline{x}_3 \notin \mathcal{X}_{0.67}$. So, the feasibility set is not convex in general.

4.2. Conditions for convexity

For a given ϵ, denote the feasibility set (from (3)) by $\mathcal{X}_\epsilon = \{\overline{x} : Pr(S(\overline{x}) > 0) < \epsilon\}$. Using existing literature,

---

[2] Note that the variables u and v depend on i; we drop the suffix i for simplicity of notation.

we derive sufficient conditions for convexity of the feasibility set $\mathcal{X}_\epsilon$.

Theorem 2. $\mathcal{X}_\epsilon$ is convex if any of the following conditions hold:

(a) The random vector $\overline{B} = [B_1 \ B_2 \ ...B_N]^T$ and the demand Q have log-concave and symmetric density functions, and $0 \le \epsilon \le 0.5$.

(b) The random vector $-\overline{B} = [B_1 \ B_2 \ ...B_N]^T$ and the demand Q have a joint normal distribution, and $0 \le \epsilon \le 0.5$.

(c) If the CDF of the demand, F_Q, is a concave function.

Proof of part a: We invoke the results from [23] to show this. From [23], we know that the function $Pr(\overline{x}^T a < b)$ is quasi-concave, if the joint density function of the random vector a and the random variable b are log-concave and symmetric. This result readily applies to our case, by rewriting the constraint (Equation 3) as $Pr(-B(\overline{x}) < -Q) \ge 1 - \epsilon$. Specifically, the function $Pr(-B(\overline{x}) < -Q)$ is quasi-concave if the joint density of the random vector $\overline{B}$ and Q are log-concave and symmetric. This implies that the feasibility set $\mathcal{X}_\epsilon = \{\overline{x} : Pr(-B(\overline{x}) < -Q) \ge 1 - \epsilon\}$, is convex.

Proof of part b: We invoke the results from [26] to show this. From Theorem 3 of [26], we know that the set $\mathcal{X}_\epsilon = \{\overline{x} : Pr(\overline{x}^T a < b) \ge p\}$ is convex, if the density function of the random vector a and the random variable b is jointly normal. This result can be applied by rewriting the DSP constraint (Equation 3) as $Pr(-B(\overline{x}) < -Q) \ge 1 - \epsilon$.

We also derive another condition for convexity, which only requires a non-increasing assumption on the distribution function of Q, and none on the bandwidth return variables B_i.

Proof of part c: Consider portfolios $\overline{y} = (y_0, y_1, y_2, ..., y_N)$, and $\overline{z} = (z_0, z_1, z_2, ..., z_N)$.

Let $\overline{x} = \lambda \overline{y} + (1 - \lambda)\overline{z}$. Now,

$$Pr(S(\overline{x}) > 0) = Pr(S(\lambda \overline{y} + (1 - \lambda)\overline{z}) > 0)$$

$$= \int_{\underline{b}} f_{\overline{B}}(\overline{b}) P(Q > x_0 + \sum_{i=1}^{N} x_i \times b_i) d\overline{b}$$

$$= \int_{\underline{b}} f_{\overline{B}}(\overline{b})(1 - F_Q(x_0 + \sum_{i=1}^{N} x_i \times b_i)) d\overline{b}.$$

Here F_Q is the distribution function of the demand Q. Now, $x_0 = \lambda y_0 + (1 - \lambda)z_0$ and $x_i = \lambda y_i + (1 - \lambda)z_i$.

$$Pr(S(\overline{x}) > 0) = \int_{\underline{b}} f_{\overline{B}}(\overline{b})(1 - F_Q(\lambda y_0 + (1 - \lambda)z_0$$

$$+ \sum_{i=1}^{N}(\lambda y_i + (1 - \lambda)z_i) \times b_i)) d\overline{b}.$$

$$= \int_{\underline{b}} f_{\overline{B}}(\overline{b})(1 - F_Q(\lambda(y_0 + \sum_{i=1}^{N} y_i b_i)$$

$$+ (1 - \lambda)(z_0 + \sum_{i=1}^{N} z_i \times b_i))) d\overline{b}.$$

If F_Q is a concave function we get the inequality,

$$Pr(S(\overline{x}) > 0) \le \int_{\underline{b}} f_{\overline{B}}(\overline{b})(1 - \lambda \times F_Q(y_0 + \sum_{i=1}^{N} y_i b_i)$$

$$+ (1 - \lambda)F_Q(z_0 + \sum_{i=1}^{N} z_i \times b_i)) d\overline{b}$$

$$= \int_{\underline{b}} f_{\overline{B}}(\overline{b})(\lambda + 1 - \lambda - \lambda \times F_Q(y_0 + \sum_{i=1}^{N} y_i b_i)$$

$$+ (1 - \lambda)F_Q(z_0 + \sum_{i=1}^{N} z_i \times b_i)) d\overline{b}$$

$$= \lambda \int_{\underline{b}} f_{\overline{B}}(\overline{b})(1 - F_Q(y_0 + \sum_{i=1}^{N} y_i b_i)) d\overline{b}$$

$$+ (1 - \lambda) \int_{\underline{b}} f_{\overline{B}}(\overline{b})(1 - F_Q(z_0 + \sum_{i=1}^{N} z_i b_i)) d\overline{b}$$

$$= \lambda Pr(S(\overline{y}) > 0) + (1 - \lambda)Pr(S(\overline{z}) > 0) \quad (7)$$

From Theorem 2 (c), it also follows that $Pr(S(\overline{x}) > 0)$ is convex if $f_Q' \le 0$ everywhere. $\square$

It can be shown that the gradient of the DSP constraint, $Pr(S(\overline{x}) > 0)$, is given by,

$$\frac{\partial Pr(S(\overline{x}) > 0)}{\partial x_0} = -f_{S(\overline{x})}(0),$$

$$\frac{\partial Pr(S(\overline{x}) > 0)}{\partial x_k} = -f_{S(\overline{x})}(0) \times E[B_k | S(\overline{x}) = 0].$$

We use the above expressions, when we solve the SPO problem numerically in Section 7.

Theorems 2(a) and 2(b) covers important distributions such as the Gaussian, log-normal, and the uniform density functions (both $\overline{B}$ and Q must follow some symmetric, log-concave distribution, although they need not be the same distribution). Theorem 2(c) covers concave and other asymmetric decreasing density functions for Q that are not included in Theorem 2(a) (the distribution of $\overline{B}$ can be arbitrary).

Remark 1: Let $N = 1$. If Q is deterministic, then the DSP constraint reduces to a linear constraint. In this case, the optimal portfolio consists of entirely primary or entirely secondary contracts. The optimal portfolio is $(Q, 0)$, if $\epsilon < F_{B_1}(p_1)$ and $(0, \frac{Q}{F_{B_1}^{-1}(\epsilon)})$, if $\epsilon >= F_{B_1}(p_1)$, where F_{B_1} is the cumulative distribution function of B_1.

Figure 1a shows the empirical distribution (cumulative) of the total daily traffic of a Verizon Wi-Fi HotSpot network from [27]. Note that the shape of the cumulative distribution function matches well with a concave distribution function also shown in the figure. The concave function used for fitting is $1 - e^{-0.4(x-7.9)}$. Similarly, Figure 1b shows the traffic distribution of a large US-based cellular network (Refer Figure 1a of [28]). Here, we see that the cellular traffic distribution can be approximated by a log-normal distribution. From Theorem 2, we know that the SPO problem is convex if the distribution function for the demand is concave or log-normal. Therefore, we can formulate the SPO problem for empirical distributions as a convex programs and study the nature of optimal portfolio, after approximating the empirical distributions with concave or log-normal distributions.

4.3. Convex Approximation

In practice, the demand and bandwidth return distributions may not satisfy the properties stated in Theorem 2, leading to the DSP constraint being non-convex. The SPO problem under the DSP constraint is also non-convex in such cases. However, several approximation techniques (inner as well as outer) have been developed in order to approximate a non-convex probability constraint to a convex contraint. In this paper, we specifically consider the Bernstein approximation technique developed in [24].

Bernstein approximation finds a convex inner approximation to the original probability constraint such that it is computationally tractable. The probability of shortage is upper bounded by the expected value of a (suitably defined) function of the shortage. The (non-convex) probability constraint is then replaced by a (convex) expectation constraint. The SPO problem under Bernstein approximation can be stated as,

$$\underset{x, t>0}{\text{minimize}} \quad C(\bar{x}) = x_0 + \sum_{i=1}^{N} x_i \times p_i$$

$$\text{subject to} \quad \underset{t>0}{\inf}[\Psi(x, t) - t\epsilon] \leq 0$$

where $\Psi(x, t) = tE\left[\psi(t^{-1}S(\bar{x}))\right]$ and $\psi : \mathcal{R} \to \mathcal{R}$ is a non-negative valued, non-decreasing convex function (called the *generating function*) such that $\psi(z) > \psi(0) = 1$ for any $z > 0$. We consider two generating functions - a piecewise linear generating function $\psi(z) = [1 + z]_+$, and the exponential generating function $\psi(z) = e^z$. Note

that ϵ is the bound on demand shortage probability (Equation 3).

5. SPO over Multiple Regions

Spectrum contracts typically come with clauses that restrict the use of the spectrum to certain geographical regions. This could be due to licensing or coverage limitations of the seller provider. For example, a seller provider may only have the license to use a part of the spectrum in certain regions (say certain counties or states in the United States), and not others. Alternatively, the base stations of the seller provider may only cover certain sub-areas of the overall area of interest to the buyer, which can span multiple regions. This adds additional complexity to the SPO problem, since the spectrum portfolio should satisfy the buyer provider's requirements for each of these regions. In this section, we formulate the SPO problem over multiple regions and argue that the results for the single region problem extend to multi-region case as well.

Let us assume that the buyer of spectrum contracts operates over a set of K disjoint geographical regions. The buyer's objective is to construct a portfolio of spectrum contracts in order to satisfy the user demand in each of the K regions. Denote the set of regions by $\mathcal{R}$, i.e, $\mathcal{R} = \{1, 2, .., K\}$. Let there be M primary and N secondary contracts in the market. Let z_i, p_j denote the unit price of ith primary contract and j^{th} secondary contract, respectively. Let $\mathcal{R}_i^p \subset \mathcal{R}, 1 \leq i \leq M$ denote the set of regions in which the ith primary contract is valid. Similarly, let $\mathcal{R}_j^s \subset \mathcal{R}, 1 \leq j \leq N$ denote the set of regions in which the jth secondary contract is valid. The user demand for each region is uncertain, denoted by the random variable $Q_k, 1 \leq k \leq K$.

The multi-region SPO problem under DSR constraint can be stated as follows:

$$\text{Minimize} \quad C(\bar{x}) = \sum_{i=1}^{M} y_i \times z_i + \sum_{j=1}^{N} x_j \times p_j, \quad (8)$$

$$E[\{Q_k - \sum_{i \in \mathcal{C}_k^p} y_i - \sum_{j \in \mathcal{C}_k^s} x_j \times B_{jk}\}^+] < \delta_k \ \forall k, \quad (9)$$

$$E[\sum_{k=1}^{K} \{Q_k - \sum_{i \in \mathcal{C}_k^p} y_i - \sum_{j \in \mathcal{C}_k^s} x_j \times B_{jk}\}^+] < \delta. \quad (10)$$

Here $\{y_1, ..., y_M, x_1, ..., x_N\}$ denotes the spectrum portfolio. $\mathcal{C}_k^p$ and $\mathcal{C}_k^s$ denote the set of primary and secondary contracts that are valid in the kth region ($1 \leq k \leq K$), respectively. $\mathcal{C}_k^p$ and $\mathcal{C}_k^s$ can be obtained from $\mathcal{R}_i^p, 1 \leq i \leq M$ and $\mathcal{R}_j^s, 1 \leq j \leq N$. Note that $\mathcal{C}_k^p \subset \{1, 2, .., M\}$ and $\mathcal{C}_k^s \subset \{1, 2, .., N\}$. The random variable B_{jk} represents the bandwidth return of the jth secondary contract in the

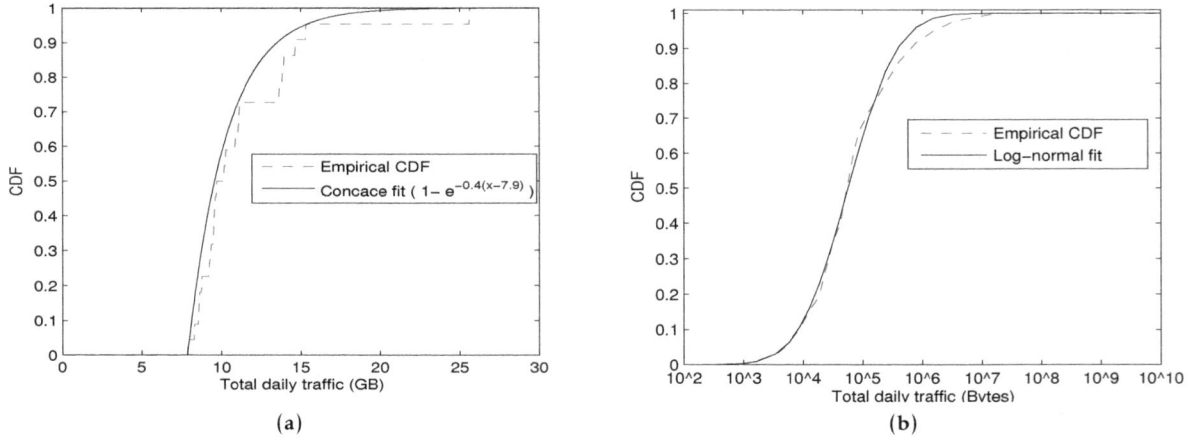

Figure 1. a) Empirical distribution of traffic from a Wi-Fi hotspot along with a concave fit, b) Empirical distribution of cellular traffic with a log-normal fit.

k^{th} region. For the multiple region problem, there are totally $K + 1$ inequality constraints; one DSR constraint for each of the K regions and one overall DSR constraint for all the regions. The LHS of the $(K + 1)^{th}$ constraint is simply the summation of the LHS of the first K constraints. However, note that $\sum_{k=1}^{K} \delta_k > \delta$, else the last constraint would be redundant; typically, the buyer provider may want to have $\delta_k > \delta/K$, for each k. The motivation of both types of constraints (per-region as well as overall) is as follows. While the buyer provider would be interested in the ensuring a certain DSR over its overall customer base, it may also want to ensure a certain DSR (possibly a smaller normalized DSR than the overall DSR) is ensured in each of its regions of operation, to avoid excessive customer dissatisfaction in each individual region. The SPO problem under DSP constraint can be defined similarly as above, but by replacing the expectation constraints with the corresponding probability constraints, and δ_k and δ by ϵ_k and ϵ, respectively.

For both the SPO problems, we see that the kth constraint ($1 \leq k \leq K$) is similar to the constraint for the single region problem ((2) and (3)) except for the presence or absence of few variables inside the two summations. First, consider the SPO problem under DSR constraint (8-(10)). Let the kth rate constraint be denoted by g_k; g_k involves only some of the y_i and x_j variables. It can be rewritten as,

$$E[\{Q_k - \sum_{1 \leq i \leq M} y_i \times I(i \in C_k^p) - \sum_{1 \leq j \leq N} x_j \times B_j'\}^+] < \delta_k,$$

(11)

where $B_j' = B_{jk}$, if $j \in \mathcal{C}_k^s$, else $B_j' = 0$. $I(i \in C_k^p)$ is the indicator function for the set C_k^p. Now, the proof technique for the single-region problem can be readily

extended to show that g_k is convex in y_i, x_j. The final constraint (g_{K+1}) is also convex, since it is the sum of several convex functions. Therefore, the feasible set for this problem is convex, since the intersection of several convex sets is convex. Similarly, the feasible set for the multiple-region SPO problem under DSP constraint is also convex, if the density functions of all the random parameters involved are log-concave and symmetric, or the demand variables Q_k have non-decreasing density functions.

6. SPO under Discrete Portfolio Constraints

In practice, the primary and secondary contracts can be bought and sold only in discrete units. In such scenarios, the SPO problem is represented as a discrete (or integer) program. Despite the discreteness (integrality) requirements in the variables, discrete (integer) programs derived from convex problem can often be solved efficiently [29]. That is however not the case with the SPO problem as we argue in this section. We will first argue that the SPO problem is NP-hard. We then argue that it is not submodular either. These results essentially imply that it is unlikely that an efficient solution to the SPO problem exists when the allocations are constrained to take a discrete set of values.

6.1. Complexity of SPO under Discrete Portfolio Constraints

NP-hardness of discrete-SPO (both under DSR and DSP constraints) can be established by reduction from the NP-hard Knapsack problem. Next, let us assume that the portfolio $\overline{x}$ is constrained to be an integer vector. Now let us consider the special case where the return from all secondary contracts, B_i, as well as the customer demand, Q, are deterministic. Then the DSR constraint (2) reduces to $Q - x_0 - \sum_{i=1}^{N} x_i \times B_i < \delta$, or $x_0 + \sum_{i=1}^{N} x_i \times B_i > Q - \delta$. In the same setting, the DSP constraint (3) becomes $x_0 + \sum_{i=1}^{N} x_i \times B_i > Q$. The problem of minimizing the objective in (1) subject to constraint (2) or (3) is then equivalent to the *minimization version* of the unbounded Knapsack problem [30]. In the minimization version of the Knapsack problem, the objective of the standard (maximization version) Knapsack problem is replaced by minimization, and the inequality in the constraint is reversed. Since the minimization version of the Knapsack problem can be transformed into an equivalent maximization version in polynomial time, and the maximization version (standard) unbounded Knapsack problem is known to be NP-hard [31], it follows that the integral versions of the DSR and DSP problems are NP-hard as well.

In the discrete domain, the equivalent of convexity is the submodularity property. For minimization of submodular functions under integrality constraints, efficient algorithms exist (that can attain a solution in pseudo-polynomial time, for example) [29]. However, we show next that the DSR and DSP constraints are not submodular.

Consider a function $g(\overline{x}) : \mathcal{Z}^n \to \mathcal{R}$. From Theorems 7.7, 7.20, and 7.21 of [29], the function g is submodular if and only it satisfies the discrete midpoint convexity defined below:

$$g(p) + g(q) \geq g\left(\left\lceil \frac{p+q}{2} \right\rceil\right) + g\left(\left\lfloor \frac{p+q}{2} \right\rfloor\right), \text{ For any p,q} \in \mathcal{Z}^n$$

(12)

Theorem 3. The probability of shortfall $Pr(S(\overline{x}) > 0)$ and the expected shortfall $E[S(\overline{x})^+]$ are not submodular.

Proof: We provide counter examples demonstrating that both the probabilty of shortfall as well as expected shortfall violate the discrete midpoint convexity property.

Consider the case where there is a single primary and a single secondary contract. Also, consider the DSR function, $g_1(\overline{x} : E(S(\overline{x}))$. Let the demand Q and the bandwidth return B_1 be deterministic. Note that the DSR constraint is convex for any probability density function. Let $Q = 5$ and $B_1 = 1$.

Consider portfolios $p = (1, 4)$ and $q = (4, 1)$. Now,

$$g_1(p) = E[max(Q - 1 - 4 \times B_1, 0)] = 0,$$
$$g_1(q) = E[max(Q - 4 - 1 \times B_1, 0)] = 0$$

But,

$$g_1\left(\left\lceil \frac{p+q}{2} \right\rceil\right) = g_1((3,3)) = E[max(Q - 3 - 3 \times B_1, 0)] = 0,$$
$$g_1\left(\left\lfloor \frac{p+q}{2} \right\rfloor\right) = g_1((2,2)) = E[max(Q - 2 - 2 \times B_1, 0)] = 1$$

Thus, we find that the discrete midpoint convexity is violated.

Next, consider the DSP function $g_2(\overline{x} : Pr(S(\overline{x}) > 0)$. As before let B_1 be deterministic with a value of 1. However, let Q be uniformly distributed $\mathcal{U}(0, 5)$. These density functions satisfy sufficient conditions for the convexity of the DSP constraint. Now, it can be quickly calculated that,

$$g_2(p) = g_2(q) = 0,$$
$$g_1\left(\left\lceil \frac{p+q}{2} \right\rceil\right) = 0, g_1\left(\left\lfloor \frac{p+q}{2} \right\rfloor\right) = 0.2$$

Thus, we find the DSP function also violates the discrete midpoint convexity condition for submodularity.

In view of the above negative results on the efficient computability of the discrete-SPO problem, we discuss a branch-and-bound algorithm for solving the problem. The same dynamic problem algorithm is used to compute solution to the SPO problem under discreteness constraints in our evaluation section (Section 7). It is worth noting that application of dynamic programming to solve similar problems have been discussed in prior literature. In particular, the integral-SP problem under the DSP constraint is closely related to the *stochastic Knapsack problem*, for which a dynamic programming solution approach is described in [32]. In the stochastic Knapsack problem as discussed in [32], objective and contraint functions are reversed: the non-linear shortage probability function is set as the objective, while the linear cost function constitutes a constraint. In the following section, we describe a branch-and-bound algorithm that is tuned to the two versions of the SPO problem that we consider in this paper.

6.2. Branch and Bound Algorithm for DSR and Convex DSP under Discrete Portfolio Constraints

We provide a branch-and-bound algorithm to solve the DSR and the convex DSP problem under discrete portfolio constraints, which was also used in Section 7 to evaluate our solutions under discreteness constraints on the spectrum portfolio. The basic idea behind branch-and-bound algorithms for solving mixed-integer non-linear programs (MINLP), is to relax the integrality restrictions on the original problem. If the solution to the relaxed problem is integral, then this is solution to original problem. However, if some variables (say y) are non-integers, then two sub-problems are created by adding bounds $y \leq [y]$ and $y \geq [y] + 1$ (where $[y]$ is the largest integer not greater than y). The process is repeated until an integer solution is found. For convex MINLPs, several global optimization algorithms have been proposed. In this paper, we implement the algorithm proposed in [33] by adapting it to our problem context. In [33], the authors solve a sequence of quadratic programming problems before branching; it is not required that the quadratic problems at the intermediate stages be solved optimally. The SPO problem (when the discreteness constraints are relaxed) can be stated as,

$$P: \text{minimize } C(\overline{x}) \tag{13}$$

$$\text{subject to: } g(\overline{x}) - \alpha \leq 0,$$
$$\overline{x} \in \mathcal{S},$$

where $C(\overline{x}) = x_0 + \sum_{i=1}^{N} x_i \times p_i$ is the porfolio cost, and $g(\overline{x})$ represents expected shortfall ($E[S(\overline{x})^+]$) or probability of shortfall ($Pr(S(\overline{x}) > 0)$), for the DSR and DSP problems, respectively. $\alpha = \delta$ for the DSR problem and $\alpha = \epsilon$ for the DSP problem. The set $\mathcal{S}$ does not take into account the discrete (integrality) assumptions on the portfolio.

Then the branch and bound algorithm requires solving the following quadratic program (QP^k) iteratively, so as to divide the original problem into sub-problems:

$$QP^k: \text{minimize } C^{(k)} + \nabla C^{(k)T} \overline{d} + \frac{1}{2} \overline{d}^T W^{(k)} \overline{d} \tag{14}$$

$$\text{subject to: } g^{(k)} + \nabla g^{(k)T} \overline{d} \leq 0,$$
$$C^{(k)} + \nabla C^{(k)T} \overline{d} \leq U_{bb} - \theta,$$
$$\overline{x}^{(k)} + \overline{d} \in \hat{\mathcal{S}},$$

where $C^{(k)} = C(\overline{x}^{(k)})$ and $W^{(k)} = \nabla^2 C^{(k)} + \lambda \nabla^2 g^{(k)}$. $\hat{\mathcal{S}}$ denotes the feasibility set after adding bounds for non-integral solution variables. U_{bb} denotes current upper bound on the objective function and θ denotes the optimality tolerance of the branch-and-bound algorithm. Intially, U_{bb} is set to ∞ and updated whenever an integer solution is found at an intermediate step. QP^k is solved to obtain the increment $\overline{d}$ from the current solution ($\overline{x}^{(k)}$).

For the SPO problem, $\nabla C^{(k)} = [1 \ p_1 \ p_2 ... \ p_N]^T$ and $\nabla^2 C^{(k)} = 0$. For the DSR constraint, the gradient ($\nabla g^{(k)}$) and the hessian ($\nabla^2 g^{(k)}$) can be found using Equations (5), (4), and (6). For the DSP constraint, the gradients can be written as,

$$\frac{\partial Pr(S(\overline{x}) > 0)}{\partial x_0} = -f_{S(\overline{x})}(0)$$

$$\frac{\partial Pr(S(\overline{x}) > 0)}{\partial x_k} = -f_{S(\overline{x})}(0) \times E[B_k | S(\overline{x}) = 0]$$

$$= -\int w \times f_{U,W}(x_0 + x_k w, w) dw,$$

where $U = Q - \sum_{j \neq k} x_j B_j$. The Hessian of the probability of shortfall, $Pr(S(\overline{x}) > 0)$, can be computed (approximately) numerically, using the above equations.

7. Numerical Evaluation

We numerically solve the SPO problems using Matlab to study the characteristics of the spectrum portfolio in a wide range of scenarios. Our goal is to examine how the parameters of the problem, namely, the price of the secondary contracts, the bandwidth return distributions, and the constraints (ϵ, δ) influence the portfolio composition. The results for the single-region SPO problems are presented in sections 7.1 and 7.2, while the results for the multiple-region problem are presented in section 7.3.

7.1. Single Primary and Single Secondary contract

We first consider the simplest case of there being a single secondary contract seller in the market. The bandwidth return B_1 and the demand Q are assumed to have truncated normal distributions. B_1 has a mean of 0.5, while the demand Q has a mean of 1.5. The distribution of Q is restricted to the interval $[0, 3]$. We obtain optimal portfolio when the key parameters of the problem (ϵ, δ, p_1) are changed.

Figure 2a shows the spectrum portfolio composition for different choices of the DSR constraint (δ) and DSP constraint (ϵ), respectively. The unit price of the secondary contract, $p_1 = 0.25$. In the figure, $\overline{x}^E = \{x_0^E, x_1^E\}$ and $\overline{x}^P = \{x_0^P, x_1^P\}$ denote the portfolios for SPO problems with DSR and DSP constraints, respectively. As expected, when $\delta = \epsilon = 0$, we observe that the portfolio consists of primary contract units only. This is due to the fact that the secondary contracts having stochastic returns introduce bandwidth shortage (or demand violation) even if they are bought in large quantities. Moreover, the number of primary contract units in both the cases is equal to the maximum possible demand (i.e 3). As the constraint (ϵ, δ) is relaxed, we find that the number of primary contract units reduces sharply until it becomes zero. On the other hand the number of secondary contract units (x_1^E, x_1^P) increases initially, but starts decreasing as soon as the number of primary contract units becomes zero. This can be explained as follows: As the constraint (ϵ, δ) is increased from zero, it becomes unnecessary to meet the demand with probability one. Therefore, total cost of the portfolio can be reduced, by reducing the number of primary contract units, while adding the requisite amount of secondary contract units to keep the demand violation below the desired value. This happens until the number of primary contract units becomes zero. Beyond this point, the only way to reduce the cost is to reduce the number of secondary contract units, which can be reduced as ϵ, δ increase.

In Figure 2a, we assume that the primary demand Q and the bandwidth return B_1 are independent. We know that a secondary contract provides access to unused or leftover channels with the seller provider.

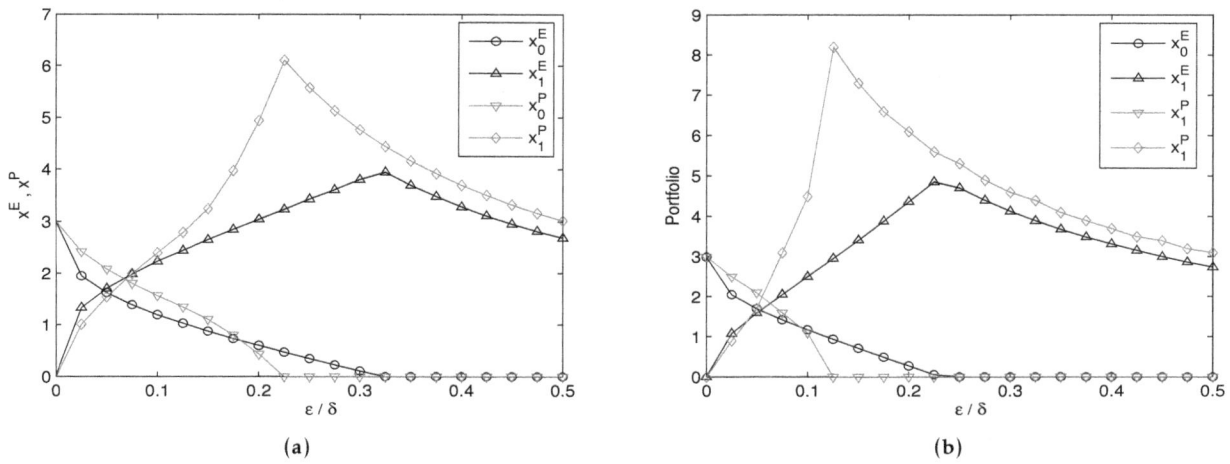

Figure 2. Number of primary (x_0^E, x_0^P) and secondary (x_1^E, x_1^P) contract units in the optimal portfolio for the SPO problem under DSR and DSP constraint. a) Demand and the secondary bandwidth return are independent, b) Demand and the bandwidth return are correlated.

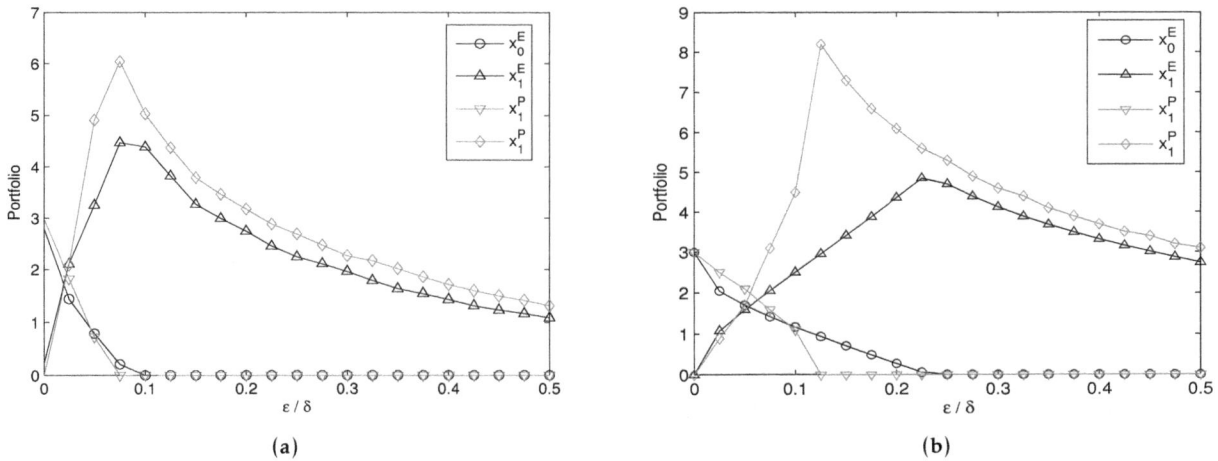

Figure 3. Primary (x_0^E, x_0^P) and secondary (x_1^E, x_1^P) contract units in the SPO when demand and bandwidth distributions are modeled as beta distributions. a) Demand and secondary bandwidth return are independent, b) Demand and bandwidth return are correlated with degree of correlation being 0.5.

Therefore, the bandwidth return of the secondary contract would be negatively correlated with the seller's own customer demand. Moreover, the traffic demand of different providers would have similar temporal characteristics. Hence, we can expect negative correlation between the buyer's own demand Q and the bandwidth return B_1. Figure 2b shows the optimal portfolio composition for the SPO problem under DSR and DSP constraint. The degree of correlation between Q and B_1 was set to 0.5. Note that the overall trend in portfolio remains similar Figure 2a. Later in this

section, we vary the degree of correlation and observe the changes in portfolio composition.

In Figures 3a and 3b, we use *beta* distribution to model the demand (Q) and bandwidth return (B_1). Some recent studies ([34], [35]) have shown that the channel occupancy probability due to primary users can be accurately modeled using *beta*, or the closely related *kumaraswamy* distributions. Therefore, next we model the demand (Q) using a beta distribution, since the channel occupancy probability is directly proportional to the primary user demand. Since

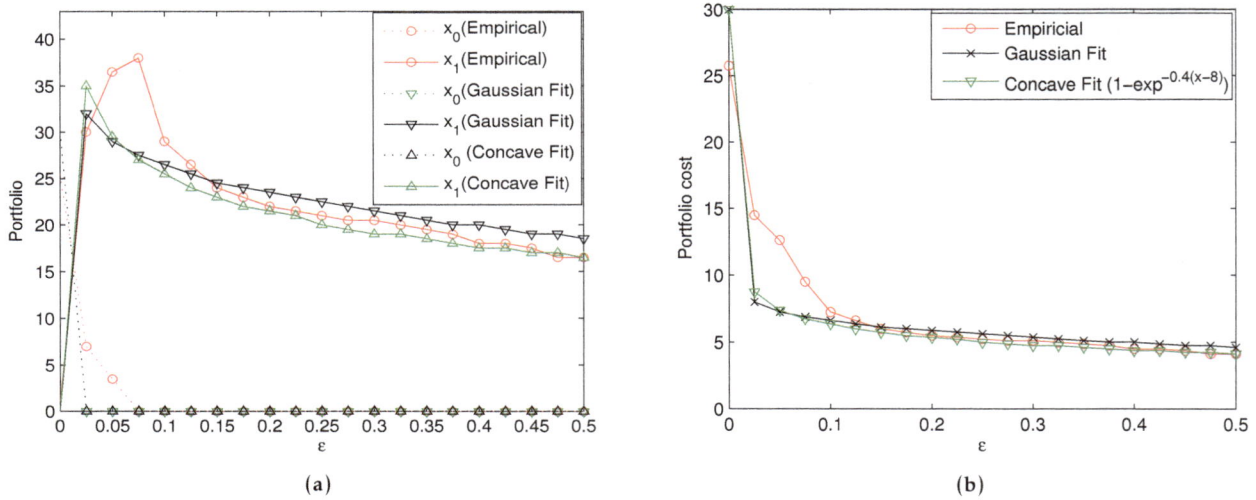

Figure 4. SPO solution for empirical and fitted distributions under DSP constraint. a) Number of primary and secondary contract units in the optimal portfolio, b) Portfolio cost.

the secondary bandwidth return is essentially the probability of the channel not being occupied, we set $B_1 = 1 - Y$, where Y is a random varible with *beta* distribution. The shape of the *beta* distribution is governed by its parameters α and β. Therefore, the DSP constraint is not convex for all choices of α and β. However, it can be shown that the DSP constraint satisfies Theorem (c), if we choose $\alpha = 1$ and $\beta > 1$ for the *beta* distribution governing the demand Q. For the results shown in Figure 3a and 3b, we chose $\alpha = 1$ and $\beta = 2$ for the beta distributions underlying both the demand as well as the bandwidth return. Morevoer, we solved the SPO problems when the demand and bandwidth return are independent as well as correlated. It can be seen that the general nature of results is similar to the gaussian distribution results.

Empirical distributions: Next, we study the sensitivity of the portfolio composition to changes in the distribution of the demand (Q) and the bandwidth return (B_1). We obtain the empirical distribution of the total daily traffic of a Verizon Wi-Fi HotSpot network from [27] (Refer to Figure 12 of [27]) and consider this distribution for the user demand Q. From this, we compute the distribution of B_1 as $f_{B_1}(b) = f_Q(\beta(1 - b))$, $0 \le b \le 1$, since bandwidth availability is related negatively to the user demand (the scaling factor β is used for normalization). For these empirical distributions, the results for SPO problem under DSP constraint are shown in Figures 4a and 4b. For comparison, we also solve and plot the SPO problem with Q modeled as gaussian and exponential concave distribution that approximate the empirical distribution. Refer Figure 1a for the concave

disctribution. In both cases, the distribution for B_1 is obtained as as $f_{B_1}(b) = f_Q(\beta(1 - b))$. The SPO problem under DSP constraint is convex when distribution of Q is concave. Some small differences notwithstanding, we see from Figure 4a that the general trend in the optimal portfolio composition for the empirical and fitted distributions is the same. Similarly, Figure 4b shows that the optimal portfolio cost for the empirical and fitted distributions are close to each other for higher values of ϵ. Additionally, we ran simulations with uniform distribution and observed similar results. Therefore, in the following we only present the results for (truncated) Gaussian distributions.

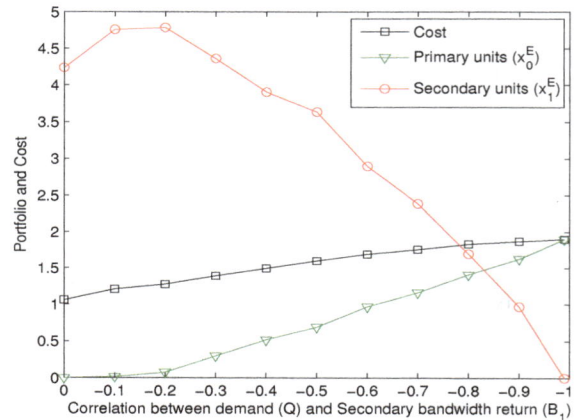

Figure 7. Correlation between demand and secondary bandwidth return modeled as beta distributions.

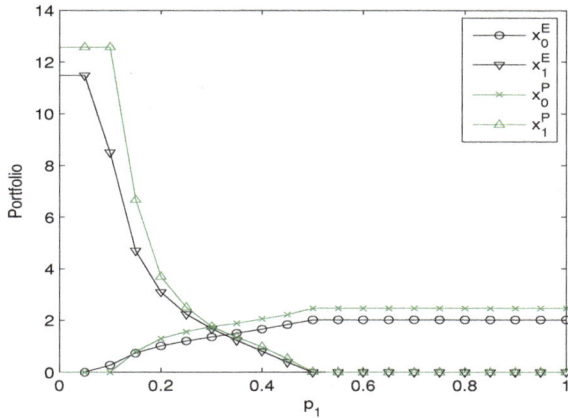

Figure 5. Optimal spectrum portfolio composition for different choices of the unit price of secondary contract.

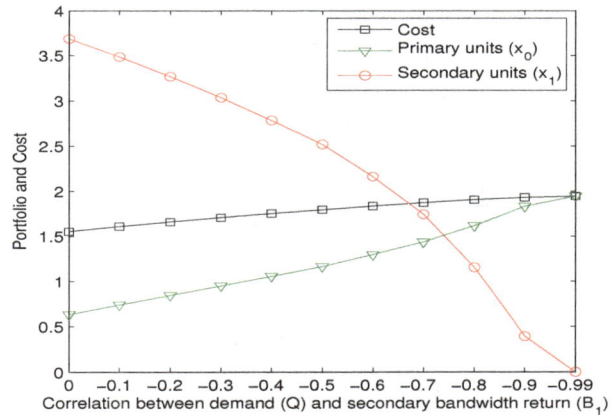

Figure 6. Correlation between demand and secondary return.

Figure 5 shows the effect of unit price of the secondary contract on the portfolio. The simulation parameters for this figure are same as the ones chosen for Figure 2a. But, we now fix ϵ and δ at 0.1, and increase the price of the secondary contract p_1. As p_1 is increased from 0 to 1, the spectrum portfolio composition gradually changes from those with entirely secondary contract units to those with entirely primary contract units. The transition in this case happens when $p_1 = 0.5$. For prices between 0.1 and 0.5, the portfolio has a mix of primary as well as secondary contract units. Since the mean bandwidth return of the secondary contract is 0.5, we find that buying secondary contract units make sense only when their price is roughly half the price of the primary contract (i.e. 1×0.5).

Effect of Correlation: Figure 6 shows the effect of correlation (negative) between the demand Q and the bandwidth return B_1 of the single secondary contract. Q and B_1 have truncated and jointly gaussian distribution. We obtain the portfolio composition for increasing negative correlation between Q and B_1. When the correlation is high, the bandwidth available from the secondary contract tends to be low with high probability when the demand Q is high. This increases the possibility of shortage. Therefore, we find that the amount of primary units in the portfolio increases, as they are risk-free. Similar results can be observed in Figure 7, where Q and B_1 are modeled using beta distributions (as discussed in Section 7.1) and the degree of correlation increased.

Convex Approximation: Next, we numerically evaluate the performance of the Bernstein approximation in cases where the DSP constraint is non-convex. For this study, we consider a single primary and a single secondary contract. The demand and the bandwidth return have empirical distributions discussed earlier. The disitributions do not satisfy Theorem 2 and hence

(a) Primary units

ϵ	0.1	0.2	0.3	0.4	0.5
x_0^A	14	12	10	12	8
x_0^O	4	10	6	2	0

(b) Secondary units

ϵ	0.1	0.2	0.3	0.4	0.5
x_1^A	0	2	4	0	6
x_1^O	16	2	8	14	16

(c) Total cost

ϵ	0.1	0.2	0.3	0.4	0.5
$Pr(S(\overline{x}) > 0)$	0	0.13	0.14	0.14	0.18
$Cost^A$	14	13	12	12	11
$Cost^O$	12	11	10	9	8
% Dev	17	18	20	33	37

Table 1. Deviation in SPO results when Bernstein approximation is used to solve SPO under non-convex DSP constraint

the DSP constraint may not be convex. We solve the Bernstein approximation (Section 4.3) to the SPO problem and obtain the optimal portfolio (x_0^A, x_1^A) and cost $(Cost^A)$. We also solve the original problem (without any approximations) through brute-force search (Equation 1 and 3) and obtain the solution $((x_0^O, x_1^O), Cost^O)$. The results are summarized in Table 1. We only present the results for the linear generating function, since we observed better approximation using the linear generating function than the exponential generating function. Tables 1a, 1b, and 1c, show respectively, the primary units (x_0), the secondary units (x_1), and the portfolio cost $(Cost)$ of the two solutions. From the percentage deviation values shown in Table 1c, we observe that the cost of the approximate solution is within $20 - 40\%$

(a) Primary units

ϵ	0.1	0.2	0.3	0.4	0.5
x_0^A	5.7	2.4	0	0	0
x_0^O	2	0.5	0	0	0

(b) Secondary units

ϵ	0.1	0.2	0.3	0.4	0.5
x_1^A	9.7	12.5	14.8	13.4	12.2
x_1^O	12.5	12.0	10.5	9.0	7.5

(c) Total cost

ϵ	0.1	0.2	0.3	0.4	0.5
$Pr(S(\overline{x}) > 0)$	0.04	0.08	0.13	0.17	0.21
$Cost^A$	11.5	9.9	8.8	8.0	7.3
$Cost^O$	9.5	7.7	6.3	5.4	4.5
% Dev	20	28	40	49	63

Table 2. Deviation in SPO results when Bernstein approximation is employed. Bandwidth return and demand are modeled as beta distributions.

(a) Demand is deterministic

$\hat{Q}$	$x_0^{Int} - x_0^*$	$x_1^{Int} - x_1^*$	Cost (% dev)
2	0	0.4552	8.2095
4	0	0.2456	1.9256
8	0.0391	0.2029	1.2545
16	0.0393	0.2022	0.5927
32	0.0393	0.2025	0.2886
64	0.0392	0.2028	0.1423
128	0.0391	0.2030	0.0707

(b) Demand is random

$\overline{Q}$	$x_0^{Int} - x_0^*$	$x_1^{Int} - x_1^*$	Cost (% dev)
2	0	0.4621	10.1825
4	0.2954	−0.7026	3.5430
8	1.1720	−3.9428	2.3836
16	−0.1282	1.3377	1.2468
32	1.2444	−4.0581	0.6808
64	0.0559	0.6854	0.3354
128	−0.0441	0.2946	0.0219

Table 3. SPO under DSR constraint and discrete portfolio assumption

of the optimum in most cases. However, Tables 1a and 1b show that the portfolio composition could be significantly different for some values of ϵ. The price of secondary contract was chosen close to the mean bandwidth return (0.61) under the empirical distribution.

We also performed a similar study after modeling the demand and bandwidth return using beta distributions. As noted earlier, the DSP constraint is not convex for all choices of its parameters α and β. Morevoer, it has been shown in literature that the beta distribution is neither log-concave or log-convex when $\alpha < 1$ and $\beta > 1$. Therefore, we choose $\alpha = 0.75$ and $\beta = 2$ for both Q and B_1, so that the DSP constraint is non-convex. We then apply Bernstein approximation as before and compare the percentage deviation between approximate solution and the optimum. The results are summarized in Tables 2a, 2b, and 2c. We find that the percentage deviation in cost as well as portfolio are slightly higher when compared to results in Table 1.

Discrete Constraints: Finally, we solve the SPO problem under discrete constraints using the branch and bound algorithm discussed in 6.2. We obtain the optimal portfolio under integral portfolio constraints and compare it with the non-integer optimum. Table 3 shows the results for the DSR constraint. For the results shown in Table 3a, the demand is assumed to be deterministic quantity ($\hat{Q}$), but the bandwidth return B_1 is a gaussian variable. And for the results shown in Table 3b, both the demand and the bandwidth return B_1 are gaussian random variables. In Table 3a, we increase $\hat{Q}$ and in Table 3b, we increase the mean demand $\overline{Q}$. Note that the portfolio increases in both cases. We did not observe any trend in the difference between the

integer and the non-integer optimal portfolio solutions for increasing values of $\hat{Q}$ or $\overline{Q}$. However, the percentage difference in cost due to integral restrictions seems to reduce in both cases. Therefore, when the portfolios are large, integral restrictions on portfolio does not affect the optimal portfolio cost significantly.

7.2. Single Primary and Two Secondary contracts

We next consider two types of secondary contracts and study how the price and bandwidth return characteristics of a contract affects the choice of the secondary contract. We only present results on the SPO problem under the DSR constraint, as the results for the DSP constraint are broadly similar in nature. As before, the demand has normal distribution between 0 and 3. The price of the single primary contract is 1. The bandwidth returns of the two secondaries, B_1 and B_2, have normal distribution between 0 and 1, but with different mean and variance.

We obtain the optimal portfolio $\overline{x} = \{x_0, x_1, x_2\}$ as the ratio of the unit prices of the two secondaries, i.e. $\frac{p_1}{p_2}$, is increased. The results are shown in Figures 8a and 8b. For the results shown in Figure 8a, B_1 and B_2 have same mean (of 0.5) but different variances. Figure 8a shows $x_1 - x_2$ as $\frac{p_1}{p_2}$ is increased from 0.5 to 2. Each of the three curves corresponds to a fixed choice of the variance (σ_1, σ_2) of the bandwidth returns. Consider the curve corresponding to the variance choice $\sigma_1 = 0.2\sigma_2$. We find that $x_1 - x_2 > 0$, until $\frac{p_1}{p_2} \leq 1.4$. This implies that the contribution of the first secondary contract units to the overall portfolio is higher than that of the second

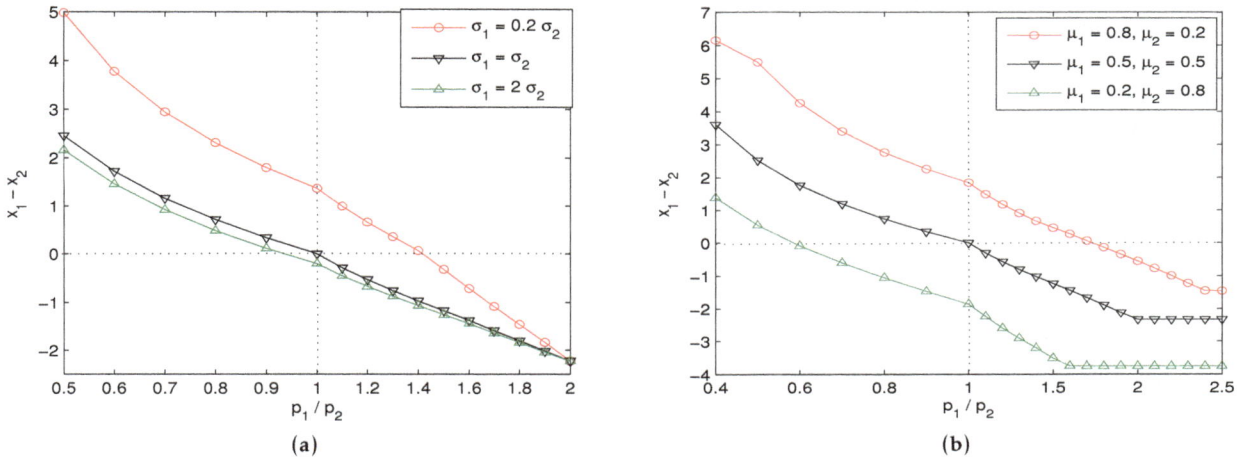

Figure 8. a) Relative contribution of the two secondary contract units as the ratio of the unit prices of the two secondary contracts is increased. Each curve corresponds to a fixed choice of the variance of B_1 and B_2, b) Relative contribution of the two secondaries. Each curve corresponds to a fixed choice of the mean of B_1 and B_2.

contract even if the unit price of the first contract is higher than the unit price of the second contract. This is clearly due to the fact that B_1 has lesser variance than B_2. However, if $\frac{p_1}{p_2} > 1.4$, the second contract units are more, since it is much lesser priced. On the other hand, when $\sigma_1 = 2\sigma_2$, the first secondary contract is preferred over second contract, only if it costs lesser than the second contract. These results suggest that secondary contracts that have lower variance of bandwidth return can be priced higher than those with higher variances, provided they have the same mean bandwidth return. Moreover, it was observed that the portfolio consisted of non-zero units of both the secondary contracts for price ratios shown, i.e. $x_1 \neq 0, x_2 \neq 0$, for $0.5 \leq \frac{p_1}{p_2} \leq 2$. This suggests that it is cost efficient to buy a mix of secondary contract units from multiple sellers, instead of just one, provided their prices are not very different. x_1 or x_2 became zero only when $\frac{p_1}{p_2}$ is either too high or too low, respectively.

Figure 8b shows $x_1 - x_2$ vs $\frac{p_1}{p_2}$, for different choices of means of B_1 and B_2, keeping the variance fixed at 0.1. When $\mu_1 = 0.8$ and $\mu_2 = 0.2$, we find that $x_1 - x_2 > 0$ as long as $\frac{p_1}{p_2} \leq 1.75$. That is, the secondary contract with 4 times higher mean bandwidth return is preferred even at 75% higher price. We also observe that the secondary contract with lesser mean is preferred only if it has lower price (For the curve with $\mu_1 = 0.2, \mu_2 = 0.8$, $x_1 - x_2 > 0$ only for $\frac{p_1}{p_2} \leq 0.6$). Figures 8a and 8b suggest that the mean as well as the variance of the bandwidth return of a secondary contract play important roles in determining the unit price of the secondary contract.

Next, consider Figure 9. Although the two secondary contracts are sold by different providers, we can expect positive correlation between their returns B_1 and B_2.

When the coustomer demand for one seller provider tends to low, it is likely that the demand for other provider is also low. Hence, the amount of secondary units available from the two providers would be comparable. The optimal portfolio and cost for different levels of correlation is shown in Figure 9. As the correlation between the secondary returns, B_1 and B_2 increase, the two secondary contracts behave like a single secondary contract, but with increased variance in the banwidth return. Due to increased riskiness in the bandwidth available from secondary contracts, the portfolio shifts to those with increased primary units.

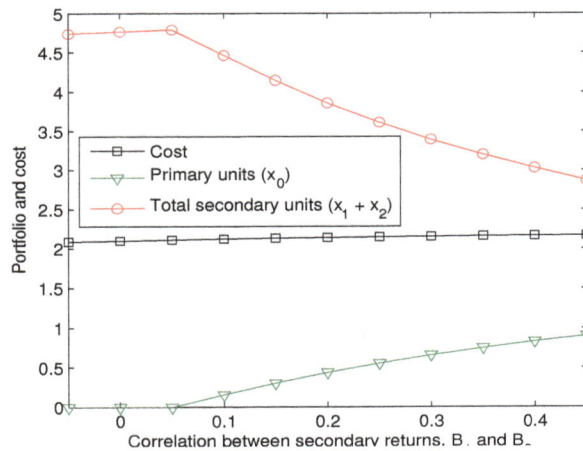

Figure 9. Correlation between secondary returns.

7.3. Multiple Regions

For the multiple-region problem, we consider two simulation scenarios. In the first scenario (Scenario A), there are totally K regions, $K + 1$ primary contracts, and $K + 1$ secondary contracts. The ith primary and secondary contract, where $1 \leq i \leq K$, is valid in the ith region only. In other words, the first K primary and secondary contracts are single-region contracts each valid in one of the K regions. However, the $K + 1^{th}$ primary and secondary contract is valid over all the K regions. The first K secondary contracts are identical in terms of their bandwidth return distributions and unit prices. The prices of all the single-region secondary contracts, $p_1, p_2.., p_K$, are set to 1. We examine the composition of secondary contract units in the optimal portfolio, when the price of the K-region secondary contract, i.e. p_{K+1}, changes. The bandwidth return variables (B_i, $1 \leq i \leq K + 1$) follow truncated normal distribution with mean 0.5 and variance 0.25. The prices of all the primary contracts is set to a large value such that the portfolio consists of only secondary contract units.

Figure 10a shows the simulation results for Scenario A when $K = 2$ (the effect of larger values of K is considered later). We only show the results for the multiple-region SPO problem under DSR constraints, since the results were similar for the DSP constraint. It was observed that the total number of primary contract units is zero as expected, i.e. $y_1 = y_2 = .. = y_{K+1} = 0$. Moreover, all the single-region secondary contracts contributed equal units to the portfolio, i.e. $x_1 = x_2 = .. = x_K$. Therefore, we plot x_{K+1} and x_1 for different price ratios $\frac{p_{K+1}}{p_1}$, where $K = 2$. When the price ratio $\frac{p_{K+1}}{p_1} < 2$, we find the portfolio consists of higher quantity of $(K + 1)^{th}$ secondary contract units compared to the single-region secondary contract, i.e. $x_{K+1} > x_1$. However, when $\frac{p_{K+1}}{p_1} \geq 2$, single-region secondary contracts are preferred over the K-region contracts ($x_{K+1} < x_1$).

In the second scenario (Scenario B), we consider four geographical regions. There are four single-region and four K-region contracts of both primary and secondary type ($K > 1$). Each K-region contract covers K regions out of the four regions, symmetrically. Unlike the previous simulation setup, we have multiple K-region contracts in this setup and there is overlap in the regions covered by these contracts. We fix the price of each single-region secondary contract (p_1) to 1. All the K-region secondary contracts have the same price denoted by p_K. We again increase the price of the K-region secondary contract p_K and observe the optimal portfolio.

The simulation results for Scenario B is shown in Figure 10b for $K = 2$. The amount of K-region secondary contract units x_K and the single-region secondary contract units x_1 in the optimal portfolio are shown. We again observe similar behaviour in x_K and x_1 when compared to Figure 10a. That is, K-region secondary contracts have higher weightage in

the optimal portfolio whenever the price ratio $\frac{p_{K+1}}{p_1} \leq 2$. Figures 10a and 10b show that to compete fairly in the market, the 2-region secondary contracts can be priced twice of that of the single-region contracts. This "pricing advantage" of the multi-region contracts is not undue however, as they cover twice the area of the single-region contracts. In general, offering spectrum contracts over a larger area implies larger licensing cost for the seller provider; moreover, the infrastructure investment and operational costs that the seller provider incurs will also be proportional to the area covered.

Next, we consider Scenario A and solve the SPO problem for higher values of K. Figure 11 shows the simulation results for $K = 2, 3, 4$. We now plot $x_{K+1} - x_1$ for different values of the price ratio $\frac{p_{K+1}}{p_1}$. For each K, when the price ratio $\frac{p_{K+1}}{p_1} < K$, we find the portfolio consists of higher quantity of $(K + 1)^{th}$ secondary contract units compared to the single-region secondary contract, i.e. $x_{K+1} - x_1 > 0$. However, when $\frac{p_{K+1}}{p_1} \geq K$, the single-region secondary contracts are preferred over the K-region contract ($x_{K+1} - x_1 < 0$, if $\frac{p_{K+1}}{p_1} \geq K$). Therefore, we find that the provider (seller) of K-region secondary contract can scale up its price upto a factor of K and still enjoy preference over the single-region contracts offered by smaller providers. This happens due to the fact that the provider can either buy one unit of the K-region secondary contract or one unit from each of the K single-region secondary contracts to provide the same service over the K regions at the same cost. The portfolio shifts completely in favor of single-region contracts only when the price, p_{K+1}, is too high. For the above choice of parameters, x_{K+1} became zero when $\frac{p_{K+1}}{p_1} > 12, 18$, and 24, respectively, for $K = 2, 3$, and 4. That is, when the price of K-region secondary contract is roughly $6K$ times (or higher) the price of the single-region secondary contract, the portfolio no longer consists of K-region secondary contract units. We obtained similar results as Figure 11 for the multi-region problem when demand and bandwidth return are modeled as the beta distribution; these results are therefore omitted for brevity.

8. Conclusion

In this paper, we have proposed a secondary spectrum market with two types of spectrum contracts – primary and secondary – and formulated the spectrum portfolio optimization (SPO) problem in this context. The two types of contracts vary in their risk-return tradeoffs, as well as their prices, and allows buyers (local or small regional providers, for example) to balance their cost with customer satisfaction level. We provide results and conditions on the convexity of the SPO problem, under both demand satisfaction rate (expectation) and

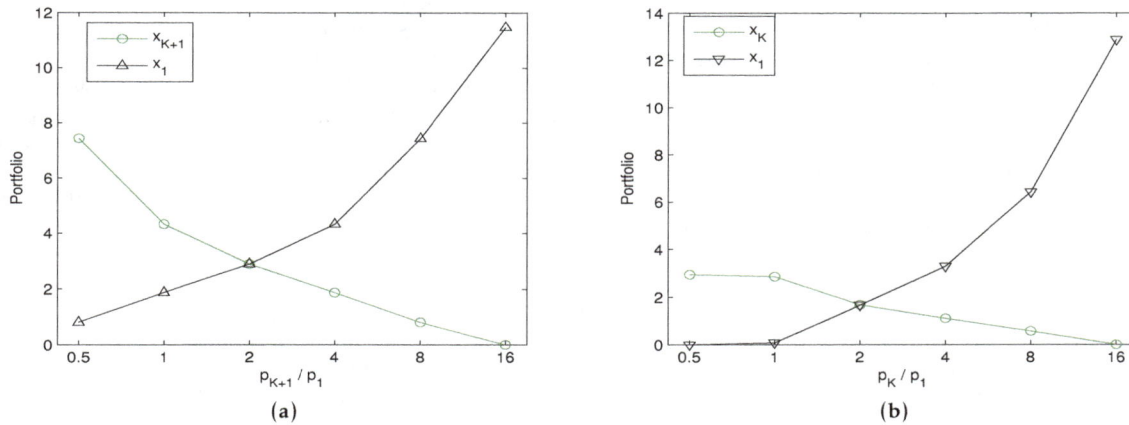

Figure 10. Optimal portfolio composition when the ratio of the unit price of the K-region secondary contract to that of the single-region secondary contract is increased; (a) Scenario A (b) Scenario B.

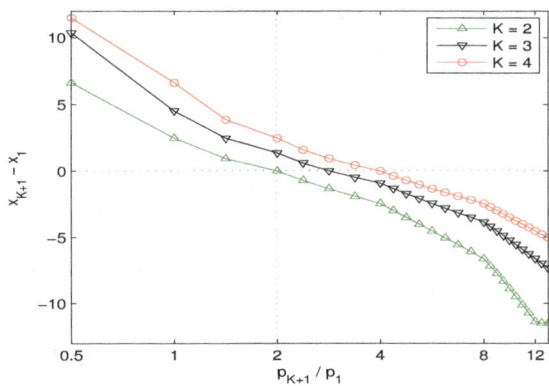

Figure 11. Optimal portfolio composition for Scenario A when the ratio of the unit price of the K-region secondary contract to that of the single-region secondary contract is increased.

demand satisfaction probability constraints. Convexity of the problems allows us to compute the optimal portfolios efficiently; we also provide expressions for the gradient that can be used for this purpose. We have also shown that convexity of the demand satisfaction constraint implies convexity of the efficient frontier, for both types of constraints. These results naturally extend to scenarios where the contracts are associated with a spatial dimension, and each contract can only provide coverage to a certain set of regions (which can differ across contracts). The convexity properties however do not extend to the integer programming formulation, where the spectrum contracts can be bought/sold only in discrete units. We have shown that the SPO problem under such constraints is not submodular.

We have used our formulation and results to compute and study the properties of optimal spectrum

portfolio in a wide range of simulation scenarios. Numerical experimentation with truncated gaussian, uniform, and empirically obtained distributions (of bandwidth availability and subscriber demand), have shown that the general nature of the variations in the optimal portfolio structure and cost, with respect to variations in key parameters like prices and customer satisfaction levels, remain similar across distributions. The composition of the optimal spectrum portfolio is also strongly influenced by the relative prices of the primary and secondary contracts, and in the multi-region case, the relative prices of the single-region and multi-region contracts. Finally, the discrete SPO problem is solved efficiently using a branch and bound algorithm to show that integrality restrictions on the portfolio does not affect the optimal portfolio cost significantly when the portfolios are large.

9. Acknowledgement

This work was supported by the National Science Foundation through awards CNS-0916958 and EARS-1247958.

References

[1] G.R. Faulhaber and D. Farber. Spectrum management: Property rights, markets, and the commons. *Telecommunications Policy Research Conference Proceedings*, 2003.

[2] E. Kwerel and J. Williams. A proposal for a rapid transition to market allocation of spectrum, opp working paper no. 38. *Office of Plans and Policy, Federal Communications Commission*, November 2002.

[3] G.R. Faulhaber. The question of spectrum: Technology, management, and regime change. *Journal on Telecommunication and High Technology Law*, 4(1):123–182, February 2005.

[4] J.M. Peha. Approaches to spectrum sharing. *IEEE Communications Magazine*, 43(2):10–12, February 2005.

[5] Z. Ji and K.J.R Liu. Collusion-resistant dynamic spectrum allocation for wireless networks via pricing. In *IEEE Symposium on New Frontiers in Dynamic Spectrum*, pages 187–190, April 2007.

[6] C.E. Caicedo and M.B.H. Weiss. The viability of spectrum trading markets. In *IEEE Symposium on New Frontiers in Dynamic Spectrum*, pages 1–10, April 2010.

[7] S. Sengupta, M. Chatterjee, and S. Ganguly. An economic framework for spectrum allocation and service pricing with competitive wireless service providers. In *IEEE Symposium on New Frontiers in Dynamic Spectrum*, pages 89–98, April 2007.

[8] A. Al Daoud, M. Alanyali, and D. Starobinski. Secondary pricing of spectrum in cellular cdma networks. In *IEEE Symposium on New Frontiers in Dynamic Spectrum*, pages 535–542, April 2007.

[9] A. Sahasrabudhe and K. Kar. Bandwidth allocation games under budget and access constraints. In *Proceedings of the CISS*, pages 761–769, March 2008.

[10] R. Etkin, A. Parekh, and D. Tse. Spectrum sharing for unlicensed bands. *IEEE Journal on Selected Areas in Communications*, 25(3):517–528, April 2007.

[11] T. Wysocki and A. Jamalipour. Pricing of cognitive radio rights to maintain the risk-reward of primary user spectrum investment. In *IEEE Symposium on New Frontiers in Dynamic Spectrum*, pages 1–8, April 2010.

[12] T.M. Valletti. Spectrum trading. *Telecommunications Policy*, 25(11):655–670, October 2001.

[13] J.M. Peha. Relieving spectrum scarcity through real-time secondary markets. *ISART Conference, Presentation Slides*, 2003.

[14] H. Xu, J. Jin, and B. Li. A secondary market for spectrum. In *IEEE International Conference on Computer Communications*, pages 1–5, March 2010.

[15] G. Kasbekar, S. Sarkar, K. Kar, P.K. Muthuswamy, and A. Gupta. Dynamic contract trading in spectrum markets. In *Allerton Conference*, pages 791–799, October 2010.

[16] L. Gao, X. Wang, Y. Xu, and Q. Zhang. Spectrum trading in cognitive radio networks: A contract-theoretic modeling approach. *IEEE Journal on Selected Areas in Communications*, 29(4):843–855, April 2011.

[17] D. Niyato, E. Hossain, and Z. Han. Dynamics of multiple-seller and multiple-buyer spectrum trading in cognitive radio networks: A game-theoretic modeling approach. *IEEE Transactions on Mobile Computing*, 8(8):1009–1022, Aug 2009.

[18] H. Markowitz. Portfolio selection. *The Journal of Finance*, 7(1):77–91, March 1952.

[19] V.S. Bawa. Safety-First, stochastic dominance, and optimal portfolio choice. *The Journal of Financial and Quantitative Analysis*, pages 255–271, June 1978.

[20] P. Bonami and M.A. Lejeune. An exact solution approach for portfolio optimization problems under stochastic and integer Constraints. *Operations Research*, 57(3):650–670, June 2009.

[21] A.D. Roy. Safety first and the holding of assets. *Econometrica*, 20(3):431–449, July 1952.

[22] D. Bertsimas, G. J. Lauprete, and A. Samarov. Shortfall as a risk measure: properties, optimization and applications. *Elsevier Journal of Economics Dynamics and Control*, 28(7):1353–1381, April 2004.

[23] C.M. Lagoa, X. Li, and M. Sznaier. Probabilistically constrained linear Programs and risk-adjusted controller design. *SIAM Journal of Optimization*, 15(3):938–951, 2005.

[24] A. Nemirovski and A. Shapiro. Convex approximations of chance constrained programs. *SIAM Journal of Optimization*, 17(4):969–996, December 2006.

[25] C. Courcoubetis and R. Weber. *Pricing Communication Networks: Economics, Technology and Modelling*. John Wiley & Sons, New York, NY, USA, 2003.

[26] A. Prekopa. A class of stochastic programming decision problems. *Math. Operationsforsch. u. Statist.*, 3:349–354, 1973.

[27] D.P. Blinn, T. Henderson, and D. Kotz. Analysis of a wifi hotspot network. In *International Workshop on Wireless Traffic Measurements and Modeling*, pages 1–6, 2005.

[28] U. Paul, A.P. Subramanian, M.M. Buddhikot, and S.R. Das. Understanding Traffic Dynamics in Cellular Data Networks. *Proc. of Infocom*, 2011.

[29] K. Murota. *Discrete Convex Analysis: Monographs on Discrete Mathematics and Applications 10*. Society for Industrial and Applied Mathematics, Philadelphia, PA, USA, 2003.

[30] S. Martello and P. Toth. An exact algorithm for large unbounded knapsack problems. *Oper. Res. Lett.*, 9(1):15–20, January 1990.

[31] G.S. Lueker. Two np-complete problems in non-negative integer programming. *Report No. 178, Computer Science Laboratory, Princeton University*, 1975.

[32] D.P. Morton and R.K. Wood. Advances in computational and stochastic optimization, logic programming, and heuristic search. chapter On a Stochastic Knapsack Problem and Generalizations, pages 149–168. Kluwer Academic Publishers, 1998.

[33] S. Leyffer. Integrating sqp and branch-and-bound for mixed integer nonlinear programming. *Computational Optimization and Applications*, pages 295–309, 2001.

[34] M.L. Benitez. Spectrum usage models for the analysis, design, and simulation of cognitive radio networks. *Dissertation, Department of Signal Theory and Communications, Universitat Politecnica de Catalunya*, 2011.

[35] M. Wellens, J. Riihijarvi, and P. Mahonen. Empirical time and frequency domain models of spectrum use. *Elsevier Physical Communication*, 2:10–32, 2009.

13

Spectrum Sensing and Primary User Localization in Cognitive Radio Networks via Sparsity

Lanchao Liu[1], Zhu Han[1,*], Zhiqiang Wu[2], Lijun Qian[3]

[1]Department of Electrical and Computer Engineering, University of Houston, Houston, TX, USA
[2]Department of Electrical Engineering, Wright State University, Dayton, Ohio, USA
[3]Department of Electrical and Computer Engineering, Prairie View A&M University, Prairie View, TX, USA

Abstract

The theory of compressive sensing (CS) has been employed to detect available spectrum resource in cognitive radio (CR) networks recently. Capitalizing on the spectrum resource underutilization and spatial sparsity of primary user (PU) locations, CS enables the identification of the unused spectrum bands and PU locations at a low sampling rate. Although CS has been studied in the cooperative spectrum sensing mechanism in which CR nodes work collaboratively to accomplish the spectrum sensing and PU localization task, many important issues remain unsettled. Does the designed compressive spectrum sensing mechanism satisfy the Restricted Isometry Property, which guarantees a successful recovery of the original sparse signal? Can the spectrum sensing results help the localization of PUs? What are the characteristics of localization errors? To answer those questions, we try to justify the applicability of the CS theory to the compressive spectrum sensing framework in this paper, and propose a design of PU localization utilizing the spectrum usage information. The localization error is analyzed by the Cramér-Rao lower bound, which can be exploited to improve the localization performance. Detail analysis and simulations are presented to support the claims and demonstrate the efficacy and efficiency of the proposed mechanism.

Keywords: Compressive sensing, decentralized spectrum sensing, localization, cognitive radio networks.

1. Introduction

Today, with rapid developments of wireless communication technologies and emergence of new wireless services, radio spectrum resource are facing the crisis of "spectrum drought" [1]. On the other hand, the spectrum resource are underutilized because of the current fixed spectrum allocation policy. Due to its capability of improving network spectrum utilization efficiency, cognitive radio (CR) [2], which enables the dynamical use of licensed spectrum bands from licensed/primary users (PUs)[3], has become a promising technique to resolve the dilemma of spectrum resource shortage versus underutilization.

The success of CR networks relies on the spectrum sensing technique [4, 5] that detects the unoccupied spectrum bands for opportunistic spectrum access. In CR networks, the CR nodes sense the spectrum and the locally collected measurements can be used collaboratively to detect the unused spectrum bands. These cooperative approaches significantly enhance the sensing reliability and efficiency. Traditionally, the CR nodes used to sense one band at a time, which results in a long sensing delay and high computational complexity in order to have the full picture of spectrums. Recent advances in compressive sensing (CS)[8, 9] make it possible to obtain the wideband signals at a sub-Nyquist sampling rate by taking random projections, which can relax the analog-to-digital (ADC) requirements as well as shorten the sensing time. The reduced number of measurements also saves the energy for communication and reduces the signalling for the CR nodes. Moreover, the

---

*Please ensure that you use the most up to date class file, available from ICST at http://icst.org/icst-transactions/
*Corresponding author. Email: zhan2@uh.edu

proliferation of CR devices has fostered the demand for lots of context-aware applications, in which the location information of PUs is viewed as one significant context. As a byproduct, the compressive spectrum sensing scheme exploits the spatial sparsity of PU locations to estimate their locations, which facilitates the spatial frequency reuse and improves spectrum resource utilization.

Despite many benefits of compressive spectrum sensing, several significant issues in this mechanism have not been fully addressed. Firstly, in order to guarantee a successful recovery of the original sparse signal, the CS sensing/sampling matrix is required to have certain properties. In particular, for the recovery to be robust to sensing noises, the matrix needs to satisfy the restricted isometry property (RIP) [10]. Does the adopted sampling mechanism in compressive spectrum sensing possess this property? Secondly, the scheme of localization via spatial sparsity imposes a requirement on the (spatial) incoherence of the channel gain matrix, which is unnecessary for the purpose of spectrum sensing. Instead of estimating the spectrum occupation and PU locations at the same time, can the two processes be separated, and can the spectrum sensing results be exploited to reduce the requirement on the channel gain matrix that is needed for the stable recovery of PU locations? Finally, what are the error characteristics of the localization errors in this mechanism, and how can we utilize those characteristics in the deployed system to control the errors? To guarantee a stable performance, all those issues in the compressive spectrum sensing scheme should be carefully discussed and analyzed.

In this paper, while we try to answer the above questions, the framework of compressive spectrum sensing is redesigned and a novel mechanism is proposed to accomplish the spectrum sensing and PUs localization tasks. Specifically, a compressive sampling mechanism is utilized by each CR node to take the measurements of the received signal spectrum. By exploiting the spectrum resource underutilization, the CS techniques can acquire a sparse signal at a sampling rate much lower than the Nyquist rate by taking a small set of random linear projections. The measurements collected at the CR nodes are processed in a collaborative way to reconstruct the original sparse signal, and we propose algorithms based on the alternating direction method of multipliers [6, 7] to recover the received signal spectrum. Then, we utilize the spectrum sensing results to localize the PUs by a Bayesian inference scheme and improve the algorithm based on analyzing the Cramer-Rao lower bound (CRLB). The main contributions of this paper are listed as follows:

1. The RIP of the cooperative compressive spectrum sensing mechanism is justified mathematically, which guarantees a successful reconstruction of the original sparse signal with high probability. Three different algorithms, Lasso, group Lasso, and distributed group Lasso with feature splitting are used to recover the spectrum usage information. Compared to Lasso, group Lasso is an improved centralized algorithm exploiting joint sparsity, and distributed group Lasso with feature splitting is an approach suitable for distributed implementation. Hence, we validate the applicability of CS to spectrum sensing both theoretically and numerically.

2. The two tasks – spectrum sensing and PU localization – are done one after another. Instead of recovering the transmitted signal spectrum at PUs, the signal spectrum at CR nodes is recovered. After the recovery, the results are used for the localization of PUs. The benefits of this treatment are two-fold: the incoherence requirement of the channel gain matrix required for stable spectrum and PU location recoveries is relaxed, and the spectrum sensing results can be exploited to reduce the computation complexity and improve the performance of the localization step.

3. The error characteristics of localization are analyzed by calculating the Cramér-Rao lower bound, which is a lower bound on the covariance of any unbiased location estimate. We analyze the localization error characteristics in detail and improve the localization algorithm design by exploiting the revealed error trends associated with the deployed system.

The rest of this work is organized as follows: Section 2 discusses the related work. The system model and problem formulation are given Section 3. The proposed approaches for spectrum sensing and PU location are detailed in Section 4 and Section 5, respectively. Section 6 presents the simulation results, and finally Section 7 concludes this paper.

2. Related Work

Closely related to this paper is the existing work on compressive spectrum sensing and PU localization taking advantages of sparsity and CS. CS has been first proposed in wideband CR in [11] for spectrum hole identification. Exploiting the sparsity of the signal spectrum, CS samples the wideband signal at a sub-Nyquist sampling rate, and thus, significantly reduces the ADC cost. [12] adopts CS in combination with decentralized recovery algorithms in cooperative CR networks, which the spectral estimates from all CR nodes are fused together to determine the spectrum availability. The spatial diversity gain induced by the distributed CR nodes greatly improves the sensing performance under fading channel environments.

Recently, CS has been used to solve the spectrum sensing problem in cooperative CR networks in [13–16]. In [13], the CS technique has been adopted for sampling wideband channel information at a sub-Nyquist rate at CR nodes. A small portion of the compressed measurements of received signals spectrum are reported to the fusion center, which recovers the spectrum usage information by matrix completion or joint sparsity reconstruction. In this centralized approach, the sampling rate at CR nodes is reduced, which saves time and energy for spectrum sensing. Also, the significantly compressed measurements save the energy consumed by transmission from the CR nodes to fusion center. [14] uses a distributed computing approach instead. Each CR node obtains its compressed individual samplings of received signal spectrum, and only communicates with the neighboring CR nodes with the one-hop range. A consensus constraint is imposed on each CR to ensure convergence, and the global convergence can be obtained by several rounds of information exchange. Upon convergence, all CR nodes obtain the fused spectrum sensing results. This approach gets rid of the fusion center at an expense of a certain amount of communication overhead. The locations of PUs are taken into consideration in [16], which exploits both the spectrum sparsity and PUs' spatial sparsity. A basis expansion model of the power spectral density (PSD) map in space and frequency is utilized to model the transmitted signals in CR networks. Each CR node collects the attenuated transmitted signals by the channel and work collaboratively to estimate the signal spectrum. The PSD of the transmitted signal is estimated by the Lasso [17] algorithm implemented via distributed online iterations, which can sense the spectrum occupation as well as recover the PU locations in the network. This kind of localization via spatial sparsity is formally defined in [18], and explicitly uses the CS theory in the localization algorithm. To guarantee a successful recovery of the source locations in the network, the requirement on the incoherence of the channel gain matrix is given therein. [19] further develops [18], which introduces a Bayesian framework for the localization problem and provides sparse approximations to its optimal solution.

Despite the early emergence and wide range of applications of the CS theory to compressive spectrum sensing, the applicability of the CS theory to this application scenario has not been rigorously justified yet. For example, the channel gain matrix is required to be sufficiently incoherent for reliable PU localization, but this requirement is often unexamined.

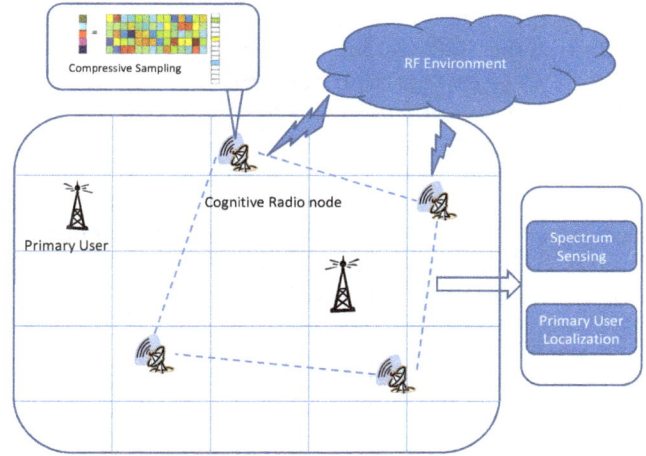

Figure 1. An illustration of system model

Furthermore, the characteristics of the localization errors have not been explicitly analyzed before. All these issues are the motivations for this paper.

3. System Model and Problem Formulation

Consider a CR network with K PUs and J CR nodes. The CR nodes locate randomly in the network with known locations $\beta = \{\beta_1, \beta_2, \ldots, \beta_J\}$. However, for PUs, neither the number of the active PUs K nor their locations are known. We define a finite position space $\alpha = \{\alpha_1, \alpha_2, \ldots, \alpha_I\}$ as the candidate positions of the PUs in the planar deployment area. Note that the PUs locate sparsely in the network and only present at K out of I positions. Each CR node in the network receives a superposition of the transmitted signals of the PUs, and our objective is to infer the spectrum occupation and locations of the PUs based on the received signals. An illustration of the system model is shown in Fig. 1.

In our model, we adopt a slotted frequency segmentation model to describe the PSD of the transmitted signal. The whole bandwidth is divided into B non-overlapping narrowband slots centered at $f_b (b = 1, \ldots, B)$ with full-width W. Suppose the PUs transmitted signals $x_i(t)$ are stationary over t and mutually uncorrelated. Here i means candidate position α_i. The PSD of the transmitted signal $x_i(t)$ for the PU at position α_i can be approximated by:

$$s_i(f) = \sum_{b=1}^{B} \theta_{i,b} \text{rect}\left(\frac{f - f_b}{W}\right), \quad i = 1, 2, \ldots, I, \quad (1)$$

where $\theta_{i,b}$ is the PSD for the corresponding narrowband slot centered at f_b. Apparently, the locations of those bases are known in our model, but the PSD levels are dynamic varying and need to be determined. If the PSD levels for certain bases are under a predefined threshold, those temporarily idle frequency bands are called spectral holes and can be opportunistically accessed by the CR nodes. We use vector $\mathbf{s_i} = [\theta_{i,1}, \theta_{i,2}, \ldots, \theta_{i,B}]$ to represent the PSD for PU at α_i. Note that only K out of I positions have PUs and many of $\mathbf{s_i}$ are all-zero vectors.

The transmitted signals are attenuated by the multipath channels between the PUs and CR nodes. The received signal of the CR node at β_j is a summation of amplitudes and delays of the multiple arriving signals from all PUs, and can be expressed as:

$$y_j(t) = \sum_{i=1}^{I} h(d_{i,j}; t) * x_i(t) + w_j(t), \quad j = 1, 2, \ldots, J, \quad (2)$$

where $*$ represents the convolution, $h(d_{i,j}; t)$ is the channel response that is a function of distance $d_{i,j}$ between α_i and β_j, and $w_j(t)$ is the noise. We assume that the signals propagate in a frequency-selective fading channel, and the channel responses are stationary w.r.t. t. Hence, we can calculate the PSD of the channel response by taking the Fourier transform of the autocorrelation of $h(d_{i,j}; t)$, i.e., $H_{i,j}(f) = \mathcal{F}(\mathbf{R}(\tau))$, where $\mathbf{R}(\tau) = \mathbb{E}[h(d_{i,j}; t)h(d_{i,j}; t - \tau)]$. The channel gain $H_{i,j}(f)$ can be obtained through extensive measurement campaigns or through the path loss model[1].

At the receiver end, the received PSD can be expressed as the product of the PSD $s_i(f)$ of the transmitted signal and channel gain $H_{i,j}(f)$:

$$r_j(f) = \sum_{i=1}^{I} H_{i,j}(f)s_i(f) + \sigma_j(f)$$

$$= \sum_{i=1}^{I} H_{i,j}(f) \sum_{b=1}^{B} \theta_{i,b}\text{rect}\left(\frac{f - f_b}{W}\right) + \sigma_j(f), \quad (3)$$

where $r_j(f)$ is the PSD of the CR nodes at position β_j, and can be estimated by the Fourier Transform of the received time domain signal $y_j(t)$ traditionally. However, since the transmitted bands occupied by the PUs are quite narrow compared with the whole available band, we can exploit the spectral sparsity and utilize CS to effectively sample the signal.

To clarify this point, we first simplify the expression of (3) to a matrix-vector form:

$$\mathbf{r_j} = \mathbf{H}_{\alpha \to \beta_j}\mathbf{s} + \sigma_j, \quad j = 1, 2, \ldots, J, \quad (4)$$

where vector $\mathbf{r_j}$ denotes the received PSD of CR node at β_j, and $\mathbf{s} = [\mathbf{s}_1^\top, \mathbf{s}_2^\top, \ldots, \mathbf{s}_I^\top]^\top$ denotes the transmitted PSD at all I candidate points. Mapping operator $\mathbf{H}_{\alpha \to \beta_j}$

stands for the corresponding channel gain matrix for all sub-bands between I candidate positions and CR node at β_j. The complete channel gain matrix from all candidate positions to all CR nodes can be written as $\mathbf{H}_{\alpha \to \beta}$.

By utilizing the CS theory, the compressed measurements at the CR node can be expressed as follows:

$$\mathbf{z_j} = \mathbf{\Phi_j}\mathcal{F}^{-1}\mathbf{r_j} + \tilde{\sigma}_j = \mathbf{\Phi_j}\mathcal{F}^{-1}\mathbf{H}_{\alpha \to \beta_j}\mathbf{s} + \tilde{\sigma}_j, \quad \forall j, \quad (5)$$

where matrix $\mathbf{\Phi_j} \in \mathbb{R}^{M \times B}, M \ll B$. Matrix $\mathbf{\Phi_j}$ can be designed sampling from a certain distribution, e.g., i.i.d. Gaussian distribution or i.i.d. Bernoulli distribution with ± 1. This matrix can be implemented by the techniques used in CS for analog signals like Xamping [20, 21], random modulator [22] and frequency-selective surface [11]. The benefits of utilizing matrix $\mathbf{\Phi_j}$ are two-fold: Firstly, it reduces the number of measurements of the CR nodes, which saves time and energy for spectrum sensing; Secondly, the reduced measurements obtained by the random projection are quite compressed compared with the original time domain signals. Since all CR nodes need to transmit the measurements back to the fusion center or exchange them with neighbors for spectrum sensing and localization, the reduction in the number of measurements saves the energy consumed by the transmission and signalling for the CR nodes.

We remark that the spectrum sensing problem is recover $\{\mathbf{r_j}\}_{j=1}^{J}$ from $\{\mathbf{z_j}\}_{j=1}^{J}$ for all CR nodes, and can be solved without the channel state information (CSI) $\mathbf{H}_{\alpha \to \beta}$. As long as the channel does not experience deep fadings, the received signals PSD $\{\mathbf{r_j}\}_{j=1}^{J}$ provide sufficient information for spectrum usage. Moreover, note that the received signals PSD $\{\mathbf{r_j}\}_{j=1}^{J}$ are also sparse. To clarify this point, we assume $\tilde{\mathbf{s}} = \sum_{i=1}^{I} \mathbf{s_i}$. The operation of summation neglects the location information of PUs and $\tilde{\mathbf{s}}$ only contains the spectrum occupation information of transmitted signals. For each $\mathbf{r}_j$ and $\tilde{\mathbf{s}}$, their supports, i.e., the positions of non-zero entries, are the same. The physical explanation is that every $\mathbf{r}_j$ is only the attenuated duplicates of $\tilde{\mathbf{s}}$ by channel effects. By exploiting these key observations, the received signals PSD are acquired by a small number of incoherent linear measurements in (5). As to localization, a Bayesian inference scheme can be adopted to localize the PUs from $\{\mathbf{r_j}\}_{j=1}^{J}$.

4. Compressive Spectrum Sensing

The spectrum sensing task is completed by recovering the received signals PSD, which can be implemented in two ways, centralized or distributed. An overview of CS is given first, and then the proposed algorithms is illustrated in detail.

---

[1] For a fading channel, the channel gain can be calculated by $H_{i,j}(f) = d_{i,j}^{-\gamma/2}|h_{i,j}(f)|$, where $h_{i,j}(f)$ is the channel fading gain that can be obtained by averaging out the effect of the channel fading and γ is the path loss factor.

4.1. Compressive Sensing Overview

CS is an emerging signal processing technique for finding sparse solutions to under-determined linear systems. By utilizing the fact that a signal is sparse or compressible in a certain domain, the CS technique can powerfully acquire a signal from a small set of randomly projected measurements with a very low sampling rate. Suppose $\mathbf{x}$ is an $N \times 1$ vector with k nonzero entries, and $\mathbf{y}$ is an $M \times 1$ vector that $k < M \ll N$. If sensing matrix $\mathbf{\Phi}$ satisfies the RIP, the solution of the noisy under-determined system of equations $\mathbf{y} = \mathbf{\Phi}\mathbf{x} + \mathbf{w}$ can be reconstructed by

$$\min_{x} \|\mathbf{x}\|_1 + \lambda \|\mathbf{y} - \mathbf{\Phi}\mathbf{x}\|_2^2, \qquad (6)$$

where $\|.\|_1$ and $\|.\|_2$ are the l_1 norm and l_2 norm, respectively. The parameter λ is chosen according to the amount of noise in the measurements. The optimization problem in (6) is known as the least-absolute shrinkage and selection operator (Lasso) [17], a.k.a. the de-nosing basis pursuit for solving a sparse linear regression problem. A detail review of traditional and advanced methods to solve (6) can be found in [23] and the references therein. In this paper, we use the ADMM [24, 25] to solve (6). In particular, to be robust again noise, we require the sensing matrix to satisfy the RIP, which is defined as follows:

Definition 1. (restricted isometry property) [10] Let $\mathbf{\Phi} \in \mathbb{R}^{m \times n}$ matrix having unit l_2-norm columns. For each integer $S \in \mathbb{N}$, we say that $\mathbf{\Phi}$ satisfies the restricted isometry property of order K with the smallest restricted isometry constant $\delta_K \in (0, 1)$, and write $\mathbf{\Phi} \in$ RIP(K, δ_K), if

$$(1 - \delta_K)\|\theta\|_2^2 \leq \|\mathbf{\Phi}\theta\|_2^2 \leq (1 + \delta_K)\|\theta\|_2^2, \quad \forall \theta : \|\theta\|_0 \leq K, \tag{7}$$

where $\|\theta\|_0$ is the number of nonzero elements in θ.

The same framework applies if signal $\mathbf{x}$ is sparse in some transformed domain $\mathbf{\Psi}$ instead of the canonical domain. In this case, $\mathbf{y} = \mathbf{\Phi}\mathbf{\Psi}\mathbf{x} + \mathbf{w}$, and we recover $\mathbf{x}$ from $\mathbf{y}$ by:

$$\min_{x} \|\mathbf{x}\|_1 + \lambda \|\mathbf{y} - \mathbf{\Phi}\mathbf{\Psi}\mathbf{x}\|_2^2, \qquad (8)$$

where $\mathbf{\Phi}\mathbf{\Psi}$ is the sensing matrix and satisfy RIP if $\mathbf{\Psi}$ and $\mathbf{\Phi}$ are incoherent.

4.2. A "Naive" Approach

The compressed measurements $\{\mathbf{z}_j\}_{j=1}^{J}$ are cooperatively used to recover the spectral estimates $\{\mathbf{r}_j\}_{j=1}^{J}$ of the received signals at all CR nodes. According to (5), $\mathbf{z}_j = \mathbf{\Phi}_j \mathcal{F}^{-1}\mathbf{r}_j + \tilde{\sigma}_j = \tilde{\mathbf{\Phi}}_j \mathbf{r}_j + \tilde{\sigma}_j$. When a fusion center is presented to collect all the compressed measurements $\{\mathbf{z}_j\}_{j=1}^{J}$, the fused measurements can be concatenated

together to form a new vector $\mathbf{z} = [\mathbf{z}_1^\top, \mathbf{z}_2^\top, \ldots, \mathbf{z}_J^\top]^\top$. Conformably, the received PSD of CR nodes can be written as $\mathbf{r} = [\mathbf{r}_1^\top, \mathbf{r}_2^\top, \ldots, \mathbf{r}_J^\top]^\top$ and the sensing matrix is

$$\tilde{\mathbf{\Phi}} = \begin{bmatrix} \tilde{\mathbf{\Phi}}_1 & 0 & \cdots & 0 \\ 0 & \tilde{\mathbf{\Phi}}_2 & \cdots & \vdots \\ \vdots & \vdots & \ddots & 0 \\ 0 & \cdots & 0 & \tilde{\mathbf{\Phi}}_J \end{bmatrix}. \qquad (9)$$

Here, we assume all CR nodes use the same random sampling matrix $\{\mathbf{\Phi}_j\}_{j=1}^{J}$, denoted as $\mathbf{\Phi}$, sampled from the i.i.d. Gaussian distribution, and thus, $\mathbf{\Phi}$ satisfies the RIP. Since the inverse Fourier transform matrix $\mathcal{F}^{-1}$ is orthonormal, the RIP of $\{\tilde{\mathbf{\Phi}}_j\}_{j=1}^{J}$ (denoted as $\tilde{\mathbf{\Phi}}$) is unaffected. Formula (9) can be write as $\tilde{\mathbf{\Phi}} = \mathbf{I}_J \otimes \tilde{\mathbf{\Phi}}$. $\otimes$ is the Kronecker product and $\mathbf{I}_J$ is a $J \times J$ identity matrix. The overall observation model can be expressed as:

$$\mathbf{z} = \bar{\mathbf{\Phi}}\mathbf{r} + \tilde{\sigma}, \qquad (10)$$

where $\tilde{\sigma} = [\tilde{\sigma}_1^\top, \tilde{\sigma}_2^\top, \ldots, \tilde{\sigma}_J^\top]^\top$. $\mathbf{r}$ can be solved by:

$$\min_{r} \|\mathbf{r}\|_1 + \lambda \|\mathbf{z} - \bar{\mathbf{\Phi}}\mathbf{r}\|_2^2. \qquad (11)$$

To guarantee a successful recovery of signal $\mathbf{r}$, the overall sensing matrix $\bar{\mathbf{\Phi}}$ must satisfy the RIP, which can be justified by the following theorem.

Theorem 1. Let $\mathbf{A}$ be an orthonormal basis, and let $\mathbf{B}$ be the matrix with restricted isomery constant $\delta_K(\mathbf{B})$. The restricted isometry constant of the Kronecker product of $\mathbf{A}$ and $\mathbf{B}$ satisfies:

$$\delta_K(\mathbf{A} \otimes \mathbf{B}) = \delta_K(\mathbf{B}), \qquad (12)$$

This theorem provides conservation of restricted isomery constants across the Kronecker product of an orthonormal matrix and a sensing matrix. The proof is given in Appendix A. For our case, $\delta_K(\bar{\mathbf{\Phi}}) = \delta_K(\mathbf{I}_J \otimes \tilde{\mathbf{\Phi}}) = \delta_K(\tilde{\mathbf{\Phi}})$, which implies that sensing matrix $\bar{\mathbf{\Phi}}$ satisfies RIP.

The centralized fusion (11) results in the global optimal solution by incorporating the measurements from all the CR nodes. This approach to recover the PSD of the received signal is straight-forward and we use it as a simple illustration. To solve (11), we first rewrite it as:

$$\min_{r, p} \|\mathbf{p}\|_1 + \lambda \|\mathbf{z} - \bar{\mathbf{\Phi}}\mathbf{r}\|_2^2, \quad s.t. \quad \mathbf{r} - \mathbf{p} = 0. \qquad (13)$$

The scaled augmented Lagrangian can be written as:

$$\mathcal{L}_\rho(\mathbf{r}, \mathbf{p}, \mathbf{u}) = \|\mathbf{p}\|_1 + \lambda \|\mathbf{z} - \bar{\mathbf{\Phi}}\mathbf{r}\|_2^2 + \rho/2\|\mathbf{r} - \mathbf{p} + \mathbf{u}\|_2^2. \qquad (14)$$

The objective function and variables are split into two parts. The separated variables $\mathbf{p}$ and $\mathbf{r}$ can be updated in an alternating fashion as follows:

$$\mathbf{r}^{k+1} = arg\min_{\mathbf{r}}(\lambda\|\mathbf{z} - \bar{\boldsymbol{\Phi}}\mathbf{r}\|_2^2 + \rho/2\|\mathbf{r} - \mathbf{p}^k + \mathbf{u}^k\|_2^2), \quad (15)$$

$$\mathbf{p}^{k+1} = arg\min_{\mathbf{p}}(\|\mathbf{p}\|_1 + \rho/2\|\mathbf{r}^{k+1} - \mathbf{p} + \mathbf{u}^k\|_2^2), \quad (16)$$

$$\mathbf{u}^{k+1} = \mathbf{u}^k + \mathbf{r}^{k+1} - \mathbf{p}^{k+1}. \quad (17)$$

The $\mathbf{r}$-update process is a quadratically regularized least-square problem, and the $\mathbf{p}$-update can be solved in closed form by soft-thresholding. In this case, the fusion center collects the compressed measurements $\{\mathbf{z}_j\}_{j=1}^J$ from all the CR nodes. The compressed measurements are computed centrally to recover the received signals PSD $\{\mathbf{r}_j\}_{j=1}^J$, and thus, the unoccupied spectrum bands are determined.

4.3. Spectrum Sensing via Joint Sparsity

Unlike the approach above, a more sophisticated approach further exploits the structure of the received signals PSD $\{\mathbf{r}_j\}_{j=1}^J$ is proposed in the sequel. Since the signal ensembles $\{\mathbf{r}_j\}_{j=1}^J$ are the received signals PSD at all the CR nodes, the sparse structures of $\{\mathbf{r}_j\}_{j=1}^J$ are identical to $\tilde{\mathbf{s}}$. In other words, the occupied frequency bands of the transmitted signals and received signals are the same as long as the channel does not experience deep fading. Note that $\{\mathbf{r}_j\}_{j=1}^J$ are the attenuated duplicates of $\tilde{\mathbf{s}}$ and share a common support. This kind of jointly sparse signals can be solved by the proposed algorithm as follows,

$$\min\sum_{b=1}^B \|\tilde{\mathbf{r}}_\mathbf{b}\|_2 + \lambda\|\mathbf{z} - \bar{\boldsymbol{\Phi}}\mathbf{r}\|_2^2, \quad (18)$$

where $\tilde{\mathbf{r}}_\mathbf{b} = \mathbf{R}(b,:)$, $\mathbf{R}$ is a matrix formed by $\{\mathbf{r}_j\}_{j=1}^J$, and $\mathbf{R} = [\mathbf{r}_1, \mathbf{r}_2, \ldots, \mathbf{r_J}]$. The solution to (18) has a natural grouping of its components, and the components within the group are likely to be either all zeros or all non-zeros. Each $\tilde{\mathbf{r}}_\mathbf{b}$ corresponds to the received signals PSD in the bth frequency band. We remark that the model in (18) incooperating joint sparsity outperforms that in (11), which recovers the spectra separately. In (18), the first term tents to select the same sparse structure of the received signals PSD, and thus, it can achieve a fast convergence and high accuracy compared with (11).

To solve (18), we first rewrite it as:

$$\min\sum_{b=1}^B \|\tilde{\mathbf{r}}_\mathbf{b}\|_2 + \lambda\|\mathbf{z} - \bar{\boldsymbol{\Phi}}\mathbf{p}\|_2^2$$
$$s.t. \quad \tilde{\mathbf{r}}_\mathbf{b} - \tilde{\mathbf{p}}_\mathbf{b} = 0, \quad b = 1, 2, \ldots, B, \quad (19)$$

where $\{\tilde{\mathbf{r}}_\mathbf{b}\}_{b=1}^B$ are the received signals PSD at each frequency bands, and $\mathbf{p}$ is the globe variable cascading

all received signals PSD. In this case, $\tilde{\mathbf{p}}_b$ is the corresponding part of $\mathbf{r}_b$ in $\mathbf{p}$. To solve (19), we first write down the scaled augmented Lagrangian:

$$\mathcal{L}_\rho(\tilde{\mathbf{r}}_\mathbf{b}, \mathbf{p}, \mathbf{u}) = \sum_{b=1}^B (\|\tilde{\mathbf{r}}_\mathbf{b}\|_2 + \rho/2\|\tilde{\mathbf{r}}_\mathbf{b} - \tilde{\mathbf{p}}_\mathbf{b} + \mathbf{u}_\mathbf{b}\|_2^2) + \lambda\|\mathbf{z} - \bar{\boldsymbol{\Phi}}\mathbf{p}\|_2^2. \quad (20)$$

Similarity, the separate variables can be updated in an alternating fashion:

$$\tilde{\mathbf{r}}_b^{k+1} = arg\min_{\tilde{\mathbf{r}}_\mathbf{b}}(\|\tilde{\mathbf{r}}_\mathbf{b}\|_2 + \rho/2\|\tilde{\mathbf{r}}_b - \tilde{\mathbf{p}}_b^k + \mathbf{u}_b^k\|_2^2), \quad (21)$$

$$\mathbf{p}^{k+1} = arg\min_{\mathbf{p}}(\sum_{b=1}^B \rho/2\|\tilde{\mathbf{r}}_b - \tilde{\mathbf{p}}_b^{k+1} + \mathbf{u}_b^k\|_2^2 + \lambda\|\mathbf{z} - \bar{\boldsymbol{\Phi}}\mathbf{p}\|_2^2), \quad (22)$$

$$\mathbf{u}_b^{k+1} = \mathbf{u}_b^k + \tilde{\mathbf{r}}_b^{k+1} - \tilde{\mathbf{p}}_b^{k+1}. \quad (23)$$

The update rule is similar to the above with the $\tilde{\mathbf{r}}_\mathbf{b}$-update replaced by block soft-thresholding. This approach exploits the structure of joint sparsity which further reduces the number of measurements and improves recovery quality. Once $\{\mathbf{r}_b\}_{b=1}^B$ are recovered, the availability of the spectrum resource in the network is determined.

4.4. Distributed Spectrum Sensing via Joint Sparsity

The centralized approaches are costly and impractical to implement when a fusion center is absent. To address this situation, a distributed spectrum sensing algorithm is proposed in the sequel.

Note that $\bar{\boldsymbol{\Phi}}$ is a block diagonal matrix, and the second term of (18) can be split into J least squares terms corresponding to the measurements form all CR nodes. The optimization problem can be equivalently expressed as:

$$\min\sum_{b=1}^B \|\tilde{\mathbf{r}}_\mathbf{b}\|_2 + \sum_{j=1}^J \lambda_j\|\mathbf{z}_\mathbf{j} - \bar{\boldsymbol{\Phi}}\mathbf{r}_\mathbf{j}\|_2^2. \quad (24)$$

The second least-square term is naturally separable across j. At the same time, the first term enforcing group sparsity of the received signals PSD and couples all CR nodes. Assume the CR nodes are connected to each other in the network. In this case, the spectrum occupation can be estimated in a cooperative way by the CR nodes through one-hop communication. To design a decentralized implementation of (24), we first define a function $f(\mathbf{R}) = \sum_{b=1}^B \|\tilde{\mathbf{r}}_\mathbf{b}\|_2$, which calculates the sum of row vector norm of matrix $\mathbf{R}$. The formula in (24) can be rewritten as:

$$\min\sum_{j=1}^J (f(\mathbf{R}) + \lambda_j\|\mathbf{z}_\mathbf{j} - \bar{\boldsymbol{\Phi}}\mathbf{r}_\mathbf{j}\|_2^2). \quad (25)$$

To distributively solve the above problem, J identical copies $\{\mathbf{R}_j\}_{j=1}^J$ of $\mathbf{R}$ are induced and the above problem can be reformulated as:

$$\min \sum_{j=1}^J (f(\mathbf{R}_j) + \lambda_j \|\mathbf{z}_j - \tilde{\mathbf{\Phi}}\mathbf{p}_j\|_2^2)$$

$$s.t. \quad \mathbf{p}_j - \mathbf{R}_j(:,j) = 0, \quad \mathbf{R} - \mathbf{R}_j = 0, \quad j = 1, 2, \ldots, J. \tag{26}$$

The basic idea of this approach is to update the received signals PSD at each CR nodes locally, and a synchronized update procedure on the coupled variable $\mathbf{R}$ is performed by using the consensus averaging technique. Specificity, the variables $\{\mathbf{R}_j\}$, $\{\mathbf{p}_j\}$ and $\{\mathbf{R}\}$ can to be grouped into two sets, one set is $\{\mathbf{R}_j\}$ and the other one is $(\{\mathbf{p}_j\}, \{\mathbf{R}\})$. After separation, we can update the two sets in an alternative fashion. When given $\{\mathbf{p}_j\}$ and $\{\mathbf{R}\}$, the updates of $\mathbf{R}_j$ are separable. When given $\{\mathbf{R}_j\}$, we can update $(\{\mathbf{p}_j\}, \{\mathbf{R}\})$ separatively. The scaled augmented Lagrangian of this problem is given as follows:

$$\mathcal{L}(\mathbf{p_j}, \mathbf{R}, \mathbf{R_j}, \mathbf{u_{1j}}, \mathbf{u_{2j}}) = \sum_{j=1}^J (f(\mathbf{R}_j) + \lambda_j \|\mathbf{z_j} - \tilde{\mathbf{\Phi}}\mathbf{p_j}\|_2^2$$
$$+ (\rho_1/2)\|\mathbf{p}_j - \mathbf{R}_j(:,j) + \mathbf{u_{1j}}\|_2^2$$
$$+ (\rho_2/2)\|\mathbf{R}(:) - \mathbf{R}_j(:) + \mathbf{u_{2j}}\|_2^2), \tag{27}$$

During the iterative updating procedure, each CR node maintains a copy $\mathbf{R}_j$ of the global variable $\mathbf{R}$ and its local variable $\mathbf{p}_j$. At a specific iteration step, $\mathbf{R}_j$ is updated locally by each CR node:

$$\mathbf{R}_j^{k+1} = arg \min f(\mathbf{R}_j) + (\rho_1/2)\|\mathbf{p}_j^k - \mathbf{R}_j(:,j) + \mathbf{u}_{1j}^k\|_2^2 +$$
$$(\rho_2/2)\|\mathbf{R}(:)^k - \mathbf{R}_j(:) + \mathbf{u}_{2j}^k\|_2^2), \tag{28}$$

where $\mathbf{R}$ is the consensus average of the $\mathbf{R}_j$ from all the CR nodes, which can be obtained by the one-hop communication between a CR node and its neighbors. The global variable $\mathbf{R}$ is updated in a synchronous fashion and disseminated throughout the network to all CR nodes, as follows:

$$\mathbf{R}^{k+1} = arg \min(1/J) \sum_{j=1}^J (\mathbf{R}_j(:)^{k+1} + \mathbf{u}_{2j}^k), \tag{29}$$

and the local received signal PSD $\mathbf{p}_j$ is updated locally at each CR node by calculating

$$\mathbf{p}_j^{k+1} = arg \min \lambda_j \|\mathbf{z_j} - \tilde{\mathbf{\Phi}}\mathbf{p_j}\|_2^2$$
$$+ (\rho_1/2)\|\mathbf{p}_j - \mathbf{R}_j(:,j)^{k+1} + \mathbf{u}_{1j}^k\|_2^2. \tag{30}$$

The parameters are updated locally at the end of each iteration step as follows:

$$\mathbf{u}_{1j}^{k+1} = \mathbf{u}_{1j}^k + \mathbf{p}_j^{k+1} - \mathbf{R}_j(:,j)^{k+1},$$
$$\mathbf{u}_{2j}^{k+1} = \mathbf{u}_{2j}^k + \mathbf{R}(:)^{k+1} - \mathbf{R}_j(:)^{k+1}. \tag{31}$$

Upon the convergence, every CR nodes in the network obtain the same global variable $\mathbf{R}$ and its own local received signal PSD. Hence, the spectrum occupancy is determined locally at each CR node.

5. Primary User Localization

This section illustrates the PU localization mechanism. The spectrum sensing results are utilized to relax the incoherence requirement on $\mathbf{H}_{\alpha \to \beta}$, and then the CRLB is provided to give a lower bound of the errors of the location estimates and to improve the performance of the proposed algorithm.

5.1. Incoherence Requirement on $\mathbf{H}_{\alpha \to \beta}$

Before analyzing the incoherence requirement on channel gain matrix $\mathbf{H}_{\alpha \to \beta}$, we first recall the definition of the coherence of a matrix as follows:

Definition 2. (Coherence)[26] Let $\psi_1, \ldots, \psi_n$ be the columns of matrix $\mathbf{\Psi} \in \mathbb{R}^{m \times n}$. The coherence of matrix $\mathbf{\Psi}$ is defined as the maximum absolute value of the cross-correlation between any two columns ψ_i, ψ_j for $\forall i, j, 1 \le i \ne j \le n$, i.e.,

$$\mu(\mathbf{\Psi}) = \max_{1 \le i \ne j \le n} \frac{\langle \psi_i, \psi_j \rangle}{\|\psi_i\|_2 \|\psi_j\|_2}. \tag{32}$$

By definition, we have $\mu(\mathbf{\Psi}) < 1$ for all $\mu(\mathbf{\Psi})$. On the other hand, if $\mathbf{\Psi}$ has orthonormal columns $\mu(\mathbf{\Psi}) = 0$. Actually, we can find the lower bound of $\mu(\mathbf{\Psi})$ for $\mathbf{\Psi}$ such that $\mu(\mathbf{\Psi}) \in [\sqrt{\frac{n-m}{m(n-1)}}, 1]$.

Intuitively, the PU location can be obtained by recovering $\mathbf{s} = \{\mathbf{s}_i\}_{i=1}^I$ from the measurements $\mathbf{z} = \{\mathbf{z}_j\}_{j=1}^J$ of all CR nodes,

$$\mathbf{z} = \bar{\mathbf{\Phi}}\mathbf{H}_{\alpha \to \beta}\mathbf{s} + \tilde{\sigma}. \tag{33}$$

Although straightforward, this approach imposes the incoherence requirement on $\mathbf{H}_{\alpha \to \beta}$, which can be explained by the following theorem.

Theorem 2. [18, 27] For a K-sparse signal θ with exact K non-zero entries. Let $\mathbf{\Psi} \in \mathbb{R}^{L \times N}$ be a matrix with coherence $\mu(\mathbf{\Psi})$, and $\bar{\mathbf{\Phi}} = \mathbf{I} \otimes \mathbf{\Phi} \in \mathbb{R}^{M \times L}$ with $\mathbf{\Phi} \in$ RIP(K, δ_K). Then with high probability, any K-sparse signal θ can be reconstructed from the measurement $y = \bar{\mathbf{\Phi}}\mathbf{\Psi}\theta$ by the l_1 minimization if

$$K \le 1 + \frac{1}{16\mu(\mathbf{\Psi})}, \tag{34}$$

and the number of measurements M obeys the relation $M = O(K \log(N/K))$.

For the inequality in (34), when there are more than one PU in the CR network, the coherence of $\mathbf{H}_{\alpha \to \beta}$ should be quite small, which limits the application scope of this algorithm. To recover multiple PUs in the CR network, channel matrix $\mathbf{H}_{\alpha \to \beta}$ should be incoherent enough to separate the PUs and recover their locations. The coherence property of $\mathbf{H}_{\alpha \to \beta}$ depends on the geographical separation of the PUs, channel frequency response, and the number of CR nodes in the network. An incoherent matrix $\mathbf{H}_{\alpha \to \beta}$ can separate multiple PUs in the network to implement the localization task. This may require the well spaced PUs in the geographical field, detailed profile of the channel information, or a large number of CR nodes in the network, which may be impractical to satisfy.

If we view the problem from a different perspective, however, the incoherence requirement on $\mathbf{H}_{\alpha \to \beta}$ can be delicately removed. We remark that the PUs are naturally separated in the frequency domain due to the avoidance of communication interference. So at different frequency band $\{f_b\}_{b=1}^{B}$, the geographical positions of the transmitted signal can be estimated. Note that there is only one transmitted signal at a specific frequency f_b, which can greatly relax the incoherence requirement on $\mathbf{H}_{\alpha \to \beta}$ for localization. Once the geographical position of the transmitted signal is obtained, the corresponding PU locations are determined. Capitalizing this key observation, a localization algorithm based on the Bayesian inference at each frequency band is proposed in the sequel.

5.2. The Localization Inference Problem

Once the spectrum sensing results $\{\mathbf{r}_j\}_{j=1}^{J}$ are obtained, the occupied spectral bands are determined. For a specific occupied frequency band f_b, the geographical location of the transmitted signal can be estimated through a Bayesian inference. Specifically, the posterior density of α_i can be determined by $p(\alpha_i|\tilde{\mathbf{r}}_b) \propto p(\tilde{\mathbf{r}}_b|\alpha_i)p(\alpha_i)$, where $p(\alpha_i)$ is the prior probability of being at location α_i. Without loss of generality, we assume $p(\alpha_i)$ subject to an uniform distribution in our problem. The conditional probability $p(\mathbf{r}_b|\alpha_i)$ is calculated as:

$$p(\tilde{\mathbf{r}}_b|\alpha_i) = \prod_{j=1}^{J} p(\mathbf{r}_b(j)|\alpha_i), \qquad (35)$$

and can be obtained during the off-line training. For a specific position α_i, the probability $p(\tilde{\mathbf{r}}_b|\alpha_i)$ for the PSD $\tilde{\mathbf{r}}_b$ at frequency band f_b subjects to $\mathcal{N}(\mathbf{u}_b, \Sigma_b)$. The localization problem can be solved as:

$$\alpha_i = arg \max p(\alpha_i|\tilde{\mathbf{r}}_b) = arg \max p(\tilde{\mathbf{r}}_b|\alpha_i). \qquad (36)$$

In this case, the α_i with the largest $p(\alpha_i|\tilde{\mathbf{r}}_b)$ is chosen as the geographical position of the transmitted signal. The centroid of K locations with the largest $p(\alpha_i|\tilde{\mathbf{r}}_b)$ can also be used as the final estimate of the transmitted signal geographical position, and this approach is called KNN (K-Nearest Neighbors).

5.3. Localization CRLB

Understanding the characteristics of the localization error is an essential step to the error control and performance improvement [28]. In CR networks, the localization error depends on variety of network configuration parameters. In this part, the CRLB is calculated to specify this dependency and understand the error characteristics of network localization.

The CRLB provides a lower bound on the error covariance of any unbiased estimator. It is independent on particular estimation methods as long as the statistical model of observations on the variable is specified. It is a benchmark to evaluate various estimators that provide estimates equal to the ground truth if averaged over enough realizations.

In our case, the parameters to be estimated include the coordinates of α_i and can be denoted as $\alpha_i = (x_i, y_i)^{\top}$. Its corresponding estimation is $\hat{\alpha}_i = (\hat{x}_i, \hat{y}_i)^{\top}$. The covariance matrix of the estimate $\hat{\alpha}_i$ can be written as

$$Cov(\hat{\alpha}_i) = E_{\alpha_i}\{(\hat{\alpha}_i - \alpha_i)(\hat{\alpha}_i - \alpha_i)^{\top}\} = \begin{bmatrix} \sigma_{\hat{x}_i}^2 & \sigma_{\hat{x}_i\hat{y}_i} \\ \sigma_{\hat{y}_i\hat{x}_i} & \sigma_{\hat{y}_i}^2 \end{bmatrix},$$
$$(37)$$

where $E_{\alpha_i}\{\cdot\}$ stands for the expectation value, and the mean square error of the location estimate $\hat{\alpha}_i$ is $var(\hat{\alpha}_i) = \sigma_{\hat{x}_i}^2 + \sigma_{\hat{y}_i}^2$.

Let $f(P|\alpha_i)$ denote the probability function of observations P conditioned on parameter α_i. The fisher information matrix (FIM) for exponential family $f(P|\alpha_i)$ is defined as:

$$J(\alpha_i) = -E\left\{\frac{\partial^2 \ln f(P|\alpha_i)}{\partial \alpha_i^2}\right\} = \begin{bmatrix} J_{x_i x_i}(\alpha_i) & J_{x_i y_i}(\alpha_i) \\ J_{y_i x_i}(\alpha_i) & J_{y_i y_i}(\alpha_i) \end{bmatrix}.$$
$$(38)$$

By definition, the CRLB is the inverse of FIM, which is a lower limit for the covariance matrix, i.e., $Cov(\hat{\alpha}_i) \geq \{J(\alpha_i)\}^{-1}$. Without loss of generality, the received signal of CR j at frequency f_b can be calculated using the shadowing path loss model, as follows:

$$\mathbf{r}_b(j) = \overline{PL(d_0)} - 10\gamma \log\left(\frac{d_{ij}}{d_0}\right) + \mathcal{X}_{\sigma_{bj}}, \qquad (39)$$

where d_0 is the reference distance. γ is the path loss factor, d_{ij} is the distance between the transmitted signal at α_i and CR node at β_j, and $\mathcal{X}_{\sigma_{bj}} \sim \mathcal{N}(0, \sigma_{bj}^2)$ stands for the shadowing effect. The lower bound of the mean square of the location estimate $\hat{\alpha}_i$ is given by the following theorem.

Theorem 3. The lower bound of the mean square error of the location estimate $\hat{\alpha}_i$ is given by:

$$var(\hat{\alpha}_i) = \frac{\sum_{j=1}^{J}(\eta_{1j}\eta_{2j})^2 + \sum_{j=1}^{J}(\eta_{1j}\eta_{3j})^2}{\sum_{j=1}^{J}(\eta_{1j}\eta_{2j})^2 \sum_{j=1}^{J}(\eta_{1j}\eta_{3j})^2 - (\sum_{j=1}^{J}((\eta_{1j})^2\eta_{2j}\eta_{3j})^2},$$

(40)

where $\eta_{1j} = \frac{10\gamma}{\sigma_{bj}\ln 10}$, $\eta_{2j} = \frac{\cos\phi_{ij}}{d_{ij}}$, and $\eta_{3j} = \frac{\sin\phi_{ij}}{d_{ij}}$. ϕ_{ij} is the angle between the PU and CR node as illustrated in Fig. 2(a).

The CRLB of the location estimate depends on many factors in the network [29], which can be classified into three categories: the system resource, the environmental parameters, and the geographical layout of the CR nodes. The system resource are the number of CR nodes that can performance the localization task in the network. The environmental parameters are σ_{bj} and γ in the path loss propagation model. The geographical layout of the CR nodes determines the value of the angle ϕ_{ij} and distance d_{ij}. A simple illustration is given in Fig. 2. In Fig. 2(a), three CR nodes, A, B and C are evenly separated around PU O. This configuration achieves the highest accuracy and called the geographic dilution of precision [30]. To see the effect of the distance d_{ij} on the CRLB, CR node A is moving towards PU O along line AO while angle ϕ is unchanged. The ratio of the new CRLB while moving A is compared to the original one. In Fig. 2(b) we can see that the CRLB decreases as the distance between CR node A and PU O deceases.

By the observation that CRLB of the location estimation decreases with distance d_{ij}, a weighted modification to (35) can improve the performance of the localization algorithm. The conditional probability $p(\mathbf{r}_b|\alpha_i)$ can be rewritten as follows:

$$p(\mathbf{r}_b|\alpha_i) = \prod_{j=1}^{J} \frac{\frac{1}{d_{ij}^\xi}p(\mathbf{r}_b(j)|\alpha_i)}{\sum_{i=1}^{I}\frac{1}{d_{ij}^\xi}},$$

(41)

where we assign a weight $\frac{1}{d_{ij}^\xi}$ to each $p(\mathbf{r}_b(j)|\alpha_i)$, and $\sum_{i=1}^{I}\frac{1}{d_{ij}^\xi}$ is the normalizing factor calculated across all the possible locations $\{\alpha_i\}_{i=1}^{I}$. During the calculation in (41), the probability $p(\mathbf{r}_b(j)|\alpha_i)$ provided by a nearer CR node is assigned to more importance, and thus, the precision of the localization mechanism is enhanced.

6. Simulation Results

In this section, the numerical simulation results are presented to demonstrate the performance of the proposed spectrum sensing and PU localization

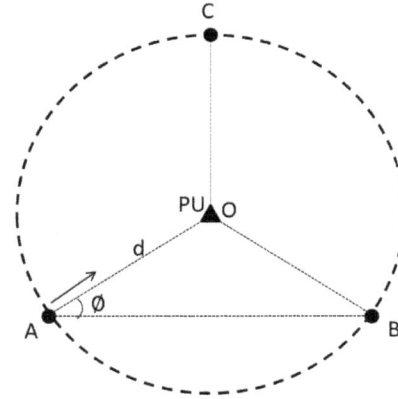

(a) An illustration of geographical dilution of precision

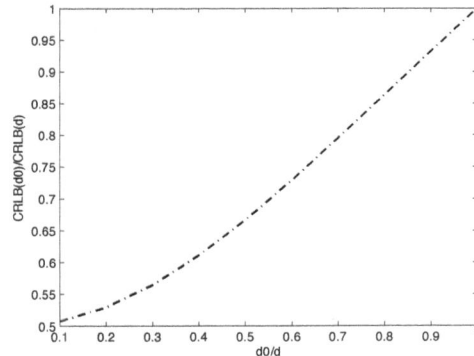

(b) CRLB will decrease as the distance between the CR node and PU decrease

Figure 2. Geographical layout of the CR nodes affects the CRLB

approaches. We first describe the simulation setup and the performance metrics. The performance of our proposed spectrum sensing mechanism is evaluated over different scenarios: different numbers of CR nodes, different numbers of PUs, and different compressed sampling ratios. Finally, the localization performance is presented.

6.1. Simulation Setup and Performance Metrics

We assume the operational space is a $1000m \times 1000m$ square field, which is divided into $I = 25$ uniformly distributed grids. K PUs locate randomly on I candidate points. The wide bandwidth of the PUs' transmitted signal is equally divided into $B = 64$ sub-channels. The transmitted signals experience frequency selective fading during transmission. Here, we assume Rayleigh channel fading in our simulation. J SUs are randomly deployed in the same field working collaboratively to implement the spectrum sensing and localization tasks. Each CR node samples M random projections of the original signal and the compressed sampling ratio is defined as M/B. The SNR is defined as the ratio of the average received signal power to the noise power over the entire bandwidth.

We evaluate the performance of our algorithms using the following metrics. Denote the $\hat{\mathbf{r}}$ as the recovered sparse vector, and $\hat{\alpha}_i$ as the recovered PU locations. D is the distance between adjacent grids. For spectrum availability detection, the receiver operating characteristic (ROC) is considered, where the true positive rate and false alarm rate are defined, respectively, as follows:

$$TPR = \frac{N_{Hit}}{N_{Hit} + N_{Miss}}, \quad FAR = \frac{N_{False}}{N_{False} + N_{Correct}}, \quad (42)$$

where N_{Hit} is the number of successful detections of occupied spectrum bands, N_{Miss} is the number of miss detections, N_{False} is the number of false alarms, and $N_{Correct}$ is the number of correct reports of unoccupied spectrum bands.

The spectrum sensing and localization performance are evaluated by their normalized MSEs, which are defined as follows, respectively,

$$\mathbf{r}_{MSE} = E\left\{\frac{\|\mathbf{r} - \hat{\mathbf{r}}\|_2}{\|\mathbf{r}\|_2}\right\}, \quad L_{MSE} = E\left\{\frac{\sum_{i=1}^{K}(\|\alpha_i - \hat{\alpha}_i\|_2)}{K \times D}\right\}.$$
$$(43)$$

6.2. Performance v.s. Different Numbers of CR nodes

In this scenario, we set the number of active PUs $K = 3$ and the compressed sampling ratio $M/B = 25\%$. The SNR is $10dB$ and the number of CR nodes J varies from 3 to 7 to evaluate the performance. The simulation results are shown in Fig. 3 and Fig. 4.

From Fig. 3, we can see that the increase in the number of CR nodes can greatly improve the performance. This may mainly due to two reasons. On the one hand, the larger number of CR nodes will provide more measurements for the fusion center using the l_1 minimization algorithm to recover the original sparse signal, which will enhance the probability of successful reconstruction. One the other hand, more CR nodes will better exploits the diversity in the measurements from different CR nodes, which will improve the detection correctness and reduce the reconstruction error. The centralized algorithm with group sparsity performs best. When the number of CR nodes is three, the "naive" approach degrades greatly, which validates the effectiveness of exploiting the joint sparsity in spectrum sensing problem. When few CR nodes present in the network, utilizing the joint sparsity in the received signals can guarantee the performance of reconstruction.

Fig. 4 shows the ROC curves at different numbers of CR nodes when sampling rate equals 25%. When the number of CR nodes is 3, the proposed mechanism

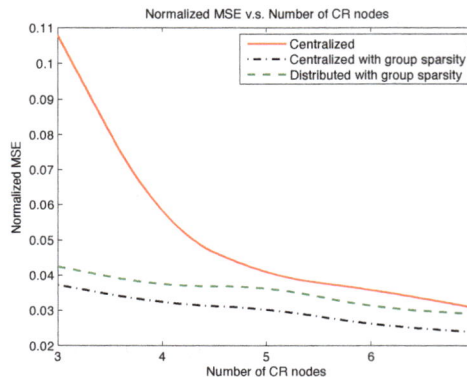

Figure 3. Normalized MSE of spectrum estimation

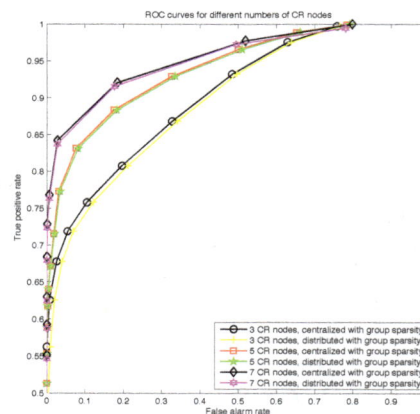

Figure 4. ROC curves at different numbers of CR nodes

can still achieve a moderate true positive rate at a low false positive rate. At the same false alarm rate, the true positive rate improves as the number of CR nodes increases. The centralized and distributed algorithms demonstrate similar performances for the spectrum occupancy detection task.

6.3. Performance v.s. Different Numbers of PUs

In this scenario, we set the number of CR nodes $J = 4$ and the compressed sampling ratio $M/N = 25\%$. The SNR is $10dB$ and the number of active PUs K varies from 3 to 7 to evaluate the performance. The simulation results are shown in Fig. 5 and Fig. 6.

From Fig. 5, we can notice that the reconstruction errors of spectrum sensing and localization increase with the number of PUs. More PUs will make the original signal less spare, which will results in the requirement of a larger number of measurements. However, the performances of the proposed mechanism are still acceptable. In the worst case, the normalized

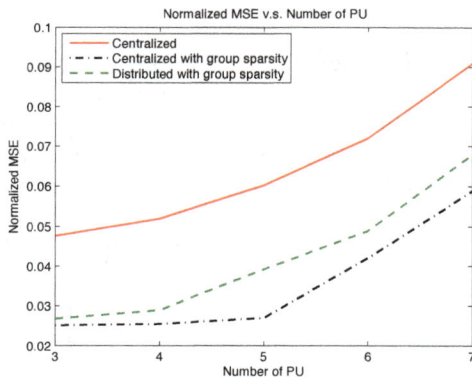

Figure 5. Normalized MSE of spectrum estimation

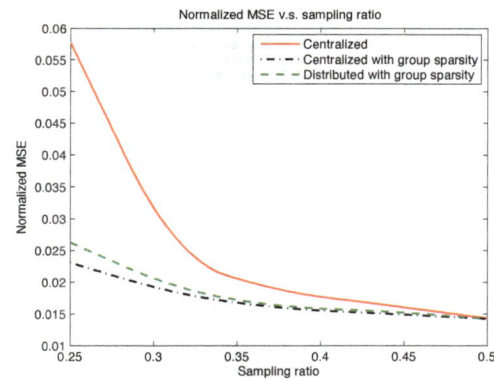

Figure 7. Normalized MSE of spectrum estimation

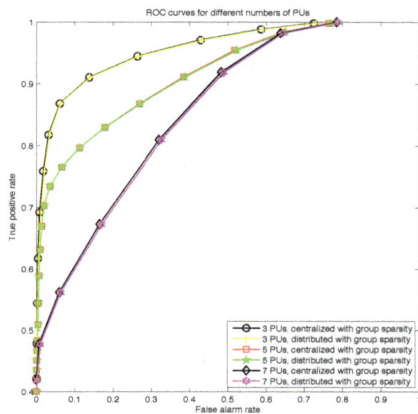

Figure 6. ROC curves at different numbers of PUs

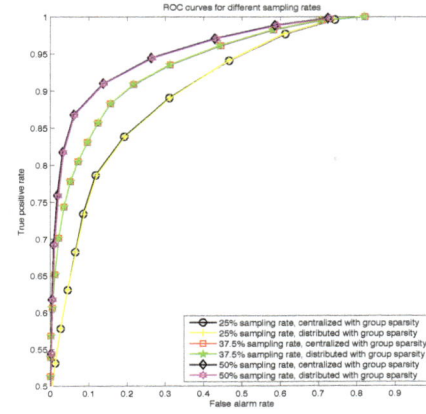

Figure 8. ROC curves at different sampling rates

MSE of spectrum sensing for 7 PUs by the centralized algorithm with group sparsity is around 0.06.

Fig. 6 shows the ROC curves at different numbers of PUs when sampling rate equals 50%. The performance of spectrum holes detection deteriorates as the number of PUs increases. However, the proposed scheme can still achieve a satisfied performance for the detection of spectrum holes when 7 PUs present in the network.

6.4. Performance v.s. Different Compressed Sampling Ratio

In this scenario, we set the number of CR nodes $J = 4$ and the number of active PUs $K = 3$. The SNR is $10dB$ and the compressed sampling ratio varies from 25% to 50% to evaluate the performance. The simulation results are shown in Fig. 7.

In Fig. 7, the normalized MSE of spectrum sensing decreases as the sampling ratio increases. A larger number of samples result in a smaller normalized MSE. The centralized algorithm with group sparsity perform best in this case. In a severe environment, the

reliable performance at the low sampling rate (25%) can significantly save the energy of the CR nodes.

Fig. 8 depicts the ROC curves of the proposed mechanism at different compressed sampling ratios. With a larger number of measurements, the performance of detecting occupied frequency band is better. When the compressed sampling ratio equals 50%, the performance is best, at the expense of more measurements taken by each CR node and processed centrally or distributively.

6.5. Localization Performance

In this scenario, we set the number of CR nodes $J = 5$, the compressed sampling ratio $M/N = 50\%$, the number of active PU $K = 3$ and the $SNR = 10dB$. We first compare the performance of the proposed algorithms with the method by reconstructing $\mathbf{s}$ in (33), and then the performance improvement of the modified Bayesian location inference is demonstrated.

Fig. 9 shows the normalized localization error for different algorithms by varying the number of CR nodes from 4 to 7. Due to the incoherence requirement

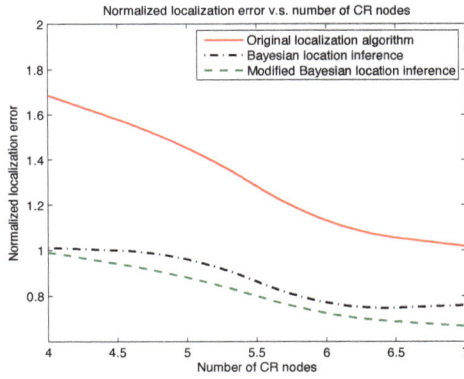

Figure 9. Comparison of localization algorithms at different numbers of CR nodes.

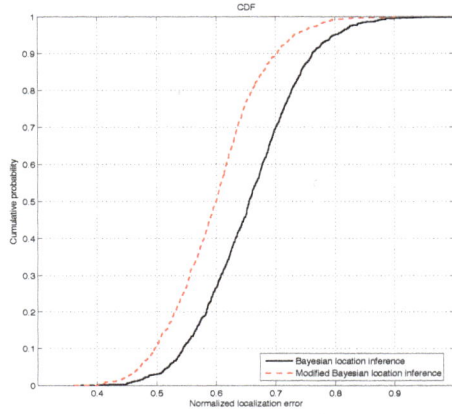

Figure 10. CDFs of normalized localization error

on $\mathbf{H}_{\alpha \to \beta}$, (33) results in significant error in location estimate when few CR nodes are present in the network. The performance is enhanced when the number of CR nodes is larger, which reduces the coherence of the channel gain matrix $\mathbf{H}_{\alpha \to \beta}$. However, the proposed Bayesian location inference approach outperforms (33) because of exploiting the spectrum sensing results. Different frequency bands naturally separate the PUs, and thus, the incoherence requirement on $\mathbf{H}_{\alpha \to \beta}$ is significantly relaxed. Moreover, the modified algorithm improves the performance by utilizing the error characteristics revealing by the CRLB. The cumulative distribution functions of the normalized localization errors are shown in Fig. 10 for the case when the number of CR nodes equals to 7. The simulation results show that the accuracy of the localization algorithm is improved.

7. Conclusion

In this paper, we redesign the framework of compressive spectrum sensing and PU localization in CR networks. The applicability of CS to cooperative spectrum sensing is justified, and the spetrum recovery performance is guaranteed. We use three different algorithms, Lasso, group Lasso, and (distributed) group Lasso with feature splitting, to recover the received signal's PSD. The locations of PUs are estimated through a Bayesian inference approach based on the results of spectrum sensing. This two-step treatment greatly relaxes the incoherence requirement of the channel gain matrix, which is required for PU localization based on spatial sparsity. The error characteristics of localization are analyzed by computing the CRLB, and the localization algorithm design is improved by exploiting the revealed error trends associated with the system. Through numerical simulations, the efficacy and efficiency of the proposed mechanism are validated.

Appendix A. Proof of Theorem 1

Proof. We follow the idea of [31] to prove the Theorem. Let matrix $\mathbf{C} = \mathbf{A} \otimes \mathbf{B}$. Suppose set Ω is of cardinality K. We denote $\mathbf{C}_\Omega$ as the sub-matrix of $\mathbf{C}$ containing the columns c_t, $t \in \Omega$. For a column $t \in \Omega$ of matrix $\mathbf{C}_\Omega$, we have $\mathbf{c}_t = \mathbf{a}_{t_1} \otimes \mathbf{b}_{t_2}$, where $\mathbf{a}_{t_1}$ is column t_1 of matrix $\mathbf{A}$, and $\mathbf{b}_{t_1}$ is column t_2 of matrix $\mathbf{B}$. $t_1 \in \Omega_1$ and $t_2 \in \Omega_2$. For each $t \in \Omega$, we can find the corresponding $t_1 \in \Omega_1$ and $t_2 \in \Omega_2$. The cardinality product $|\Omega_1||\Omega_2| \leq K^2$. We have $\Omega \subset \Omega_1 \cup \Omega_2$. Since the range of singular values of a sub-matrix are interlaced inside those of the original matrix, we have:

$$\sigma_{min}(\mathbf{A} \otimes \mathbf{B}) = \sigma_{min}(\mathbf{A}_{\Omega_1})\sigma_{min}(\mathbf{B}_{\Omega_2}) \leq \sigma_{min}(\mathbf{C}_\Omega), \quad (A.1)$$

$$\sigma_{max}(\mathbf{A} \otimes \mathbf{B}) = \sigma_{max}(\mathbf{A}_{\Omega_1})\sigma_{max}(\mathbf{B}_{\Omega_2}) \geq \sigma_{max}(\mathbf{C}_\Omega). \quad (A.2)$$

By definition of RIP, the $\delta_K(\mathbf{A}_{\Omega_1})$, $\delta_K(\mathbf{B}_{\Omega_1})$ and $\delta_K(\mathbf{C}_\Omega)$ are the smallest constants that make the following inequalities holds:

$$(1 - \delta_K(\mathbf{A}_{\Omega_1})) \leq \sigma_{min}(\mathbf{A}_{\Omega_1}) \leq \sigma_{max}(\mathbf{A}_{\Omega_1}) \leq (1 + \delta_K(\mathbf{A}_{\Omega_1})),$$
$$(A.3)$$
$$(1 - \delta_K(\mathbf{B}_{\Omega_1})) \leq \sigma_{min}(\mathbf{B}_{\Omega_1}) \leq \sigma_{max}(\mathbf{B}_{\Omega_1}) \leq (1 + \delta_K(\mathbf{B}_{\Omega_1})),$$
$$(A.4)$$
$$(1 - \delta_K(\mathbf{C}_\Omega)) \leq \sigma_{min}(\mathbf{C}_\Omega) \leq \sigma_{max}(\mathbf{C}_\Omega) \leq (1 + \delta_K(\mathbf{C}_\Omega)).$$
$$(A.5)$$

Substitute (A.3) and (A.4) into (A.1) and (A.2), we have:

$$(1 - \delta_K(\mathbf{A}_{\Omega_1}))(1 - \delta_K(\mathbf{B}_{\Omega_1})) \leq \sigma_{min}(\mathbf{C}_\Omega),$$
$$(1 + \delta_K(\mathbf{A}_{\Omega_1}))(1 + \delta_K(\mathbf{B}_{\Omega_1})) \geq \sigma_{max}(\mathbf{C}_\Omega). \quad (A.6)$$

Note that $\delta_K(\mathbf{C}_\Omega)$ is the smallest constant that satisfies (A.5)

$$(1 + \delta_K(\mathbf{C})) \leq (1 + \delta_K(\mathbf{A}))(1 + \delta_K(\mathbf{B})). \quad (A.7)$$

$\mathbf{A}$ is an orthonormal basis, and $\delta_K(\mathbf{A}) = 0$. So we have $\delta_K(\mathbf{C}) \leq \delta_K(\mathbf{B})$. By Theorem 3.7 in [32], we have $\delta_K(\mathbf{C}) \geq \delta_K(\mathbf{B})$. Hence, $\delta_K(\mathbf{C}) = \delta_K(\mathbf{B})$. $\qquad\square$

Appendix B. Proof of Theorem 2

The joint probability density distribution $f(P|\alpha_i)$ can be expressed as,

$$f(P|\alpha_i) = \prod_{j=1}^{J} \frac{1}{\sqrt{2\pi}\sigma_{bj}} \times \exp\left\{ -\frac{\mathbf{r}_b(j) - \overline{PL(d_0)} + 10\gamma \log(\frac{d_{ij}}{d_0})}{2\sigma_{bj}^2} \right\}. \tag{B.8}$$

After obtaining the log-likelihood of $f(P|\alpha_i)$, the entries of the FIM can be calculated according to (38), which can be expressed as follows:

$$J_{x_i x_i}(\alpha_i) = \sum_{j=1}^{J} \left(\frac{10\gamma}{\sigma_{bj} \ln 10} \right)^2 \left(\frac{\cos\phi_{ij}}{d_{ij}} \right)^2, \tag{B.9}$$

$$J_{x_i y_i}(\alpha_i) = J_{y_i x_i}(\alpha_i) = \sum_{j=1}^{J} \left(\frac{10\gamma}{\sigma_{bj} \ln 10} \right)^2 \left(\frac{\cos\phi_{ij}}{d_{ij}} \right)\left(\frac{\sin\phi_{ij}}{d_{ij}} \right), \tag{B.10}$$

$$J_{y_i y_i}(\alpha_i) = \sum_{j=1}^{J} \left(\frac{10\gamma}{\sigma_{bj} \ln 10} \right)^2 \left(\frac{\sin\phi_{ij}}{d_{ij}} \right)^2. \tag{B.11}$$

The definition of the angle ϕ_{ij} is illustrated in Fig. 2(a). Since $Cov(\hat{\alpha}_i) \geq \{J(\alpha_i)\}^{-1}$, the lower bound of the location estimation variance $var(\hat{\alpha}_i)$ can be calculated as:

$$\begin{aligned} var(\hat{\alpha}_i) &= \sigma_{\hat{x}_i}^2 + \sigma_{\hat{y}_i}^2 \geq \frac{J_{x_i x_i}(\alpha_i) + J_{y_i y_i}(\alpha_i)}{|J(\alpha_i)|} \\ &= \frac{J_{x_i x_i}(\alpha_i) + J_{y_i y_i}(\alpha_i)}{J_{x_i x_i}(\alpha_i)J_{y_i y_i}(\alpha_i) - J_{x_i y_i}(\alpha_i)J_{y_i x_i}(\alpha_i)}. \end{aligned} \tag{B.12}$$

By Substituting (B.9), (B.10) and (B.11) into (B.12), we can obtain the lower bound of the mean square error of the location estimate $\hat{\alpha}_i$.

References

[1] G. Staple and K. Werbach, "The end of spectrum scarcity," *IEEE Spectrum Archive*, vol. 41, no. 3, pp. 48-52, Mar. 2004.

[2] S. Haykin, "Cognitive radio: brained-empowered wireless communications," *IEEE Journal on Selected Areas in Communications*, vol. 23, no. 2, pp. 201-220, Feb. 2005.

[3] E. Hossain, D. Niyato, and Z. Han, "Dynamic Spectrum Access in Cognitive Radio Networks," *Cambridge University Press*, UK, 2009.

[4] S. Haykin, D. J. Thomson, and J. H. Reed, "Spectrum sensing for cognitive radio," *Proceedings of the IEEE*, vol. 97, no. 5, pp. 849-877, May. 2009.

[5] I. F. Akyildiz, B. F. Lo, and R. Balakrishnan, "Cooperative spectrum sensing in cognitive radio networks: a survey," *Physical Communication*, vol. 4, no. 1, pp. 40-62, Mar. 2011.

[6] S. Boyd, N. Parikh, and E. Chu, "Distributed optimization and statistical learning via the alternating direction method of multipliers," *Now Publishers Inc*, Boston, 2011.

[7] D. P. Bertsekas and J. N. Tsitsiklis, "Parallel and distributed computation: numerical methods," *Prentice Hall*, New Jersey, 1989.

[8] D. Donoho, "Compressed sensing," *IEEE Transactions on Information Theory*, vol. 52, no. 4, pp. 1289-1306, Apr. 2006

[9] E. J. Candès and T. Tao, "Near optimal signal recovery from random projections: Universal encoding strategies?" *IEEE Transactions on Information Theory*, vol. 52, no. 12, pp. 5406-5425, Dec. 2006.

[10] E. J. Candès, J. K. Romberg, and T. Tao, "Stable signal recovery from in-complete and inaccurate measurements,"Âİ *Commun. Pure Appl. Math.*, vol. 59, no. 9, pp. 1207-1223, Aug. 2006.

[11] Z. Tian and G. Giannakis, "Compressed sensing for wideband cognitive radios," *IEEE International Conference on Acoustics, Speech, and Signal Processing*, Honolulu, HI, 2007.

[12] Z. Tian, "Compressed wideband sensing in cooperative cognitive radio networks," *IEEE Global Telecommunications Conference*, New Orleans, LA, 2008.

[13] J. Meng, W. Yin, H. Li, E. Hossain, and Z. Han, "Collaborative spectrum sensing from sparse observations in cognitive radio networks," *IEEE Journal on Selected Topics on Communications, special issue on Advances in Cognitive Radio Networking and Communications*, vol. 29, no. 2, pp. 327-337, Feb. 2011.

[14] F. Zeng, C. Li, and Z. Tian, "Distributed compressive spectrum sensing in cooperative multi-hop wideband cognitive networks," *IEEE Journal of Selected Topics in Signal Processing, Special Issue on Signal Processing in Cooperative Cognitive Radio Systems*, vol. 5, no. 1, pp. 37-48, Feb. 2011.

[15] L. Liu, Z. Han, Z. Wu, and L. Qian, "Collaborative compressive sensing based dynamic spectrum sensing and mobile primary user localization in cognitive radio networks," *IEEE Global Telecommunications Conference*, Houston, TX, 2011.

[16] J. A. Bazerque and G. B. Giannakis, "Distributed spectrum sensing for cognitive radio networks by exploiting sparsity," *IEEE Transations on Signal Processing*, vol. 58, no. 3, pp. 1847-1862, Mar. 2010.

[17] R. Tibshirani, "Regression shrinkage and selection via the lasso," *J. Roy. Statist. Soc. Ser. B*, vol. 58, no. 1, pp. 267-288, 1996.

[18] V. Cevher, M. F. Duarte, and R. G. Baraniuk, "Distributed target localization via spatial sparsity," *European Signal Processing Conference*, Lausanne, Switzerland, Aug. 2008.

[19] V. Cevher, P. Boufounos, R. G. Baraniuk, A. C. Gilbert, and M. J. Strauss, "Near-optimal bayesian localization via incoherence and sparsity," *Int. Conf. on Information Processing in Sensor Networks*, San Francisco, CA, Apr. 2009.

[20] M. Mishali, Y. C. Eldar, and A. Elron, "Xampling: signal acquisition and processing in union of subspaces," *IEEE Transactions on Signal Processing*, vol. 59, no. 10, pp. 4719-4734, Oct. 2011.

[21] M. Mishali and Y. C. Eldar, "Sub-Nyquist sampling: bridging theory and practice," *IEEE Signal Processing Magazine*, vol. 28, no. 6, pp. 98-124, Nov. 2011.

[22] J. Tropp, M. Wakin, M. Duarte, D. Baron, and R. Baraniuk, "Random filters for compressive sampling and reconstruction," *IEEE Int. Conf. on Acoustics, Speech, and Signal Processing*, Toulouse, France, May 2006.

[23] Z. Han, H. Li, and W. Yin, "Compressive sensing for wireless networks," *Cambridge University Press*, UK, 2012.

[24] J. Yand and Y. Zhang, "Alternating direction algorithms for l_1-problems in compressive sensing," *SIAM journal on Scientific Computing*, vol. 33, no. 1, pp. 250-278, Feb. 2011.

[25] W. Deng, W. Yin, and Y. Zhang, "Group sparse optimization by alternating direction method," *Rice CAAM Report*, TR11-06, 2011.

[26] E. J. Candès and Y. Plan, "Near-ideal model selection by l_1 minimization," *Ann. Statist.*, vol. 27, no. 5, pp. 2145-2177, 2007.

[27] H. Rauhut, K. Schnass, and P. Vandhergheynst, "Compressed sensing and redundant dictionaries," *IEEE Trans. Info. Theory*, vol. 54, no. 5, pp. 2210-2219, May 2008.

[28] Y. Liu and Z. Yang, "Location, localization and localizability: location-awareness technology for wireless networks," *Springer*, NY, 2011.

[29] A. M. Hossian and W. Soh, "Cramér-Rao bound analysis of localization using signal strength difference as location fingerprint," *in Proceesings of IEEE Conference on Computer Communications*, San Diego, CA, USA.

[30] M. A. Spirito, "On the accuracy of cellular mobile station location estimation," *IEEE Transactions on Vehicular Technology*, vol. 50, no. 3, pp. 674-685, May. 2001.

[31] M. F. Duarte and R. G. Baraniuk, "Kronecker compressive sensing," *IEEE Transactions on Image Processing*, vol. 21, no. 2, pp. 494-504, Feb. 2012.

[32] S. Jokar and V. Mehrmann, "Sparse representation of solutions of Kronecker product systems," *Linear Algebr. Appl.*, vol. 431, no. 12, pp. 2437-2447, Dec. 2009.

14

Spectrum Hole Identification in IEEE 802.22 WRAN using Unsupervised Learning

V.Balaji[1,*], S.Anand[1], C.R.Hota[1], G.Raghurama[2]

[1]Department of Computer science and Information Systems, BITS Pilani, Hyderabad Campus, 500078, India

[2]Department of Electrical and Electronics Engineering, BITS Pilani, Goa Campus, 403726, India

Abstract

In this paper we present a Cooperative Spectrum Sensing (CSS) algorithm for Cognitive Radios (CR) based on IEEE 802.22 Wireless Regional Area Network (WRAN) standard. The core objective is to improve cooperative sensing efficiency which specifies how fast a decision can be reached in each round of cooperation (iteration) to sense an appropriate number of channels/bands (i.e. 86 channels of 7MHz bandwidth as per IEEE 802.22) within a time constraint (channel sensing time). To meet this objective, we have developed CSS algorithm using unsupervised K-means clustering classification approach. The received energy level of each Secondary User (SU) is considered as the parameter for determining channel availability. The performance of proposed algorithm is quantified in terms of detection accuracy, training and classification delay time. Further, the detection accuracy of our proposed scheme meets the requirement of IEEE 802.22 WRAN with the target probability of falsealrm as 0.1. All the simulations are carried out using Matlab tool.

Keywords: Cognitive radio, Dynamic Spectrum Access, Cooperative Sensing, TV white space, Machine Learning

1. Introduction

A Cognitive Radio (CR) is a key technology [1] that allows wireless devices to dynamically access the available spectrum opportunities. Cognitive radio is a software defined radio [2] with the capability of identifying unused spectrum in a particular time, frequency and geographic location and utilizing it in opportunistic manner. The cognition capability of a CR is defined as the ability of CR transceiver to sense the surrounding radio environment, analyze the captured information and decide the best course of action in order to decide which spectrum bands are to be used and best transmission strategy to be adopted. CR is capable of making intelligent decisions and it's actions are based on observing the wireless connections and then using intelligent algorithms and computational learning to optimize their behavior. From the Definition of CR by Simon Haykin [3], it is clear that a CR device must have the attributes: awareness, intelligence, learning, adaptivity, reliability

and efficiency. CR should intelligently sense the unused spectrum bands and learn without interfering with primary users. The experience gained through learning makes the CR to optimally reconfigure RF operating parameters and improve its decision. To perform this, CR must support the following functionalities [4]:

Spectrum awareness: It involves sensing the available spectrum bands and monitoring the activities of primary user with the help of spectrum sensing algorithms. These algorithms are used to identify the spectral activity pattern and estimate the characteristics of spectrum holes.

Learning: This phase acts as a knowledge base between spectrum sensing and decision phase. The gathered knowledge through learning can then be exploited to improve decision capability of CR.

Decisions and actions: The decision phase helps to choose appropriate spectrum band according to spectrum characteristics and user information. The actions are performed by effectively utilizing spectrum holes. The knowledge gathered during the learning phase acts as input to this module. The reconfiguration actions on RF operating parameters are performed

*Corresponding author.Email: p2011040@hyderabad.bits-pilani.ac.in

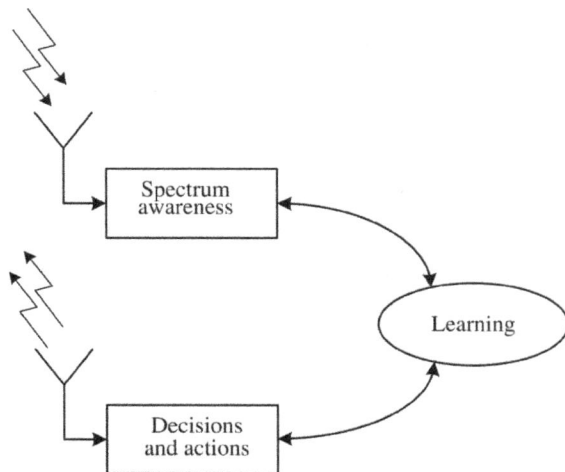

Figure 1. Functional model of CR

during this phase. The above sequence of operations by CR is schematically shown in Fig. 1.

1.1. Motivation

In CR, the most important step is to obtain necessary observations about its surrounding RF environment, such as the presence of primary users and the appearance of spectrum holes. Spectrum sensing enables the detection capability of CR to measure, learn and be aware of the radio's operating environment. The spectrum sensing problem in CR consists of three sub-problems [5]:

- Decide which channel to sense (Channel sensing is a decision making problem)

- Decide whether the channel is idle/busy based on local observations of the sensed channel (Primary signal detection or channel-state detection problem)

- Decide collaboratively whether to access the channel or not if it is indeed idle (Cooperative decision making problem)

Cooperative communication in wireless networks addresses the problem of channel impairments (i.e. multipath fading/shadowing) and improves the spatial diversity gain [6] of wireless receivers. The wireless nodes can make collaborative decision strategies to access channels with the help of cooperative communication techniques. The idea of CSS has been adopted from this cooperative communication technique. These CSS schemes greatly improve the received Signal-to-Noise ratio (SNR) under deep fading. Through this cooperation, the SUs can share their locally observed information about spectrum holes and make more accurate collaborative decision. It is

noted that the cognitive capability [7] of CR enhances the decision quality of CSS algorithms and improves cooperative sensing accuracy. Recent advances in spectrum exploration and exploitation are discussed in [8].

1.2. Related work

Although cooperative communication [9] has lot of benefits to cognitive radios, there are still numerous theoretical and technical problems that remain unsolved. In [10], author's discuss various cooperative sensing techniques with their emphasis on spectrum sensing and access based cooperation, interference constraint based adaptive cooperative feedback, cooperative transmission based on rate-less network coding and interference coordination based on limited cooperation. Cooperative sensing is proposed in the literature [11] and its performance has been investigated extensively. In a widely studied form of cooperative spectrum sensing, the Secondary Users (SU) provide locally-sensed information on the primary users activity to a decision-making fusion center (FC) which can be an access point or base station or one of the SUs [12]. The FC analyzes the information and determines the activity status of primary user. The cooperative sensing can be categorized based on the type of fusion scheme used at the FC. Hard decision combining schemes such as AND, OR and k-out-of-N rule are considered in [13]. A cooperative sensing scheme based on linear combination of the local test statistics was proposed in [14] where the combining weights were optimized to improve the detection performance. Relay based CSS schemes are studied in [15].

As already discussed, cooperative learning [16] can help a cognitive radio to learn the surrounding environment and improve its sensing accuracy. However, in recent years there has been a growing interest in applying machine learning algorithms to CSS. In [17], the author has proposed CSS scheme based on supervised learning approach such as Support Vector Machine (SVM) and K-Nearest Neighbour (KNN) classification algorithms. The same author in another paper [18] has done a comparative study of Supervised (i.e. SVM and weighted KNN) and Unsupervised learning techniques (i.e. K-means clustering and Gaussian Mixture Model) for CSS schemes. The comparison of various CSS classifiers has been carried out based on training duration, classification delay and Receiver Operating Characteristics (ROC) performance. The result concludes that unsupervised K-means clustering is a promising approach for CSS due to its high ROC performance with low classification delay and training duration. However, in [18] the author assumed that the SUs are immobile and the SNR of each SU has been normalized to Gaussian distribution. In our work, we deployed SUs randomly with mobility in a grid topology and the SNR values are changed during iterations according to distance coordinates from the primary transmitter.

A survey on state-of-the art machine learning techniques and role of learning in cognitive radio is presented in [19]. In our recent work [20] we have used Perceptron learning module in which the fusion centre collects local sensing results of each SU and makes the final decision based on soft combination of the local decisions (weighted average method). The weights corresponding to each SU is computed using energy values captured by individual SU. The weight assigned to every secondary user is multiplied to the local decision value and the cumulative sum obtained from all the secondary users is used to determine the final decision of the FC. These weighted linear combinations of the local decision vectors produce the Target Output. Then, the hard-limit function determines the final decision of FC about availability of primary channel. Due to the dynamic channel environment, feature vectors are scattered in decision boundary which affects the detection accuracy of FC. To overcome this, we have developed in this work unsupervised K-means clustering approach which partitions set of training energy vectors into K disjoint clusters. This unsupervised K-means clustering is a promising approach due to its higher detection accuracy and less training and classification delays.

1.3. Contribution

This paper discusses a framework of CSS scheme using unsupervised K-means clustering algorithm to meet the functional requirement of IEEE 802.22 WRAN standard. The key contributions of this paper are as follows:

- The simulation scenario of CSS scheme has been formulated using machine learning techniques to meet the requirements of IEEE 802.22 WRAN standard.

- Local sensing phase is carried out using energy detection to scan the complete available channel set from 54MHz-682MHz with channel bandwidth of 7MHz.

- The Cooperative Spectrum Sensing (CSS) phase is based on unsupervised K-means clustering classification algorithm. The reason for adopting learning algorithm in CSS is because of its ability to dynamically adapt and train at any time, ability to 'learn' features and attributes of the system which is often difficult to formulate analytically. The performance of our proposed algorithms are evaluated using training duration, classification delay and detection accuracy.

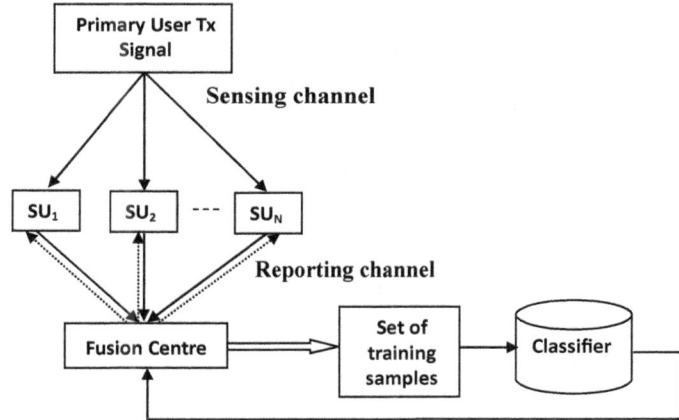

Figure 2. Working model of proposed CSS scheme

2. System model

2.1. Working model of the proposed CSS scheme

The Fig.2 shows the working model of the proposed CSS scheme. Each SU senses the primary user transmitted signal through sensing channel using energy detection scheme. The operation of energy detector is based on received signal power and noise power and comparing with local threshold to decide presence (H_1)/absence (H_0) of PU. The statistical inference drawn from energy detector which acts as local decisions $X_i(n)$ of ith SU at sample index n is given by,

$$H_0 : x_i(n) = w_i(n)$$
$$H_1 : x_i(n) = h_i(n) \times s(n) + w_i(n) \tag{1}$$

where $w_i(n)$ is the additive white-Gaussian noise (AWGN), s(n) is the primary user signal and $h_i(n)$ is the gain of the sensing channel between PU and SU.

The decision metric for the energy detector can be written as,

$$M_i = \sum_{n=0}^{N} | x_i(n) |^2 \tag{2}$$

where N is the observation vector. The performance of energy detector can be evaluated by using two probabilities: Probability of detection $'P_d'$ and Probability of false alarm $'P_f'$. The probability of detection is to decide the presence of primary user when it is truly present. In contrary, the $'P_f'$ is to decide the presence of PU when it is actually not present. It can be formulated as,

$$P_d = P_r(M_i > \lambda/H_1)$$
$$P_f = P_r(M_i > \lambda/H_0) \tag{3}$$

where $'\lambda'$ is decision threshold which can be selected for finding the optimum balance between $'P_d'$ and $'P_f'$. By setting a desired probability of false alarm and calculating the variance of a data set, the system sets a threshold to indicate signals above the noise level.

Each SU processes the received energy and compares it with a local threshold. The received signal strength of each SU depends based on its distance from Primary transmitter. The collection of energy vectors of each SU is represented using a matrix shown below. In this matrix, the row vectors and column vectors are considered as secondary users and number of channels respectively. Each secondary user has an array of values specifying the availability of each of the 92 channels. The local decision observations of all SUs are denoted as $Y_i(t)$ and represented in matrix form as,

$$Y_i(t) = \begin{pmatrix} x_1(n) & x_1(n) & \cdots & x_1(n) \\ x_2(n) & x_2(n) & \cdots & x_2(n) \\ \vdots & \vdots & \ddots & \vdots \\ x_N(n) & x_n(n) & \cdots & x_N(n) \end{pmatrix} \quad (4)$$

Based on the local decisions of the N SUs, the fusion center will take a final decision. The local sensing phase is described in Algorithm 1. First, the primary user signal is added with noise according to the distance from the primary user. This noise added signal, 'signal_at_node' acts as input to different SUs. For each of the 10 secondary users, periodograms are calculated for 'signal_at_node', and based on that a Power Spectral Density (PSD) graph is obtained. The frequency range is considered as (54-698MHz) and divided into chunks of 7MHz channel bandwidth which is scanned in steps of channel width giving around 92 channels whose status can be either 'occupied' or 'available'. The average energy values at each channel are compared to a threshold value based on a random probability of false alarm. If the energy value of the channel is greater than the threshold, the channel is specified as 'occupied', otherwise it is 'available'.

2.2. Unsupervised learning algorithm for proposed CSS scheme

Learning ability is important in cognitive radios for effective decision making. Learning algorithms are implicitly built into spectrum knowledge acquisitions and decision-making algorithms in the sense that they convert information (current and past observations) in to decisions and actions. As mentioned in [3], a CR is an intelligent wireless communication system using the attributes of intelligence and cognitive abilities that enables self-learning and self-awareness.

Learning algorithms can broadly be categorized as either Supervised or Unsupervised learning. In the recent literature on CR [15], both supervised and unsupervised techniques have been proposed for various learning tasks. Unsupervised learning may particularly be suitable for diverse RF environment to make decisions and actions without prior knowledge. In this framework, we propose to use unsupervised

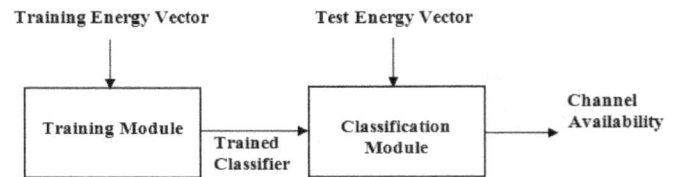

Figure 3. Schematic representation of learning module for proposed CSS scheme

K-means clustering algorithm to make cooperative decisions about channel availability. Before discussing the algorithm, it is necessary to look into the schematic representation of the learning module shown in Figure.3. It consists of training module and classification module. The training energy samples are fed into the training module which provides trained energy vectors to the classification module.

Generally, the training procedure of machine learning takes long time. To overcome this, the training module can be activated only during the initial CR deployment and any changes in primary network radio configurations. The classification module helps to determine the channel availability with the help of test energy vector. In order to achieve low classification delay, it is necessary to choose suitable classification algorithm with low complexity.

K-means clustering is an iterative, data partitioning algorithm that assigns number of observations to exactly one K clusters defined by centroids, where K is chosen before the algorithm starts. It partitions data into K mutually exclusive clusters, and returns the index of the cluster to which it has assigned each observation. It finds a partition in which objects within each cluster are as close to each other as possible, and as far from objects in other clusters as possible. Each cluster in the partition is defined by its member objects and by its Centroid. The centroid for each cluster is the point to which the sum of distances from all objects in that cluster is minimized. The Centroid of each cluster is used for classification. Once the classifier is trained, it is ready to receive test energy vectors for classification. K-means clustering aims to partition the observed energy vectors into K clusters (c1, c2,..ck) so as to minimize the distance of vectors within cluster by using distance measure. The partitioned clusters are passed using 'argmin' function as mentioned in equation (5).

$$\underset{c_1,c_2,\dots c_k}{argmin} \sum_{k=1}^{K} \sum_{Y^L \in C_k} \left\| Y^L - \alpha_k \right\|^2 \quad (5)$$

where Ck is the set of training energy vectors that belong to cluster K, Y^L is complete training

---

Algorithm 1: Local Sensing Based on Energy Detection

No_of_Nodes N;
Data: *energyDetection()*
Result: *energy vector*
begin
    for *user ← 1 to N* **do**
        Signal_at_node ← Primary_user_Signal + AWGN
        L ← size(Primary_User_Signal)
        Threshold ← qfuuncinv(Pf(user))/sqrt(L)+1
        Periodogram_at_node ← periodogram(Signal_at_node)
        Occupied[length(Periodogram_at_node)] ← 0
        while *i < lengthPeriodgram_at_node* **do**
            if *Periodgram_at_node i > Threshold* **then**
                occupied(i) ← 1
            i = i +1
        Channel_width ← 7 MHz
    Energy ← 0
    Sum ← 0
    if (*occupied* == 1) **then**
        Sum ← Sum + 1
        Energy ← Energy + Periodgram_at_node (freq)
    if *Sum > width/2* **then**
        Channel ← 1
    else
        Channel ← 0
Data: *changeVelocities(velocity_i, velocity_j, X)*
begin
    if (*mod(x, 4)* == 0) **then**
        Reverse the $Velocity_i$
    if (*mod(x, 2)* == 0) **then**
        Reverse the$Velocity_i$
    else if (*mod(x, 4)* == 1) **then**
        Reverse the $Velocity_j$
        if (*mod(x, 2)* == 0) **then**
            Reverse the $Velocity_j$
Data: *changeDistances(velocity_i, velocity_j, X, Y)*
begin
    $X ← X + Velocity_i$
    $Y ← Y + Velocity_j$
    if $(X, Y) > (100, 100)$ **then**
        $(X, Y) ← (X, Y) - 2 \times (Velocity_i, Velocity_j)$
    if $(X, Y) > (0, 0)$ **then**
        $(X, Y) ← (X, Y) + 2 \times (Velocity_i, Velocity_j)$

---

energy vectors, α_k is called Centroid of cluster K and $\|.\|^2$ is known as Square of Euclidean distance. After training, the classifier receives test energy vector for classification. The classifier classifies based on the following condition,

$$\frac{\|Y^* - \alpha_1\|}{min_{k=1,2,...K} \|Y^* - \alpha_k\|} \geq \beta \qquad (6)$$

where Y^* is known as test energy vector received by classifier, α_k is the Centroid for cluster K and β

Algorithm 2: Proposed CSS scheme using k-means clustering algorithm

---

Input energy(i,j) //Stores energy values of jth SU
for the ith band;
Initialize local decision(j,i);
$Y^L \leftarrow$ Training energy vectors;
$Y^{L,k} \leftarrow C_k$ //partitions training vectors into K
disjoint clusters (C);
$\alpha_k \leftarrow \mu_i$ //Initialize centre of cluster to determine
Centroid α_k, where $i = 1, 2, .., k$;

for *each cluster k* **do**
$\quad Y^{L,K} \leftarrow |\alpha_k|^- 1 \sum_{Y^L} \epsilon \alpha_k Y^L, \forall_k = 1, 2, .., k$
$\quad$ //calculating mean of all training energy
$\quad\quad$ vectors in cluster k
$\quad Distmeasure \leftarrow$ Euclidean $\|$ Cityblock
$\quad$ // for minimizing distance of energy vectors to
$\quad\quad$ local minima
$CH \leftarrow H_0|H_1$
// each SU reports its sensing decision to FC
$\quad CH \rightarrow$ global decision
// FC declares final decision based on suboptimal
solution through convergence

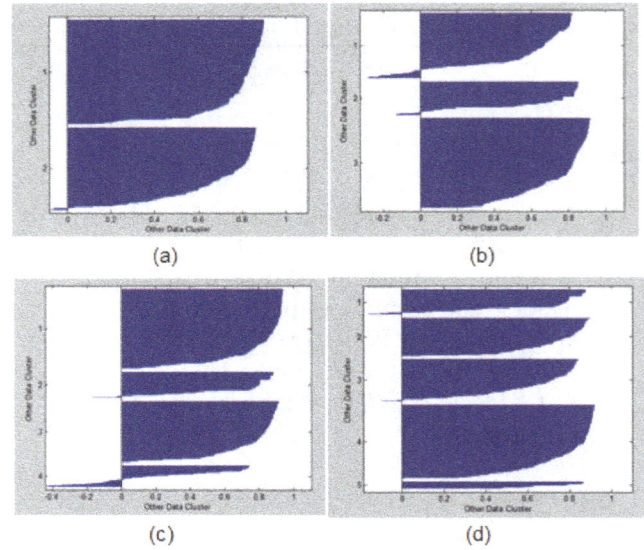

Figure 4. Silhouette plot for Euclidean distance measure: a)K=2 b)K=3 c)K=4 d)K=5

is called threshold to control trade-off between false alarm and detection probabilities. The algorithm works as follows. First, it Partitions the set of energy data into k disjoint clusters. The Centroid of first cluster (for which the class is available) is the mean of the data for which class is available. All the other data is divided into separate K clusters such that within squares sum of distances is minimized for all these K clusters. For the given training energy vectors, the data is first divided into two parts. One is for those for which the class is available, and the other for those for which class is unavailable. All the other data is divided into K clusters, where K varies from 1 to 10. For classification of test energy vectors, the classifier determines if the test energy vector belongs to cluster 1 or other clusters, based on the distance of the test energy vector to the centroids. We have considered two distance measures namely Euclidean and Cityblock. The Euclidean distance examines the root of square differences between coordinates of a pair of objects. Similarly, the cityblock distance examines the absolute differences between coordinates of a pair of objects. The classifier classifies the test vector as channel unavailable if the distance d is greater than β which is a tunable parameter. The value of this tuning parameter varies from 0.1 to 0.3 which indicates the permissible value of 'P_f' as per IEEE 802.22. The steps involved in unsupervised K-means clustering based CSS scheme is shown in Algorithm 2.

3. Simulation Setup and Results

The performance of the unsupervised K-means clustering algorithm for CSS has been analyzed by calculating delay of training as well as testing energy vectors and detection accuracy. We consider a CR simulation scenario with one primary transmitter and 10 SUs which operate in the frequency range of (54-698)MHz divided into 7MHz of channel bandwidth. Multiple secondary users are randomly deployed in a grid topology of area 120×120 Sq.km, using one FC. The distance coordinates of each SU varies during each iteration. The value of SNR for each SU changes based on the distance from the primary transmitter. The other important simulation parameters are as follows: Primary transmitter power is 200 MW, Primary signal type is BPSK modulated signal, Noise model is Additive White Gaussian Noise (AWGN). The simulation scenarios are performed using MATLAB 7.14 (R2012a) in a 64-bit computer with core i3 processor, clock speed 2.4 GHz, and 4GB RAM.

The performance of unsupervised K-means clustering algorithm for CSS scheme has been summarized on Table.1 and 2 using Euclidean and Cityblock distance metrics. The following observations can be made from the above summary. The variation of training delay with respect to number of clusters is less under Cityblock than Euclidean. There is less deviation on delay time for test energy vectors under both distance metrics. It is important to note that the detection accuracy remains same under both distance metrics. Also, the rate of detection accuracy satisfies the permissible limit given by IEEE 802.22.

Table 1. Performance Summary of k-means clustering based CSS scheme using Euclidean Distance metric

No of clusters	Training observation	Training Delay	Test observation	Test delay	Detection accuracy
2	1634	0.0097	86	0.0289	69.76
3	1634	0.0117	86	0.0470	69.76
4	1634	0.0157	86	0.0543	69.76
5	1634	0.0168	86	0.0750	69.76

Table 2. Performance Summary of k-means clustering based CSS scheme using Cityblock Distance metric

No of clusters	Training observation	Training Delay	Test observation	Test delay	Detection accuracy
2	1634	0.0116	86	0.0369	60
3	1634	0.0119	86	0.0506	60
4	1634	0.0121	86	0.0661	60
5	1634	0.0129	86	0.0757	60

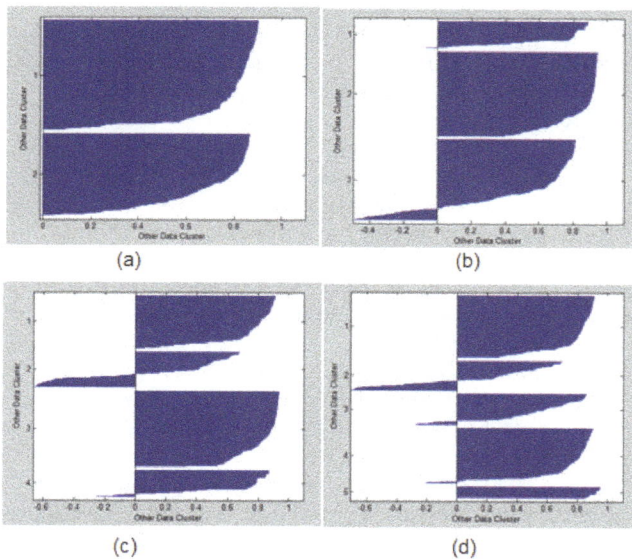

Figure 5. Silhouette plot for Cityblock distance measure: a)K=2 b)K=3 c)K=4 d)K=5

To get an idea of how well-separated the resulting clusters are, we can make a silhouette plot using the cluster indices output from K-means. The silhouette plot displays a measure of how close each point in one cluster is to points in the neighboring clusters. This measure ranges from +1, indicating points that are very distant from neighboring clusters called 'well-formed clusters', through 0, indicating points that are not distinctly in one cluster or another called 'ill-formed clusters', to -1, indicating points that are probably assigned to the wrong cluster called 'outliers'. Silhouette returns these values in its first output. The Silhouette plots using Euclidean distance metric for different values of K are shown in Fig.4.

The Silhouette graph shows two well-formed clusters, with a little fraction of data points as outliers. Since, most of the data points from both the clusters have their index greater than 0.6, this shows that the data points are tightly bound in the two clusters. With K = 3, clusters are formed from the data for which the channel is unavailable; the data points are scattered into 3 clusters as shown above. The Silhouette Graph show that the cluster 3 formed from above data is a well-formed cluster. Cluster 2 and Cluster 3 can also be thought of as well-formed, however the number of data points that are classified as outliers are more in these cases. With K = 4 clusters are formed from the data for which the channel is unavailable, the data points are scattered into 4 clusters as shown above. The Silhouette Graph shows that the cluster 4 formed from above data are an ill-formed data cluster. Cluster 1, 2 and 3 are well formed data clusters; however, cluster 2 has some ill classified points. With K = 5 clusters are formed from the data for which the channel is unavailable, the data points are scattered into 4 clusters as shown above. The Silhouette Graph shows that all the clusters formed with K = 5 are well-formed data clusters. Cluster 1 and 3 have some outliers classified under them, but all the other clusters have the index of most of the data-points above 0.6, which makes all of them distinct.

Similarly, the Silhouette plots for Cityblock distance measure are shown in Fig.5 for various cluster values. With K = 2 clusters are formed from the data for which the channel is unavailable, the data points are scattered into two clusters as shown above. The Silhouette Graph shows that both clusters are very well-formed with no outliers classified. Also, majority of data points in each of the clusters have their index greater than 0.6, which shows that the clusters have been formed tightly by the data points. With K = 3 clusters are formed from the data for which the channel is unavailable, the data points are scattered into 3 clusters as shown above. The Silhouette Graph shows that cluster 1 and 2 are well-formed, with cluster 1 having some outlier data points. However, cluster 3 has a large number of outliers, hence cannot be classified as well-formed. With K = 4 clusters are formed from the data for which the channel is unavailable, the data points are scattered into 4 clusters as shown above.

The Silhouette Graph shows that only cluster 1 and cluster 3 can be thought as well-formed. However, cluster 2 has large number of data points specified as outliers. Also, cluster 4 is not well-formed because of the outliers shown in the Fig.5. With K = 5 clusters are formed from the data for which the channel is unavailable, the data points are scattered into 5 clusters as shown above. The Silhouette Graph shows that only cluster 1 and cluster 5 are well-formed. However, cluster 5 is small as it contains less number of data points. All other clusters have outliers classified within them and hence cannot be thought of as well-formed.

4. Conclusion

In this paper, we presented cooperative spectrum sensing (CSS) scheme using unsupervised k-means clustering algorithm. The proposed CSS scheme has the capability to learn from the radio environment to achieve cognitive tasks. The received signal strength of each SU is measured using energy detection scheme and considered as feature input to the classifier module to determine channel availability. The simulation scenario has been formulated to meet the requirements of IEEE 802.22 WRAN standard. The simulation results show that the unsupervised k-means clustering algorithm significantly improves detection accuracy with training and testing delay of 16.8 and 75 milliseconds respectively. As future work, it can be extended further to various cooperation scenarios to support different wireless standards and specifications which will help to improve the cognition capability and cooperative sensing accuracy.

References

[1] Marcus M, Burtle J, Franca B, Lahjouji A, McNeil N. Federal communications commission spectrum policy task force. Report of the Unlicensed Devices and Experimental Licenses Working Group. 2002;.

[2] Mitola III J, Maguire Jr GQ. Cognitive radio: making software radios more personal. Personal Communications, IEEE. 1999;6(4):13–18.

[3] Haykin S. Cognitive radio: brain-empowered wireless communications. Selected Areas in Communications, IEEE Journal on. 2005;23(2):201–220.

[4] Akyildiz IF, Lee WY, Vuran MC, Mohanty S. NeXt generation/dynamic spectrum access/cognitive radio wireless networks: a survey. Computer Networks. 2006;50(13):2127–2159.

[5] Liang YC, Chen KC, Li GY, Mahonen P. Cognitive radio networking and communications: An overview. Vehicular Technology, IEEE Transactions on. 2011;60(7):3386–3407.

[6] Yücek T, Arslan H. A survey of spectrum sensing algorithms for cognitive radio applications. Communications Surveys & Tutorials, IEEE. 2009;11(1):116–130.

[7] Wang B, Liu K. Advances in cognitive radio networks: A survey. Selected Topics in Signal Processing, IEEE Journal of. 2011;5(1):5–23.

[8] Lunden J, Koivunen V, Poor HV. Spectrum Exploration and Exploitation for Cognitive Radio: Recent Advances. Signal Processing Magazine, IEEE. 2015;32(3):123–140.

[9] Akyildiz IF, Lo BF, Balakrishnan R. Cooperative spectrum sensing in cognitive radio networks: A survey. Physical communication. 2011;4(1):40–62.

[10] Chen X, Chen HH, Meng W. Cooperative communications for cognitive radio networksâĂŤfrom theory to applications. Communications Surveys & Tutorials, IEEE. 2014;16(3):1180–1192.

[11] Ghasemi A, Sousa ES. Collaborative spectrum sensing for opportunistic access in fading environments. In: New Frontiers in Dynamic Spectrum Access Networks, 2005. DySPAN 2005. 2005 First IEEE International Symposium on. IEEE; 2005. p. 131–136.

[12] Lee WY, Akyildiz IF. Optimal spectrum sensing framework for cognitive radio networks. Wireless Communications, IEEE Transactions on. 2008;7(10):3845–3857.

[13] Zhang W, Mallik RK, Letaief K. Optimization of cooperative spectrum sensing with energy detection in cognitive radio networks. Wireless Communications, IEEE Transactions on. 2009;8(12):5761–5766.

[14] Peh EC, Liang YC, Guan YL, Zeng Y. Cooperative spectrum sensing in cognitive radio networks with weighted decision fusion schemes. Wireless Communications, IEEE Transactions on. 2010;9(12):3838–3847.

[15] Ganesan G, Li Y. Cooperative spectrum sensing in cognitive radio, part I: Two user networks. Wireless Communications, IEEE Transactions on. 2007;6(6):2204–2213.

[16] Bkassiny M, Li Y, Jayaweera SK. A survey on machine-learning techniques in cognitive radios. Communications Surveys & Tutorials, IEEE. 2013;15(3):1136–1159.

[17] Thilina KM, Choi KW, Saquib N, Hossain E. Pattern classification techniques for cooperative spectrum sensing in cognitive radio networks: SVM and W-KNN approaches. In: Global Communications Conference (GLOBECOM), 2012 IEEE. IEEE; 2012. p. 1260–1265.

[18] Thilina KM, Choi KW, Saquib N, Hossain E. Machine learning techniques for cooperative spectrum sensing in cognitive radio networks. Selected Areas in Communications, IEEE Journal on. 2013;31(11):2209–2221.

[19] Abbas N, Nasser Y, El Ahmad K. Recent advances on artificial intelligence and learning techniques in cognitive radio networks. EURASIP Journal on Wireless Communications and Networking. 2015;2015(1):1–20.

[20] Balaji V, Kabra P, Saieesh P, Hota C, Raghurama G. Cooperative Spectrum Sensing in Cognitive Radios Using Perceptron Learning for IEEE 802.22 WRAN. Procedia Computer Science. 2015;54:14–23.

15

Cell Selection in Wireless Two-Tier Networks: A Context-Aware Matching Game

Nima Namvar[1], Walid Saad[2], Behrouz Maham[3]

[1,3]Electrical and Computer Engineering Department,University of Tehran,Tehran, Iran.
[2]Wireless@VT, Bradley Department of Electrical and Computer Engineering, Virginia Tech, Blacksburg, VA.

Abstract

The deployment of small cell networks is seen as a major feature of the next generation of wireless networks. In this paper, a novel approach for cell association in small cell networks is proposed. The proposed approach exploits new types of information extracted from the users' devices and environment to improve the way in which users are assigned to their serving base stations. Examples of such *context* information include the devices' screen size and the users' trajectory. The problem is formulated as a matching game with externalities and a new, distributed algorithm is proposed to solve this game. The proposed algorithm is shown to reach a stable matching whose properties are studied. Simulation results show that the proposed context-aware matching approach yields significant performance gains, in terms of the average utility per user, when compared with a classical max-SINR approach.

Keywords: Small Cell Networks, Context Information, User-Cell Association, Stable Matching

1. Introduction

Owing to the introduction of smartphones, tablets, and bandwidth-intensive wireless applications, the demand for the scarce radio spectrum has significantly increased in the past decade [1]. The concept of small cell networks (SCNs) is seen as a cost-effective and promising approach to cope with such an increasing demand. Indeed, the dense deployment of small cells, powered by low power, low cost base stations (BSs), is seen as a promising technique to improve the coverage and capacity of wireless cellular systems [2-4]. However, due to the presence of different categories of cells with diverse power, capacity, and range, the introduction of such heterogeneous SCNs leads to many technical challenges such as resource allocation, network modeling, interference mitigation, and network economics [5].

One important challenge in SCNs is that of cell association and handover [6]. Indeed, developing approaches to assign mobile users to their preferred small cell while also handling prospective handovers is necessary to achieve efficient SCN operation. Due to the diversity of coverage-range of the cells in SCNs, applying traditional approaches for user-cell association (UCA) in an SCN can lead to undesirable network performance and possibly increased handover failures [7].

In [7], a user association algorithm based on traffic transfer is introduced which aims at pushing the users onto the more lightly loaded cells in order to improve load balancing in small cell networks. This is achieved by proposing a novel sub-optimal solution for optimizing the long-term rate that each user experiences. The authors in [8] propose a novel UCA strategy by joint optimization of channel selection and power control for the purpose of minimizing the delay. The authors use an approach that is related to the sum of per-user SINR. The work in [9] proposes a flexible UCA method which aims at reducing the outage probability of the network. This is done by analyzing the received SINR form each tier, when the tiers are distributed randomly according to Poisson process. A new approach for UCA in the downlink of small cell networks is introduced in [10] for increasing the minimum average users' throughput which is based on an iterative algorithm that exploits the feedback information of the users. The authors in [11] and [12] proposed a load-aware cell association strategy which, by adjusting the transmit power, dynamically modifies the coverage area of the cells depending on their current load. This approach aims at balancing the load over neighboring macrocells. However, in small cell networks, one must balance the load over the various network tiers. A simple approach for user-cell association in small cell networks is proposed in [13]. In this approach, the authors use biasing factors for the transmit power of different tiers and attempt to distribute the traffic among the cells more fairly.

Strategies based on channel borrowing from lightly-loaded cells are studied in [14-16]. In these works, some resources of lightly-loaded cells will temporarily be used for servicing the users in a neighboring cell. However such channel-borrowing strategies have been proposed for cell association in macrocell-only networks and are not effective in small cell networks. Other related works can be found in [17-20].

Most of this existing literature assumes that the network makes resource allocation and cell association decisions based solely on physical layer parameters. Indeed, the current state-of-the-art often ignores the fact that the users can have different mobility patterns and diverse quality-of-service (QoS) demands. However, an effective and optimum UCA approach must be able to distinguish the individual properties of the users and, thus, be able to prioritize them based on their traffic type (i.e. urgent real-time traffic and delay tolerant traffic), QoS demands, and trajectory. For instance, a fast-moving user that is using a video application should be treated differently from a semi-static user who is downloading a file. Here, the QoS of the first user could be dramatically impeded by the slightest of delays, while the latter is relatively delay tolerant. We refer to such additional information about the users or the network as *context information*.

Thus, our main goal is to introduce a self-organizing approach for cell association in small cell networks, using which users and the network's cells can interact to decide on their preferred UCA in a way to optimize the overall network QoS. In particular, we propose a load-ware, application-aware approach for UCA which accounts for a plethora of context information including user mobility. Indeed, by exploiting context information from different network layers, we can develop a more efficient cell-association strategy which can lead to an improved network performance.

The main contribution of this paper is to introduce a novel context-aware UCA approach which employs useful information from different features of the network in order to optimize the network-wide QoS. In our proposed model, we explore a combination of several context information which, to best of our knowledge, have not been used by any other work for user association in small cells: trajectory and speed of the users, cells' load, quality of service requirements of the users, and the hardware specification of the user equipments. We show that by utilizing the mentioned combination of context information, the network can better decide on which user should be assigned to which cell. We model the UCA problem as a many-to-one matching game with externalies. To do so, we introduce novel and well-defined utility functions to capture the preferences of the users and cells. To solve the proposed matching game, we propose a novel iterative algorithm that converges to a stable matching between

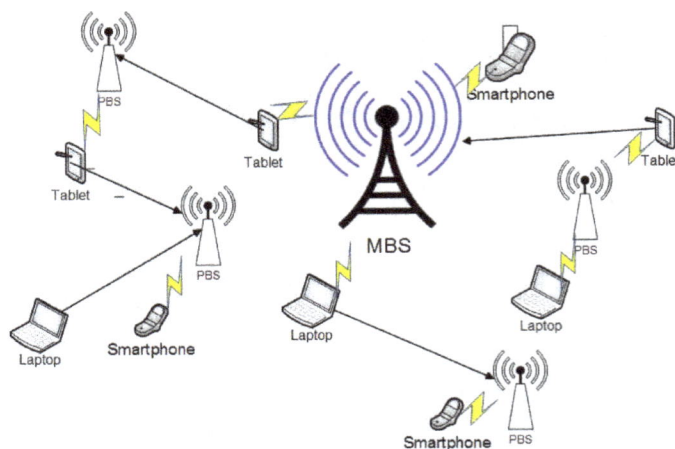

Figure 1. Users' mobility scenario in consideration

the set of users and the set of the network's cells. Simulation results show that the proposed matching-based approach yields considerable QoS improvement relative to classical, context-unaware UCA approaches. The results also show that the proposed algorithm converges in a reasonable number of iterations.

The rest of this paper is organized as follows: The system model is presented in Section 2. In Section 3, we formulate the user assignment problem in the framework of matching game with externalities and propose a novel algorithm to solve it. The performance of the proposed algorithm is assessed via simulations in the Section 4, and, finally, the conclusions are drawn in Section 5.

2. System Model

Consider the downlink of a two-tier wireless small cell network consisting of macrocells and picocells. Let $\mathcal{M}$, $\mathcal{P}$, and $\mathcal{N}$ denote the set of M macrocells, the set of P picocells, and the set of N users, respectively. Each small cell can serve a *quota* of up to q users simultaneously. We assume a wireless channel having slow multipath fading. Users are moving at low speeds and request service from the different small cells that they meet during their travel in the network. Figure 1 shows a typical small cell network in which the users are mobile. As shown in Figure 1, the communication sessions should be handed over between the neighboring cells.

Each user in the network has its own performance indicators such as the urgency of data, and the QoS demand which depends on the hardware specification of a user's device and the application type. Thus, as a first step toward developing the proposed model, we will explicitly discuss all the user context information that will be accounted for.

Screen Size: The screen size of the user equipment will affect the QoS perception of the user, especially for video-oriented applications. Indeed, user equipments with large screens have more sensitive QoS perception to a video's resolution than the smaller user equipments. We capture the impact of the screen size of each user $i \in \mathcal{N}$ using a parameter L_i that reflects the diameter length of each user's device. Devices with bigger screen size, are capable of showing the pictures with higher resolution which requires greater amount of network resources. Therefore, to satisfy the QoS demand of the devices with higher L_i, such as laptops or tablets, the network should allocate more resources to them relative to the smaller equipments such as smartphones.

Data Urgency: The resource requirements of the users naturally depend on their traffic patterns and application requirements. For example, the QoS of a live video streaming vitally depends on the delivery time since a small amount of delay could decrease the QoS dramatically. In contrast, the download of an Internet file may not be too susceptible to delay. By prioritizing the users based on their QoS needs, we are able to improve the average QoS for the users while also distributing the traffic among the cells more reasonably.

The QoS that each user experiences depends on the urgency of the user's data. Hence, we consider the QoS to be a function of delivery-time t. Naturally, for highly urgent data, the QoS will decrease more drastically as time elapses. Some suggestions to quantitatively model such behavior are presented in [21]. Consequently, for any user $i \in \mathcal{N}$, the QoS that reflects the data urgency can be given by:

$$Q_i(t) = \frac{1}{1 + e^{t - \tau_i}}, \qquad (1)$$

where τ_i is a parameter that reflects the urgency of the data. A smaller τ_i implies a more urgent data. This function shows that, within an interval of $2\tau_i$, the QoS drops to approximately $e^{-\tau_i}$ times of its initial value. This implies that only delivering the data before τ_i could be acceptable, and after that, the QoS becomes relatively small.

Handover Process: Due to the mobility of the users, the active communication sessions must be handed over between the cells. Figure 1 shows the handover scenario in consideration. A handover (HO) process cannot occur immediately when a user enters to the boundary of the cell as it requires some initial preparation time. Prior to that, no data could be handed over between two neighboring cells. To guarantee the connection of the users to the cells, the network must avoid risky HOs that could potentially incur a signal loss or erroneous communication. A handover failure occurs when the received signal to noise and interference ratio (SINR) drops under a certain threshold [19]. Therefore, one can use received SINR to determine the handover-failure circles. In particular, we will use the typical

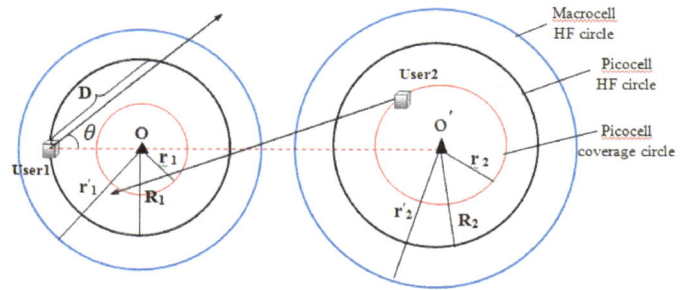

Figure 2. The handover failure and coverage regions

value of -6 dB as the threshold of the received SINR for the handover-failure circle [20]. Here, we study the probability of handover failure (HF) considering the users' speed and trajectory. It is assumed that all cells are equipped with omnidirectional antennas. We assume a circular coverage area for tractability. We note that the matching approach presented in Section 3 can easily accommodate other forms of coverage areas and mobility models.

In a two-tier network, one must consider two handover types: 1) from macrocell to picocell and, 2) from picocell to picocell. Assume that a user that has previously been served by a macrocell enters a picocell submits a request for handover. When user $i \in \mathcal{N}$ enters a picocell $j \in \mathcal{P}$, the total possible time of interaction between the user and the picocell, t_T^{ij}, could be computed as:

$$t_T^{ij} = \frac{2R_j \cos(\theta_i)}{V_i}, \qquad (2)$$

where R_j represents the radius of the coverage area, and θ_i is the angle of the user's direction with respect to the imaginary line connecting it to the center of the cell as shown in Figure 2. V_i is the user's average speed. Indeed, the numerator of (2) represents the length of the chord of the coverage circle that the user takes when it passes through the coverage area of the cell. Hereinafter, we assume that V_i is small enough that channel conditions remains constant during the handover and that the users have *low* to *medium* mobility. A successful HO process necessitates a certain preparation time of duration T_p before it could be initiated. Thus, based on the values of t_T^{ij} and T_p, we distinguish two different scenarios: 1) If $t_T^{ij} > T_p$, the user is considered as a *candidate* to be served; 2) If $t_T^{ij} < T_p$, the user is called a *temporary guest* and no HO would be initiated.

The users enter the picocell at an arbitrary direction. Therefore, θ is a random variable which is distributed uniformly in $(-\frac{\pi}{2}, \frac{\pi}{2})$. Assume D to be the length of the chord that the user takes. The cumulative distribution function (CDF) of D, $Pr(D < d)$, is equal to $2Pr(\theta > \cos^{-1}\left(\frac{d}{2R}\right))$. Therefore, given that θ has uniform distribution, the probability density function (pdf) of D, $f_D(d)$, can be given by:

$$f_D(d) = \frac{1}{\pi R \sqrt{1 - \frac{d^2}{4R^2}}}. \qquad (3)$$

A handover process fails when the user's path intersects with the handover failure (HF) circle. When the path is the tangent of the HF circle (with the radius r), D is equal to $2\sqrt{R^2 - r^2}$. Therefore, when $D \geq 2\sqrt{R^2 - r^2}$, the user's path intersects with the HF circle and the handover fails. Using (3), the probability of HF when a user enters from macrocell to picocell (M2P) can be derived as follows:

$$Pr_{HF}^{M2P} = \int_{2\sqrt{R^2-r^2}}^{2R} f_D(x)dx = \frac{2}{\pi}\cos^{-1}\left(\sqrt{1 - (\frac{r}{R})^2}\right). \qquad (4)$$

(4) shows that the probability of a handover failure is a function of $\frac{r}{R}$. Therefore, $\frac{r}{R}$ can be used as an indicator of the handover reliability. For example, assume that a handover could be initiated only if $Pr_{HF}(\frac{r}{R}) \leq 0.05$; then the next cell must hold this condition: $\frac{r}{R} \leq 0.08$. If the cell does not satisfy this condition, then, no handover should be initiated. Indeed, he HO process becomes more reliable as r becomes smaller relatively to R. The ratio of r to R varies from cell to cell and therefore, the different cells guarantee different levels of reliability during the handover process.

Now, assume that a user exits from picocell $j_1 \in \mathcal{P}$ and enters to another neighboring picocell $j_2 \in \mathcal{P}$ and sends a request for data handover. The handover process could be initiated once the user leaves j_1. However, it must be terminated before the user's distance from j_1 exceeds $r_1' > R_1$ and also before it enters the coverage of picocell j_2 to a distance of r_2. Let O and O' represent the centers of j_1 and j_2 respectively. Thus, OO' represents the distance between the two picocell base stations. To ensure a reliable and successful handover, only those cells which satisfy the inequality $R_1 + r_2 \leq OO' \leq r_1' + R_2$, must be considered for the handover.

The speed of the users can vary between two extremes V_{min} and V_{max}. In practice, as the small cells often do not have all the information on the mobility distribution, then, it would be reasonable to assume that the users' speed varies uniformly between these two extents [24]. The probability of handover failure when a user enters from picocell to another picocell (P2P) can be computed by subtracting the probability of successful handover from (1). For a successful handover, two independent conditions must be satisfied. First, the user should move slowly enough so that the handover in the first cell could be triggered. The probability of this event is given by $Pr(V < \frac{r_1'-R_1}{t_{m_1}})$.

Second, the path of the user should be in such a way that it does not intersect with the HF circle of the destination cell. Therefore, given that users' speed has a uniform distribution, the probability of handover failure is given by:

$$Pr_{HF}^{P2P} = 1 - \frac{\frac{r_1'-R_1}{T_{p_1}} - V_{min}}{V_{max} - V_{min}}\left(1 - \frac{2}{\pi}\cos^{-1}(\sqrt{1 - (\frac{r_2}{R_2})^2})\right). \qquad (5)$$

Now, considering the defined context information, in the next section, we formulate the UCA problem as a context-aware many-to-one matching game.

3. Cell Association as a Matching Game with Externalities

Originally introduced by Gale and Shapley in their seminal work [25], matching games are seen as a powerful and efficient framework to model conflicting objectives between two sets of players. Players of each set have a ranking, or preference, over the players in the opposite set. These preferences capture the objectives of players and the purpose of a matching game is to match the players of these two sets according to their preferences [26].

Among different types of matching games, the many-to-one matching scenario is especially suitable for the studied cell association problem because in this game, several players of one set can be matched with a single player of the other set. As an analogy to the many-to-one matching game, in the cell association problem several users can be assigned to a single cell. Here, using the context information introduced in the previous section, we can define proper utility functions to capture the preferences of users and small cells. Once this is done, the many-to-one matching model could be employed to assign the users to the cells based on each player's individual preferences and goals. In other words, using many-to-one matching games, we aim at maximizing the utility functions of users and small cells and thereby, optimizing the network-wide performance.

In the classical matching game introduced in [25-27], it is assumed that the preferences of the players are independent. However, this assumption does not hold in our model since the QoS metrics of the players are interdependent. In other words, as we can see from (6) and (7), the prospective utilities of the cells and users must depend on the current matching which itself depends on the preferences of the players. In such situations in which externalities affect the preferences of the players, the many-to-one matching game model with externalities is a promising approach to study the problem [28], [29]. However, there is no general solution for matching games with externalities as the general approach of Gale and Shapley cannot be generalized to this case. Therefore, introducing a novel approach which is tailored to specific nature of the proposed game is required. Indeed, the unique properties of our problem requires the introduction of a novel solution to the matching game which is tailored to the specific nature of the UCA problem.

Formally, the outcome of the UCA problem is a *matching* between two sets $\mathcal{N}$ and $\mathcal{P}$ which is defined as follows:

Definition 1. A *matching* μ is a function from $\mathcal{N} \cup \mathcal{P}$ to $2^{\mathcal{N} \cup \mathcal{P}}$ such that $\forall n \in \mathcal{N}$ and $\forall p \in \mathcal{P}$: (i) $\mu(n) \in \mathcal{P} \cup \emptyset$ and $|\mu(n)| \leq 1$, (ii) $\mu(p) \in 2^{\mathcal{N}}$ and $|\mu(p)| \leq q_p$, and (iii) $\mu(n) = p$ if and only if n is in $\mu(p)$.

The users who are not assigned to any member of $\mathcal{P}$, will be assigned to the nearest macrocell. Members of $\mathcal{N}$ and $\mathcal{P}$ must have strict, reflexive and transitive preferences over the agents in the opposite set. In the next subsections, exploiting the context information we introduce some properly-defined utility functions to effectively capture the preferences of each set.

3.1. Users' Preferences

Each user seeks to maximize its QoS requirements. Indeed, the users prefer those cells that are able to provide a reasonable delay while also meeting the QoS requirements as dictated by the application type and the screen size of each user's device. Users require a target rate $\hat{C}$ that reflects the type of applications which fits their screen size. Therefore, for each user $i \in \mathcal{N}$ with screen size L_i, we assign a target rate $\hat{C}_i(L_i)$ which quantifies the QoS requirement of the user. Moreover, the users seek to optimize their transmission rate which depends on the received power and the interference caused by neighboring small cells. Hence, those cells that are less congested and have higher transmission rate are prioritized by the users. In fact, the available amount of resources in a cell depends on the number of its current users, in such a way that the less congested the cell is, the more resources could likely be available. For each user i serviced by a small cell j, the utility function can be given by:

$$U_i^{user}(\mu, j, L_i) = \begin{cases} \left(\frac{C_i - \hat{C}_i(L_i)}{K_i}\right)^{\alpha_i} - \gamma_i(q_j - m_j) \\ \text{if } \hat{C}_i(L_i) \leq C_i, \\ \\ -\lambda_i \left(\frac{\hat{C}_i(L_i) - C_i}{K_i}\right)^{\beta_i} - \gamma_i(q_j - m_j) \\ \text{if } \hat{C}_i(L_i) > C_i, \end{cases} \quad (6)$$

where q_j is the quota of the small cell j, and m_j is the total number of users being served by it. L_i is the screen size of user i and $\hat{C}_i$ is the its target rate.

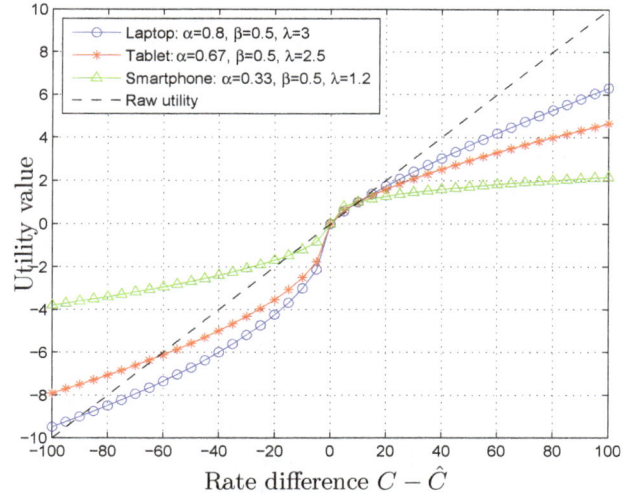

Figure 3. Utility of the users with different screen size

C_i represents the received rate of the user i which is equal to $W \log_2(1 + \frac{P_j c_{ij}}{\sum_{k \neq j} P_k c_{ik} + \sigma^2})$, where P_j is the power of small cell base station (SCBS) j, c_{ij} is the channel coefficient between user i and SCBS j, σ^2 is the power of additive noise, and W is the bandwidth. γ_i is the cost per unit traffic and α_i, β_i, λ_i and K_i are the coefficients that shape the utility function.

Figure 3 shows an example of the utility of a user for $\gamma = 0$. This illustrative example will show how each user, having different screen size, can perceive the rate gains. As we can see, for large-screen devices, such as laptops, the utility of the users is very sensitive to the received rate since a large screen allows users to better discern the quality of the application being used (e.g. video or multimedia). In contrast, the utility of the users with small screen size is not too susceptible to the received rate. Therefore, users on smartphones will overweight low rates (with respect to the reference $\hat{C}$), since the quality might be perceived as good, even though in reality it is below par. Moreover, because they are not capable of showing the pictures with extremely high resolution, receiving rates that are much higher than the target rate cannot change the utility of users with small screens significantly.

The value of m_j depends on the current matching, because it is the current matching that determines how many users are assigned to a specific small cell. As a result, the utility of each user is a function of current matching μ, as shown in (6). The first term in (6) captures the user's natural objective to maximize its transmission rate and the second term accounts for the fact that the users seek to find lightly loaded small cells to achieve more resources.

In fact, this utility function encourages the user to select lightly loaded cells and consequently, helps to offload the heavily-loaded cells by pushing the users to more lightly-loaded cells. Using (6), the users can rank the SCBSs in their vicinity based on the defined utility.

3.2. Small Cells' Preferences

The main goal of each small cell is to increase the network-wide capacity by offloading traffic from the macrocells while providing satisfactory QoS for the users. To decrease the number of total handovers, the small cells prefer the users which stay longer in the cell. The possible interaction time between the user and the cell depends on the speed and direction of the user. Clearly, users with lower mobility and a trajectory close to the cell's diameter would stay longer in the cell. On the other hand, to increase the network-wide QoS, the small cells must prioritize users having more urgent requests compared to those with less urgent ones.

By prioritizing the users coming from congested cells, the small cells could offload the heavily-loaded cells. To encourage the cells to prioritize the users coming from congested cells, we assume that each user is carrying a potential utility as a function of the pervious cell j' load, $f(\frac{m_{j'}}{q_{j'}})$. This utility depends on the current matching which determines the number of users in neighboring cells. We define the following utility that each SCBS $j \in \mathcal{P}$ obtains by serving an acceptable UE $i \in \mathcal{N}$:

$$U_j^{SCBS}\left(\mu, i, m_{j'}, q_{j'}\right) =$$
$$\frac{\cos(\theta_i)}{V_i}\left[1 + \log\left(\frac{\max(1, m_{j'})}{q_{j'}}\right)\right]\frac{1}{\tau_i}. \quad (7)$$

The first term in (7) allows to prioritize the users that stay longer in the cell. The second term accounts for the offloading concept, and the third term is the utility achieved by the SCBS j when serving a specific application. This utility function is well matched with the fact that a given small cell gains more utility by giving service to the users that are moving slower, having more urgent data, and coming from more congested cells. Thus, by doing so, the network could provide higher QoS and distribute the load more effectively.

From (6) and (7), we can see that the utilities depend on the current matching μ and consequently, the preferences of the players are interdependent. Under this condition, the preferences of players are not solely based on individuals, but some *externalities* affect the preferences and matching as well.

Definition 2. The preference relation $>_i$ of the user $i \in \mathcal{N}$ over the set of matchings $\Psi(\mathcal{N}, \mathcal{P})$ is a function that compare two matchings $\mu, \mu' \in \Psi$ such that:

$$\mu$$
$$\mu\,\mu' \Leftrightarrow U_i^{user}(\mu, j, L_i) > U_i^{user}(\mu', j, L_i). (8)$$

Table 1. Proposed Algorithm For The Matching Game

Input: context-aware utilities and the preferences of each set
Output: Stable matching between the users and SCBSs

Initializing: All the UEs are assigned to the nearest macro-BS

Stage I: Preference Lists Composition

- UEs and SCBSs exchange their context information
- UEs(SCBS) sort the set of acceptable candidate SCBSs(UEs) based on their preference functions

Stage II: Matching Evaluation
  while: $\mu^{(n+1)} \neq \mu^{(n)}$

- Update the utilities based on the current matching μ
- Construct the preference lists using preference relations $>_i$ and $>_j$ for $\forall i \in \mathcal{N}$ and $\forall j \in \mathcal{P}$
- Each user i applies to its most preferred SCBS
- Each SCBS j accept the most preferred applicants up to its quota q_j and create a waiting list while rejecting the others

  Repeat
    ● Each rejected user applies to its next preferred SCBS
    ● Each SCBS update its waiting list considering the new applicants and the pervious awaiting applicants up to its quota
  Until: all the users assigned to a waiting list
end

The preference relation for an SCBS j, $>_j$, is defined similarly. Users and SCBSs rank the members of the opposite set based on the defined preference relations. Our purpose is to match the users to the small cells so that the preferences of both side are satisfied as much as possible; thereby the network-wide efficiency would be optimized.

To solve a matching game, one suitable concept is that of a stable matching. In a matching game with externalities, stability has different definitions based on the application. Here, we consider the following notion of stability:

Definition 3. A matching μ is blocked by the user-SCBS pair (i,j) if $\mu(i) \neq j$ and if $j >_i \mu(i)$ and $i >_j i'$ for some $i' \in \mu(j)$. A many-to-one matching is *stable* if it is not blocked by any user-SCBS pair.

In the next section, we propose an efficient algorithm for solving the game that can find a stable matching between users and small cells.

3.3. Proposed Algorithm

The deferred acceptance algorithm, introduced in [26], is a well-known approach to solving the standard matching games. However, in our game, the preferences of the players as shown in (7) and (9), depend on externalities through the entire matching, unlike classical matching problems. Therefore, the classical approaches such as deferred acceptance cannot be used here because of the presence of externalities [28],[29]. To solve the formulated game, we propose a novel

Table 2. Typical values of data rate for different devices

Device type	Average screen size	Typical Data rate
Laptop	17"	1000 kbps
Tablet	10"	600 kbps
Smartphone	4.5"	400 kbps

algorithm shown in Table I. Assume that all the users are initially associated to the nearest macro base station (MBS). Each user sends its profile information (V, α, τ) to the neighboring SCBSs. Each SCBS, on the other side, only keeps the users satisfying (8) and ranks them based on their utilities (9). After ranking the acceptable UEs, the SCBS sends to the currently waiting users its own context information including its rate over load defined in (6) and its corresponding coverage and HF circle radii R and r.

Each user makes a ranking list of the available SCBSs and applies to the most preferred one. The SCBSs rank the applying users and keep the most preferred ones up to their quota and reject the others. The users who have been rejected in the former phase, would apply to their next preferred SCBS and the SCBSs modify their waiting list accordingly. This procedure continues until all the users are assigned to a waiting list.

However, since the preferences depend on the current matching μ, an iterative approach should be employed. In each step, the utilities would be updated based on the current matching. Once the utilities are updated, the preference lists would be updated accordingly as well. Therefore, in each iteration, a new temporal matching arises and based on this matching, the interdependent utilities are updated as well. The algorithm initiates the next iteration based on the modified preferences. The iterations will continue until two subsequent temporal matchings are the same and algorithm converges.

The proposed algorithm will lead to a stable matching when it converges, since by contradiction, the "deferred acceptance" in Stage II would not converge if the matching is not stable. Although a formal analytical proof of convergence for the proposed algorithm is difficult to derive, we make several observations that can help in establishing such a convergence. First, we note that in each iteration the "deferred acceptance" method in Stage II yields a temporary matching between the users and cells for any initial preferences [25], [26]. Following each iteration, the preferences are updated according to (5) and (6) which are functions of three main variables: the topology and speed of users, the channel conditions, and the current matching.

Second, in view of the fact that users have low mobility and experience a wireless channel with slow fading, we can assume that the network's topology and channel conditions remain almost constant during an algorithm run. As a result, we can conclude that in each

iteration the preferences are updated solely based on the current temporary matching. Therefore, since there is only a finite number of possible matchings between the users and their neighboring cells, the updating the preferences is not an endless process. In other words, there would be a limited number of iterations which beyond that, updating the preferences will either converge to a final, stable matching or cycle between a number of temporary matchings. However, here, we note two things: a) based on our thorough simulation results in Section 4, the case in which there is a cycling behavior only rarely occurs and b) under this case, we assume that the players can detect a cycle and stop the algorithm.

4. Simulation Results

For our simulations, we consider a single MBS with radius 1 km and overlaid by P uniformly deployed picocells. The transmit power of each picocell is 30 dBm and its bandwidth is $W = 200$ kHz. The small cells' quota is supposed to be a typical value $q = 4$ for all SCBSs [30]. The channels experience a Rayleigh fading with parameter $k = 2$. Noise level is assumed to be $\sigma^2 = -121$ dBm and the minimum acceptable SINR for the UEs is 9.56 dB [31]. There are N users distributed uniformly in the network. The QoS parameter τ_i in (1) is chosen randomly from the interval $[0.5, 5]$ ms. The users have low mobility and can be assumed approximately static during the time required for a matching. The speed of users varies between $20 km/h$ and $40 km/h$. Utility parameters in (6) are chosen in line with Figure 3. γ_i and K_i, are assumed to be 1 and 10 respectively, for all the users $i \in \mathcal{N}$. All the statistical results are averaged via a large number of runs over the random location of users and SCBSs, the channel fading coefficients, and other random parameters. The performance is compared with the max-SINR algorithm which is a well-known context-unaware approach used in wireless cellular networks for the UCA. In this approach, each user is associated to the SCBS providing the strongest SINR.

Figure 4 shows the average received rate per user for different number of SCBSs. As the number of SCBSs increases, the interference between the different cells increases. Therefore, the average rate that each user achieves will decrease. Figure 4 demonstrates that the proposed algorithm can lead to higher average rate per user in comparison to max-SINR approach reaching up to 66.7% gain for a network size of $P = 36$ SCBSs.

Figure 5 shows the average utility per different types of devices, for different number of SCBSs when the number of users is $N = 60$. According to (6), each user has a specific target rate tailored to its screen size. Typical values used for the target rates for three different types of devices are shown in Table

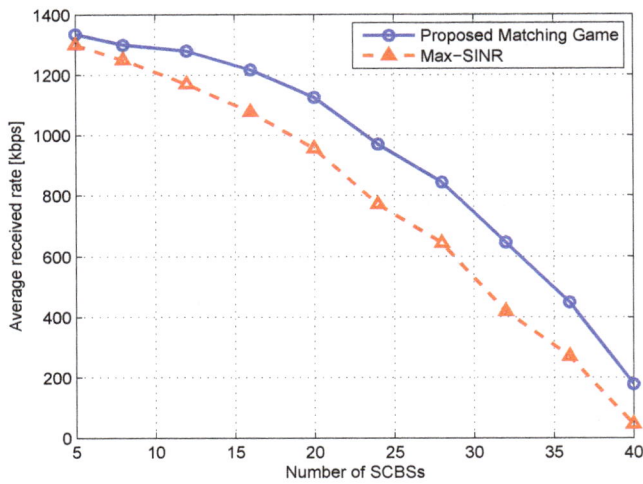

Figure 4. Average received rate per user for different number of SCBSs with $N = 60$ users.

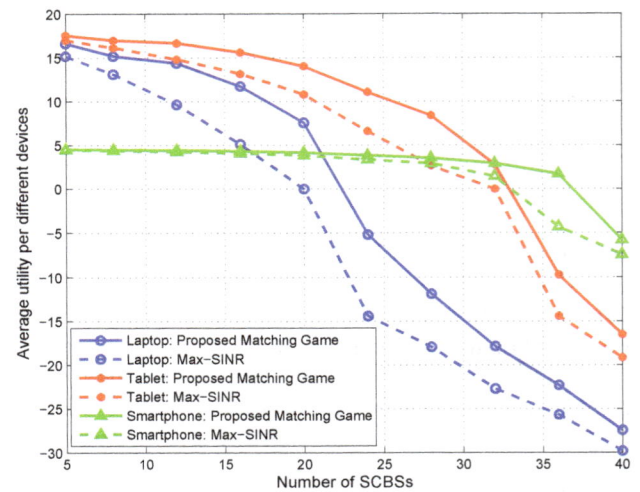

Figure 5. Average utility per different types of devices with $N = 60$ users.

2. Figure 5 shows that, for small-screen devices such as smartphones, the perceived utility of the user will not change dramatically if it receives a rate that is higher than its target rate. However, this utility for larger devices such as tablets and laptops is more sensitive to the received rate. From Figure 5, we can see that, when the number of SCBS is small and the average received rate is high, the utility of the laptops and tablets is greater than that of the smartphones because they are more sensitive to the received rate. However, as the number of the SCBSs increases and the network becomes more congested, the average received rate decreases and the utility of laptops and tablets decreases considerably, while the utility of the smartphones decreases very slowly. In Figure 5, we can see that, in general, for all types of devices, the proposed approach outperforms the conventional max-SINR approach.

Figure 6 shows the average utility per user for different number of SCBSs for $N = 60$ users. As the number of SCBSs increases, the average utility per user will decrease because the received rate will decrease due to the stronger interference. Although the cost for the traffic will also decrease (second term in (6)) when the number of SCBSs increases, but its effect is less than the effect of rate (first term in (6)). Figure 6 shows that the proposed algorithm outperforms the max-SINR algorithm for all network sizes. This performance advantage reaches up to 194% gain over to max-SINR criterion for a network with 24 SCBSs.

Figure 7 shows the average utility per user for different types of devices and, for different number of users when the number of SCBSs is $P = 15$. In Figure 7, we can see that, as the number of users increases, the average received rate per user will also increase. Therefore, the utility of the devices which is a function

of the received rate will increase as well. However, when the average received rate is small, devices with smaller screens have more utility relative to the ones with large screens. This is due to the fact that the small devices are not so sensitive to the rate since they are incapable of handling higher resolutions. Similar to Figure 5, in Figure 7, we can see that devices with larger screen size are more susceptible to the received rate, i.e. the distance from the BS. In fact, as the rate increases, we can see that the devices with large screen size such as laptop, achieve more utility in comparison to the small devices, since they are so sensitive to the rate and an increase in the received rate can increase their QoS considerably. We can see from Figure 7 that the proposed algorithm has noticeable gain over the max-SINR approach and can reach up to 4%, 32%, and 87.5% gain over the max-SINR criterion for the smartphones, tablets, and laptops respectively.

Figure 8 shows average utility per user for different number of users with $P = 15$ SCBSs. As the number of users increases, the average received rate will also increase which leads to an increase in the average user's utility. Figure 8 demonstrates that at all network sizes, the proposed approach has a performance advantage over max-SINR. The average gain of the proposed approach over the max-SINR scheme is 39.4%.

Figure 9 shows the average utility per user for different percentage of the smartphones for a network size of $N = 60$ users and $P = 20$ SCBSs. As the percentage of the smartphones increases from 50% to 100%, the gain of the proposed approach relative to max-SINR scheme decreases from 113% to 9%. This is directly related to the features of the smartphones. In fact, devices with small screen size are not very sensitive to the received rate, therefore, the proposed context-aware UCA algorithm which aims

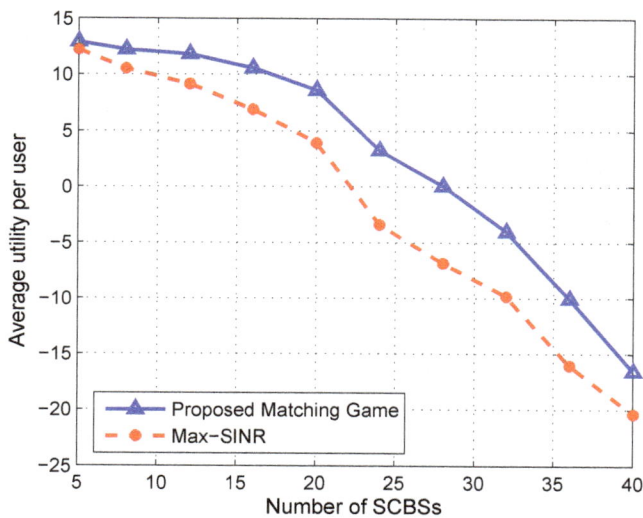

Figure 6. Average utility per user for different number of SCBSs with $N = 60$ users.

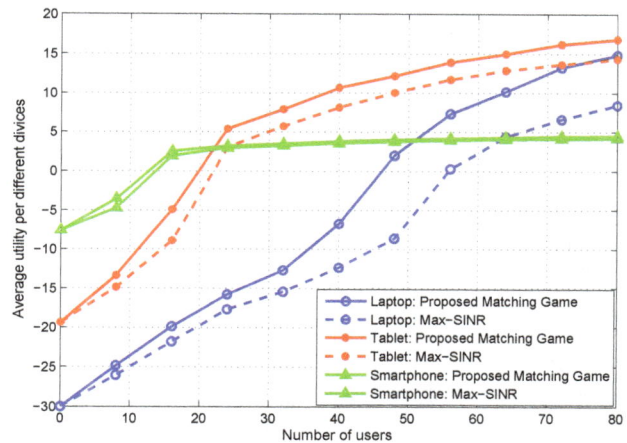

Figure 7. Average utility per different types of devices with $P = 15$ SCBSs.

at optimizing the received rate of the devices will not have considerable gains over the context-unaware max-SINR approach when the network encompasses devices with small screens only. Conversely, when the network has considerable percentage of laptops and tablets which are very sensitive to the received rate, then the proposed context-aware approach yields significant gain over the max-SINR because the proposed algorithm prioritize the devices based on their QoS demands and requirements.

In Figure 10, we show the average utility achieved by each SCBS as a function of the number of users for $P = 15$ SCBSs. As the number of users N increases, the network becomes more congested, and the probability that a new user who applies for an SCBS is coming from a congested BS increases. Therefore, it is more likely for the SCBSs to gain more utility by offloading the network. However, when the network is considerably congested, the new users that arrive to the network would be mostly assigned to the MBS, since many of SCBSs have already reached their maximum capacity. Figure 10 shows that, at all network sizes, the proposed algorithm achieves significant gains over the max-SINR approach that reach up to 72.8% gain for a network size of 40.

Figure 11 shows the average number of iterations per user required for the algorithm to converge to a stable matching for two different network sizes, as the number of users varies. In this figure, we can see that the number of algorithm iterations is an increasing function of the number of users and the number of SCBSs. Figure 10 shows that the average number of iterations varies from 1.09 and 1.1 at $N = 3$ to 8.3 and 9.7 at $N = 80$, for the cases of 15 SCBSs and 20 SCBSs, respectively. Clearly, Figure 11 demonstrates that the proposed

algorithm converges within a reasonable number of iterations and scales well with the network size.

5. Conclusions

In this paper, we have proposed a new context-aware user association algorithm for the downlink of wireless small cell networks. By introducing well-designed utility functions, our approach accounts for the trajectory and speed of the users as well as for their heterogeneous QoS requirements and their hardware specifications. We have modeled the problem as a many-to-one matching game with externalities, where the preferences of the players are interdependent and contingent on the current matching. To solve the game, we have proposed a novel algorithm that converges to a stable matching in a reasonable number of iterations. Simulation results have shown that the proposed approach yields considerable gains compared to max-SINR approach.

6. Acknowledgements

This work was supported by the U.S. National Science Foundation under Grant CNS-1253731.

References

[1] Cisco, "Cisco visual networking index: Global mobile data traffic forecast update, 20102015," Whitepaper, Feb. 2011.

[2] A. Damnjanovic, J. Montojo, Y. Wei, T. Ji, T. Luo, M. Vajapeyam, T. Yoo, O. Song, and D. Malladi, "A survey on 3gpp heterogeneous networks," IEEE Wireless Communications, vol. 18, p. 1021, June 2011.

[3] J. G. Andrews, H. Claussen, M. Dohler, S. Rangan, and M. C. Reed, "Femtocells: Past, Present, and Future," IEEE

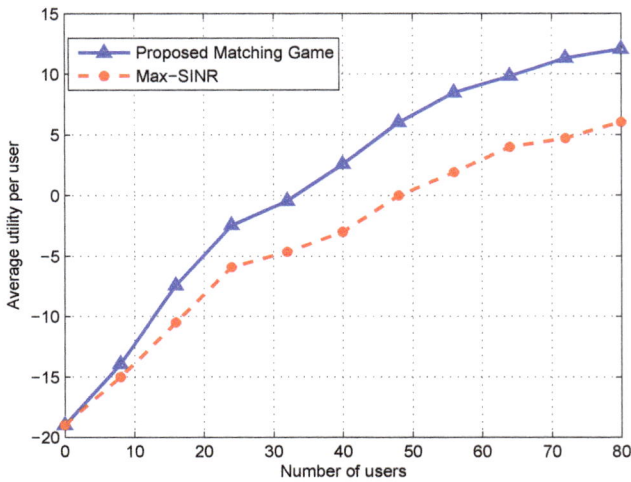

Figure 8. Average utility per user for different number of users with $P = 15$ SCBSs.

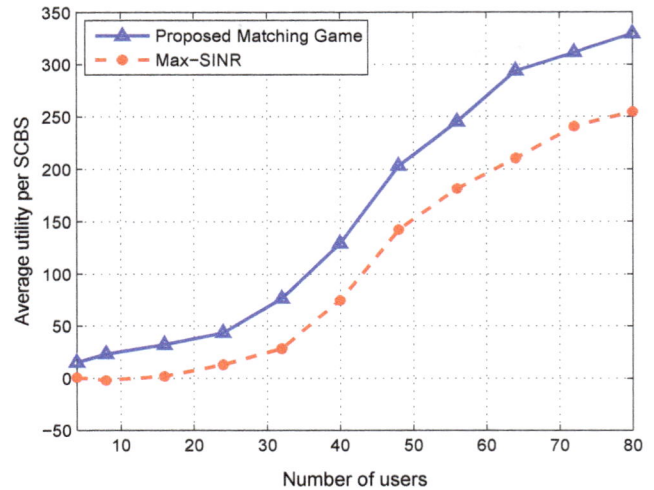

Figure 10. Average utility per SCBS for different number of users with $P = 15$ SCBSs.

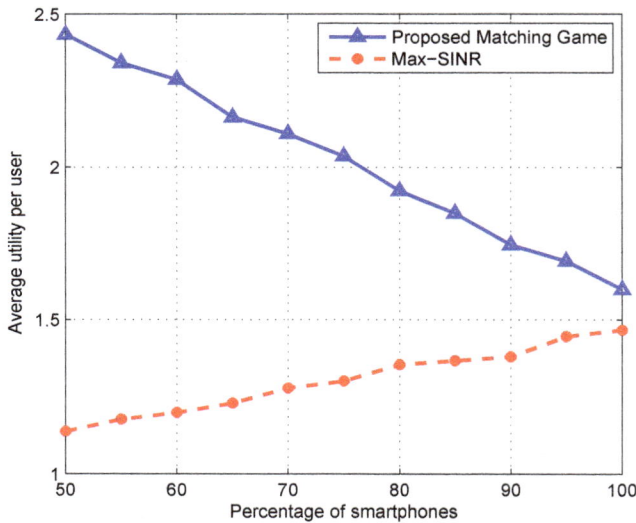

Figure 9. Average utility per user for different percentage of the smartphones with $N = 60$ users and $P = 20$ SCBSs.

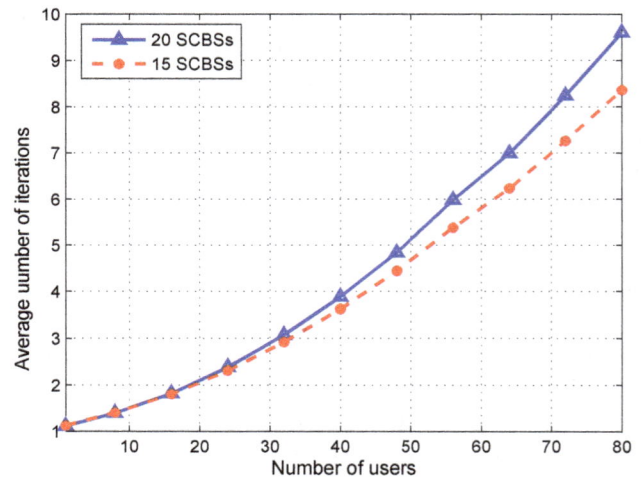

Figure 11. Average number of algorithm iterations for reaching a stable matching, for different number of users with $P = 15$ and $P = 20$ SCBSs.

Journal on Selected Areas in Communications, vol. 30, no. 3, pp. 497- 508, 2012.

[4] T. Q. S. Quek, G. de la Roche, I. Guvenc, , and M. Kountouris, "Small Cell Networks: Deployment, PHY Techniques,and Resource Management", London, England: Cambridge University Press, 2013.

[5] A. Ghosh, N. Mangalvedhe, R. Ratasuk, B. Mondal, M. Cudak, E. Visotsky, T. A. Thomas, J. G. Andrews, P. X. H. S. Jo, H. S. Dhillon, and T. D. Novlan, "Heterogeneous cellular networks: From theory to practice", Communications Magazine, IEEE, vol. 50, no. 6, pp. 54-64, Jun. 2012.

[6] H. S. Dhillon, R. K. Ganti, F. Baccelli, , and J. G. Andrews, "Modeling and analysis of k-tier downlink heterogeneous cellular networks," IEEE Journal on Selected Areas in

Communications, vol. 30, pp. 550-560, Apr. 2012.

[7] Q. Ye, B. Rong, Y. Chen, M. Al-Shalash, C. Caramanis, and J. G. Andrews, "User association for load balancing in heterogeneous cellular networks," Communications Magazine, IEEE.

[8] C. Chen, F. Baccelli, and L. Roullet, "Joint optimization of radio resources in small and macro cell networks", in Proc., IEEE Veh. Technology Conf., pp. 1-5, IEEE, May 2011.

[9] H. S. Jo, Y. J. Sang, P. Xia, and J. G. Andrews, "Heterogeneous cellular networks with flexible cell association: A comprehensive downlink sinr analysis.", IEEE Transactions on Wireless Communications, vol. 11, p. 10, Oct. 2012.

[10] T. Koizumi and K. Higuchi, "A simple decentralized cell association method for heterogeneous networks", International Symposium on Wireless Communication Systems (ISWCS), pp. 256-260, 2012.

[11] S. Das, H. Viswanathan, and G. Rittenhouse, "Dynamic load balancing through coordinated scheduling in packet data systems", in Proc., IEEE INFOCOM, vol. 1, pp. 786âĂŞ796, Apr. 2003.

[12] Y. Bejerano and S. J. Han, "Cell breathing techniques for load balancing in wireless LANs", IEEE Transactions on Mobile Computing, vol. 8, pp. 735âĂŞ749, June 2009.

[13] A. B. Saleh, O. Bulakci, S. Redana, B. Raaf, and J. Hamalainen, "Enhancing LTE-advanced relay deployments via biasing in cell selection and handover decision", in 2010 IEEE 21st International Symposium on Personal Indoor and Mobile Radio Communications, pp. 2277âĂŞ2281, Sep. 2010.

[14] T. Kahwa and N. Georganas, "A hybrid channel assignment scheme in large-scale, cellular-structured mobile communication systems", IEEE Trans. on Communications, vol. 26, pp. 432âĂŞ438, Apr. 1978.

[15] H. Jiang and S. Rappaport, "CBWL: A new channel assignment and sharing method for cellular communication systems", IEEE Trans. on Veh. Technology, vol. 43, pp. 313âĂŞ322, May 1994.

[16] S. K. Das, S. K. Sen, and R. Jayaram, "A dynamic load balancing strategy for channel assignment using selective borrowing in cellular mobile environment", Wireless Networks, vol. 3, pp. 333âĂŞ347, Oct. 1997.

[17] R. Madan, J. Borran, A. Sampath, N. Bhushan, A. Khandekar, and T. Ji, "Cell Association and Interference Coordination in Heterogeneous LTE-A Cellular Networks," Selected Areas in Communications, IEEE Journal on, vol. 28, no. 9, pp. 1479- 1489, 2010.

[18] X. Wu, B. Mukherjee, and S. H. G. Chan, "Maca-an efficient channel allocation scheme in cellular networks," in Global Telecommunications Conference (GlobeCom), pp. 1385-1389, 2000.

[19] E. Yanmaz and O. K. Tonguz, "Femtocells: Past, Present, and Future," IEEE Journal on Selected Areas in Communications, vol. 22, pp. 862-872, Jun 2004.

[20] D. Cavalcanti, D. Agrawal, C. Cordeiro, B. Xie, and A. Kumar, "Issues in integrating cellular networks wlans, and manets: a futuristic heterogeneous wireless network," IEEE Wireless Communications Magazine, vol. 12, pp. 30-41, Jun 2005.

[21] M. Proebster, M. Kaschub, and S. Valentin, "Context-aware resource allocation to improve the quality of service of heterogeneous traffic," in IEEE International Conference on Communications (ICC), pp. 1-6, Jun 2011.

[22] D. Lopez-Perez, I. Guvenc, and C. Xiaoli, "Theoretical analysis of handover failure and ping-pong rates for heterogeneous networks" in IEEE International Conference on Communications (ICC), pp. 6774-6774, Jun 2012.

[23] TR 36.839, "Mobility Enhancements in Heterogeneous Networks," 3GPP Technical Report, June 2011.

[24] A. Papoulis, S. U. Pillai, "Probability, Random Variables and Stochastic Processes" Mc Graw Hill, 2002.

[25] D. Gale and L. S. Shapley, "College Admissions and the Stability of Marriage" Amer. Math. Mon., vol. 69, no. 1, pp. 9-14, 1962.

[26] A. E. Roth and M. A. O. Sotomayo, "Two-Sided Matching: A Study in Game-Theoretic Modeling and Analysis," Cambridge University Press, 1992.

[27] H. Xu, B. Li, "Seen As Stable Marriages," IEEE INFOCOM conference, 2011.

[28] A. Salgado-Torres, "Many to one matching: Externalities and stability," Universidad Carlos III de Madrid, 2011.

[29] K. Bando, "Many-to-one matching markets with externalities among firms," Journal of Mathematical Economics, vol. 48, no. 1, pp. 14-20, 2012.

[30] D. Knisely, T. Yoshizawa, and F. Favichia, "Standardization of femtocells in 3gpp," IEEE Communication Magazine, vol. 47, p. 6875, Oct. 2009.

[31] S. Sesia, I. Toufik, and M. Baker, "LTE - the UMTS long term evolution," UK: John Wiley and Son publication., 2009.

16

Distance Based Method for Outlier Detection of Body Sensor Networks

Haibin Zhang[1], Jiajia Liu[1,*], Cheng Zhao[1]

[1]School of Cyber Engineering,hbzhang@mail.xidian.edu.cn, Xidian University, Xi'an 710071, P.R. China

Abstract

We propose a distance based method for the outlier detection of body sensor networks. Firstly, we use a Kernel Density Estimation (KDE) to calculate the probability of the distance to k nearest neighbors for diagnosed data. If the probability is less than a threshold, and the distance of this data to its left and right neighbors is greater than a pre-defined value, the diagnosed data is decided as an outlier. Further, we formalize a sliding window based method to improve the outlier detection performance. Finally, to estimate the KDE by training sensor readings with errors, we introduce a Hidden Markov Model (HMM) based method to estimate the most probable ground truth values which have the maximum probability to produce the training data. Simulation results show that the proposed method possesses a good detection accuracy with a low false alarm rate.

Keywords: outlier detection; body sensor networks; sliding window

1. Introduction

The improvement of living standards, unreasonable diet, excess energy, environmental pollution and other factors enable chronic diseases developing more quickly. This leads to a shortage of qualified healthcare professionals and equipments to treat the sick and needy persons. The wireless body sensor network (BSN) is one of solutions to this problem. BSNs use wireless devices attached to or implanted in the body to collect various vital signs such as heart rate (HR), oxygen saturation (SpO2), blood pressure (BP), etc, and transmit collected data to a central device for processing. This allows real-time monitoring and early detection of clinical deterioration, and greater freedom and mobility while maintaining the quality of medical care [9].

Wireless devices are restricted by resources. In addition, they are frequently susceptible to environmental effects, vulnerable to the malicious, which lead to unreliability sensor data. However, medical applications have strict requirements for reliability to avoid false alarm, so outlier detection is extremely important to ensure the reliability and accuracy of sensor data before the decision-making process [9].

Outlier detections in wireless sensor networks have been studied for many years [1–4]. They estimated sensor readings or probabilities of sensor readings using spatial correlation in measurements at different sensors for the outlier detection of wireless sensor networks. These methods have an assumption that there are a large number of sensors used for the same events detection, a sensor value can be deduced from other sensors' values. However, they are not suitable for body sensor networks, because it is difficult to put too many sensors on body and it is usual that different sensors are used to collect different vital signs.

Kim *et al*. in [5] gave an approach for motion outlier detection in body sensor networks, they used history data to train Gaussian Mixture Model to generate clusters of data in similar motion groups, these cluster of data are used to estimate a Gaussian distribution to compute fault probability for new node reading. These methods can detect faults that are deviated largely to the normal data. For those faults whose readings are normal values without a low probability these methods may have a bad performance.

Chen *et al*. in [6] diagnosed abnormal data of time series of multivariate variables by two steps. Firstly, they suppose that there is a relationship between the variables represented by an expression. Then a data is diagnosed as faulty if its deviation to the estimated value by this expression is larger than a threshold.

Similarly, Salem *et al.* in [8] tried to use a linear regression to estimate the reading of a sensor, e.g. HR, by values of its neighbour sensors. However, there may not be relationship between variables, and it can not decide which data is faulty when some fault occurs.

Salem *et al.* in [9] used a Kernel function to estimate the the distribution of the distance between a sensor reading to the mean of training data. A data instance is diagnosed as faulty if the probability of the calculated distance is very low. Rajasegarar *et al.* in [13],[12] used a naive Bayes based method for outlier detection. This method simply calculated the frequency of each attribute to estimate the probability. However, the Bayesian model is established on the basis of independence hypothesis, which assumed that the attributes of data are independent. This is not always true, and may affect the results of the classification.

In this paper, we propose a distance based outlier detection method for BSNs. Firstly, we calculate the average distance to k nearest neighbors of training data to estimate a KDE. For any diagnosed data, we calculate the probability of its distance to k nearest neighbors using the KDE, if this probability is less than a threshold, then the diagnosed data may be an outlier. Then we check the distances of the diagnosed data to its left and right neighbors, if both of these distances are great, then we decide the diagnosed data as an outlier.

In some conditions like that the sensor readings have successive outliers, then the previous method may have a poor performance. We introduced a sliding window to this issue. Similarly, we calculate the probability of the distance to k nearest neighbors for a diagnosed sliding window with the estimated KDE, if this probability is less than a threshold, then we decide that there are some outliers in this window. Then we check the distance to the left and right neighbors to locate the outlier.

Estimating the KDE with training data is the key issue of the proposed method for outlier detection. However, the history training data containing errors can disrupt the estimated values of the KDE. We use the Hidden Markov Model (HMM) to estimate the most probable ground truth values which have the maximum probability to produce the training sensor readings.

The rest of this paper is organized as follows. Section II introduces system models and some definitions. In section III, the distance based outlier detection method is formalized. In section IV, experiments are carried out to test the performance of the proposed method. Conclusions are drawn in Section V.

2. System Models and Definitions

Fig. 1 shows the network architecture of our considering medical deployment scenario. We use three sensors to monitor heart activities, blood pressure, respiration rate and saturation of oxygen in the arterial

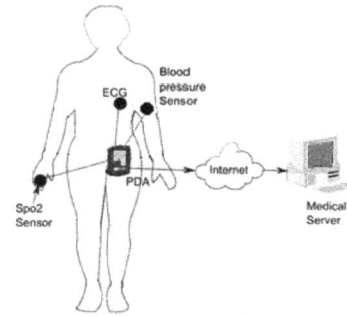

Figure 1. The Architecture of BSNs.

blood. These sensors monitor vital signs and transmit the collected data periodically at every discrete time instance to neighboring personal server devices, such as a smartphone. Then by wireless and wired connection, these data are streamed remotely to a medical doctor's site for real time diagnosis, to a medical database for record keeping, or to the corresponding equipment that issues an emergency alert.

Time series of vital sign data: It is a sequence of vital sign data arranged in time order $X = (X_a, X_{a+1}, \cdots, X_c)$, where $X_i = (x_{ih}, x_{ib}, x_{is})$ is the set of sensor readings of HR, BP and SpO2. The main purpose of time series analysis is to diagnose the current sensor readings based on the existing historical data. At any time t, suppose the ground truth values of HR, BP and SpO2 are $G_t = (g_{th}, g_{tb}, g_{ts})$, the measured values transmitted from sensors are $X_t = (x_{th}, x_{tb}, x_{ts})$, the outlier detection process decides whether x_t accords with G_t. However, for outlier detection process, the difficult is that we have no way to know the ground truth values G_t.

2.1. Kernel Density Estimation

Kernel density estimation is a non-parametric way to estimate the probability density function of a random variable. Let $(y_1, y_2, \cdots, y_n)$ be an independent and identically distributed sample obtained from some distribution with an unknown density function f. The shape of function f can be estimated as

$$\widehat{f_h}(y) = \frac{1}{n} \sum_{i=1}^{n} K_h(y - y_i) = \frac{1}{nh} \sum_{i=1}^{n} K(\frac{y - y_i}{h}) \quad (1)$$

where $K(\bullet)$ is a non-negative function called the kernel that integrates to one and has mean zero, $h > 0$ is a bandwidth. $K_h(x)$ is a kernel with subscript h is given as

$$K_h(x) = \frac{1}{h} K(\frac{x}{h}) \quad (2)$$

2.2. Hidden Markov Model

A hidden Markov model is a 5-tuple $\lambda = (Q, V, \Pi, A, B)$, where $Q = \{q_1, \cdots, q_n\}$ is a set of hidden states with s_t denoting the state at time t, $V = \{v_1, \cdots, v_m\}$ is a set of observation symbols with o_t denoting the symbol at time t, $\Pi = \{\pi_1, \cdots, \pi_n\}$ is a vector of initial probabilities with $\pi_i = P(s_1 = q_i)$, A is a matrix $(a_{ij})_{(n \times n)}$ of transition probabilities with each $a_{ij} = P(s_{t+1} = q_j s_t = q_i)$, $1 \leq i, j \leq n$, B is matrix $(b_{ij})_{(n \times m)}$ of observation probabilities with each $b_{ij} = P(o_t = v_j s_t = q_i)$. We also use π_{s_1}, $a_{s_t s_{t+1}}$ and $b_{s_t o_t}$ to denote π_i, a_{ij} and b_{ij} respectively.

3. Outlier Detection Based on Distance

3.1. Simple Distance to Neighbors

We use the Euclidean distance to calculate the distance between two multivariate data. Let $X_i = (x_{ih}, x_{ib}, x_{is})$ and $X_j = (x_{jh}, x_{jb}, x_{js})$ be sensor readings on time i and j, the Euclidean distance between X_i and X_j is

$$d_{ij} = \sqrt{(x_{ih} - x_{jh})^2 + (x_{ib} - x_{jb})^2 + (x_{is} - x_{js})^2} \quad (3)$$

The k nearest distance of X_i is calculate as follows.

$$d_i^k = \sum_{i-\lfloor k/2 \rfloor \leq j < i+\lceil k/2 \rceil} w_j d_{ij} \quad (4)$$

where w_j is a weight. The distance of X_i to its left and right neighbors (the nearest distance for short) is

$$d_i^2 = d_{i(i-1)} + d_{i(i+1)}$$

Given a time series of history training sensor readings, we firstly calculate the k nearest distance of each data X_i, then the univariate KDE is used to estimate the probability distribution of these k nearest neighbors distance. For a recently produced sensor readings X_t, the k nearest neighbors distance d_t^k of X_t and the probability p of d_t^k is calculated by the KDE obtained by history training data. If p is less than a threshold, the the nearest distance d_t^2 of X_t is calculated, if d_t^2 is greater than a pre-defined value, then X_t is diagnosed as an outlier.

If some error occurs at time t, then the k nearest distance data at time closing to t, e.g. time $t-1$ or $t+1$, may have a small probability. The similar condition exists in the nearest distance. This leads to that the average rate of outliers newly introduced will be high using the k nearest distance or the nearest distance based method alone for outlier detection. Thus, we use the combination of these two method for outlier detection in this paper.

Fig. 4 shows a simulation result of 5000 HR data from a real medical dataset using the proposed outlier

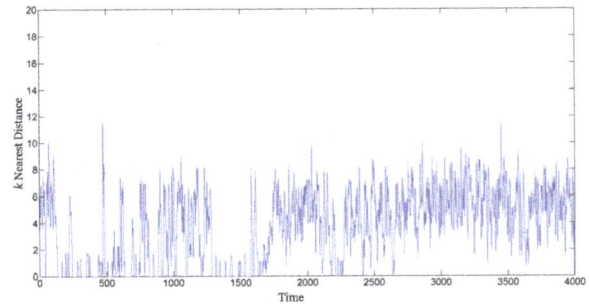

Figure 2. k Nearest Distance of Training Data.

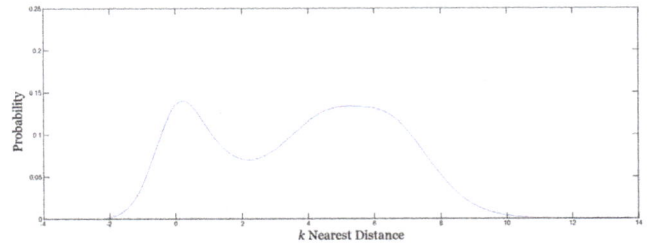

Figure 3. Probability Density Distribution of The k Nearest Distance.

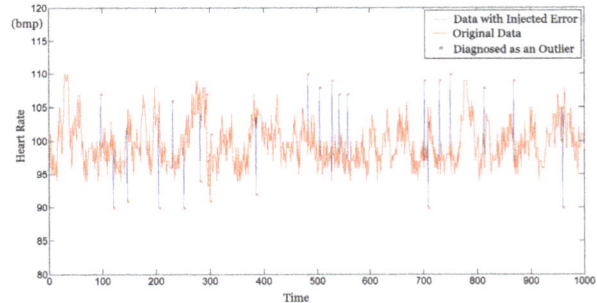

Figure 4. Outlier Detection.

detection method. We use the first 4000 data to estimate the KDE. The k nearest distance of the 4000 training data is shown in Fig. 2, where k is selected as 2, and the kernel density estimation of the distance in Fig. 2 is depicted in Fig. 3. Then we inject 5% errors to the rest of 1000 data. The original data, the data with injected errors, and the data diagnosed as outlier are all marked in Fig. 4.

3.2. Distance to Neighbors of Sliding Window

For some condition like that the time series of sensor readings has too much successive outliers, then the detection method in the above subsection may have a poor performance. The reason is that the k nearest neighbors of an outlier X_t may be contained in the successive outliers. This leads to the k nearest distance of X_t having a normal probability, and X_t is diagnosed as a normal data. To this problem, we improve the distance based outlier detection method on a sliding window.

A sliding window with width m at time t contains X_t and its left $m-1$ nearest neighbors $X_{t-m+1}, X_{t-m+2}, \cdots, X_t$, that is

$$B_t = \begin{pmatrix} x_{(t-m+1)h}, & x_{(t-m+1)b}, & x_{(t-m+1)s} \\ x_{(t-m+2)h}, & x_{(t-m+2)b}, & x_{(t-m+2)s} \\ & \cdots & \\ x_{th}, & x_{tb}, & x_{ts} \end{pmatrix}$$

We define a distance D_{ij} between sliding window B_i and B_j as

$$D_{ij} = \frac{1}{m} \sum_{l=0}^{m-1} d_{(i-l)(j-l)} \qquad (5)$$

where $d_{(i-l)(j-l)}$ is the Euclidean distance between X_{i-l} and X_{j-l}, and the weighted average distance D_t^k of sliding window B_t to its k nearest neighbors:

$$D_t^k = \sum_{t-\lfloor k/2 \rfloor \le l < i + \lceil k/2 \rceil} w_l D_{tl} \qquad (6)$$

where w_j is a weight. We can see that the weighted average distance D_t^k of sliding window B_t to its k nearest neighbors is the average of the k nearest distance for all X_i in B_t.

Similarly, given a time series of sensor readings of history training data, we firstly calculate the k nearest distance of each sliding window to estimate a univariate KDE. For a recently produced sensor reading widow B_t, the k nearest distance D_t^k of B_t is calculated. If the probability of D_t^k calculated by the KDE is less than a pre-defined threshold, then we decide that there are some outliers in the window B_t. The next thing is to locate the outlier sensor reading. Since the sliding window B_t contains three time series—B_{th} for HR, B_{tb} for BP and B_{ts} for SpO2, for each sensor reading y_i in B_{th}, B_{tb} or B_{ts}, if the nearest neighbor distance of y_i is greater than a pre-defined M, then y_i is diagnosed as an outlier.

3.3. Handling Error Training Data With HMMs

It is impossible to ensure that the history training data are all correct. The outliers in the training data can disrupt the estimated value of the KDE, and influence the performance of the proposed outlier detection method. To this issue, we use the HMM to estimate the most probable ground truth values which have the maximum probability to produce the training sensor readings.

Given a time series $\{y_1, \cdots, y_n\}$ of training sensor readings and a sensor error probability p, we select the first half data $O = \{y_1, \cdots, y_{T=\lfloor n/2 \rfloor}\}$ to estimate the parameters of an HMM λ by improving the Baum-Welch algorithm [?]. For HMM λ and the rest training sensor readings $O' = \{y_{T+1}, \cdots, y_n\}$, we can use the Viterbi algorithm [?] to find the most likely ground truth vital sign values G of all possible G' that can produce O'.

Given the sensor reading sequence O, the Baum-Welch algorithm finds a local maximum

$$\zeta = (\Pi, A, B) = \max_{\zeta'} P(O|\zeta') \qquad (7)$$

for sequence O with random initial conditions. However, through our experiments, we find that the performance using the HMM with parameters being estimated by the Baum-Welch algorithm directly is poor. We improve Baum-Welch algorithm as the following steps.

Forward procedure: let

$$\alpha_t(i) = P(o_1 = y_1, \cdots, o_t = y_t, s_t = q_i|\zeta) \qquad (8)$$

then

$$\alpha_1(i) = \pi_{q_i} b_{q_i o_1} \qquad (9)$$

$$\alpha_{t+1}(i) = b_{q_i o_{t+1}} \sum_{j=1}^{n} \alpha_t(j) a_{ji} \qquad (10)$$

Backward procedure: let

$$\beta_t(i) = P(o_{t+1} = y_{t+1}, \cdots, o_T = y_T, s_t = q_i|\zeta) \qquad (11)$$

then

$$\beta_T(i) = 1 \qquad (12)$$

$$\beta_t(i) = \sum_{j=1}^{n} \beta_{t+1}(j) a_{ij} b_{q_j o_{t+1}} \qquad (13)$$

Update: we can now calculate the temporary variables:

$$\gamma_t(i) = P(s_t = q_i|O, \zeta) = \frac{\alpha_t(i)\beta_t(i)}{\sum\limits_{j=1}^{n} \alpha_t(j)\beta_t(j)} \qquad (14)$$

$$\begin{aligned} \xi_t(ij) &= P(s_t = q_i, s_{t+1} = q_j|O, \zeta) \\ &= \frac{\alpha_t(i)a_{ij}\beta_{t+1}(j)b_{q_j o_{t+1}}}{\sum\limits_{k=1}^{n} \beta_t(k)\alpha_t(k)} \end{aligned} \qquad (15)$$

ζ can now be updated:

$$\pi_i^* = \gamma_1(i) \qquad (16)$$

let $a_{ij}' = \dfrac{\sum\limits_{t=1}^{T-1} \xi_t(ij)}{\sum\limits_{t=1}^{T-1} \gamma_t(i)}$, then

$$a_{ij}^* = \begin{cases} a_{ij}^* = \varepsilon \times a_{ij}' + (1-\varepsilon)/n & if\ P(q_i) < \Theta \\ a_{ij}^* = a_{ij}' & otherwise \end{cases} \qquad (17)$$

where $0 < \varepsilon \leq 1$ is a weighting, and Θ is a preselected threshold.

$$\text{let } b'_{ik} = \frac{\sum\limits_{t=1, O_t = v_k}^{T} \gamma_t(i)}{\sum\limits_{t=1}^{T} \gamma_t(i)}, \text{ then}$$

$$b^*_{ik} = \begin{cases} 1 - p & if \ q_i = v_k \\ p * \dfrac{b'_{ik}}{\sum\limits_{v_k \neq q_i} b'_{ik}} & otherwise \end{cases} \quad (18)$$

These steps are now repeated iteratively until a desired level of convergence.

4. Simulation Results

In order to examine the performance of the proposed outlier detection method, we carry out some experiments on medical datasets from the PhysioNet database [**?**]. The dataset contains 7 attributes: BPmean, systolic BP, diastolic BP, HR, pulse, respiration rate, and SpO2. We only focus on three attributes: BPmean, HR, and SpO2. We use a data sequence of 5000 data, in which the first 4000 data are selected as the training data, and the last 1000 data with injecting faults as diagnosed data.

4.1. Performance Without Sliding Window

Since the 4000 training data are not injected faults, so we can estimate the KDE of the k nearest distance for each sensor reading. Given sensor error probabilities, we inject faults into the 1000 diagnosed data with the position and the value of the injected error all selected by random numbers. The simulation results are the average performance of 300 times randomized experiments.

Table I shows the performance of the k nearest distance based outlier detection method. In this simulation, if the probability of the k nearest distance calculated by the estimated KDE is less than a threshold $\delta = 0.001$, then the diagnosed data is decided as an outlier. From Table I, we can see that the k nearest distance based method has a good outlier detection, but the false alarm rate is high, which leads to that the error rate after executing this method is enormous greater than the original error probability.

Table II shows the performance of the nearest distance based outlier detection method. If the nearest distance is greater than 4, then the diagnosed data is determined as an outlier. From Table II, we can see that the number of errors can be reduced by 50% approximately using this method.

Table III shows the performance of the combination of the k nearest distance and the nearest distance based outlier detection method. From this table, we can see the the performance is better than that of using k nearest distance and the nearest distance based method alone.

Table 1. k Nearest Distance Based Simulation Result (p Indicates Prior Error Probability), OD denotes Outlier Detection, FA denotes False Alarm and EP denotes Error Probability.)

p (%)	5	10	15	20	30	40
OD (%)	84	86	98	97	97	97
FA (%)	9.5	21.4	30.9	38	53.7	67.8
EP (%)	9.4	19.7	26.5	31	38.3	41.6

Table 2. The Nearest Distance Based Simulation Result (p Indicates Prior Error Probability), OD denotes Outlier Detection, FA denotes False Alarm and EP denotes Error Probability.)

p (%)	5	10	15	20	30	40
OD (%)	80	70	70	53	37	31
FA (%)	1.3	1.2	1.2	1.3	3.1	1.9
EP (%)	2.2	4.0	5.6	10.5	21.0	28.7

Table 3. The Combination Method Simulation Result (p Indicates Prior Error Probability, OD denotes Outlier Detection, FA denotes False Alarm and EP denotes Error Probability.)

p (%)	5	10	15	20	30	40
OD (%)	78	69	70	53	37	31
FA (%)	0.3	0.4	0.2	0.7	2.7	1.8
EP (%)	1.4	3.5	4.6	10.1	20.0	28.7

Table 4. The MD In [9] Based Simulation Result (p Indicates Prior Error Probability), OD denotes Outlier Detection, FA denotes False Alarm and EP denotes Error Probability.)

p (%)	5	10	15	20	30	40
OD (%)	72	71	69	72	70	71
FA (%)	0.95	0.93	0.98	0.96	0.83	0.92
EP (%)	2.3	3.8	5.5	6.4	9.6	12.2

As a comparison, Table IV gives the performance of the Mahalanobis distance (MD) and KDE based approach in [9] on the same dataset. In this method,

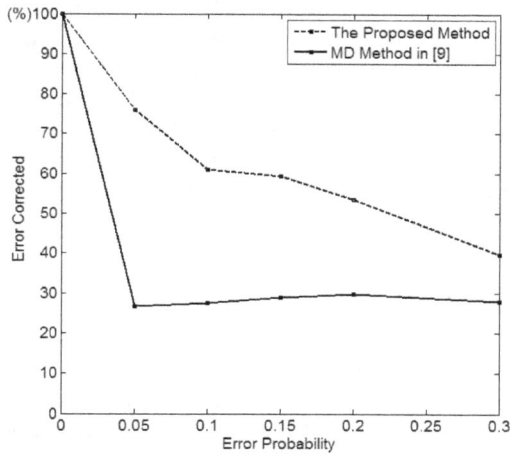

Figure 5. Comparison of Proposed Method and The Method in [9] On Probability of Errors Corrected.

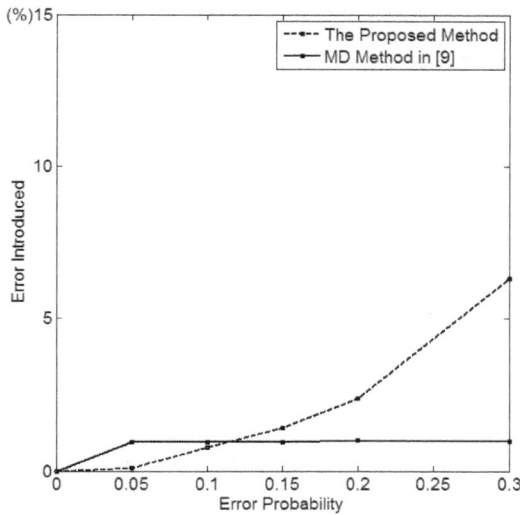

Figure 6. Comparison of Proposed Method and The Method in [9] On Probability of Errors Introduced.

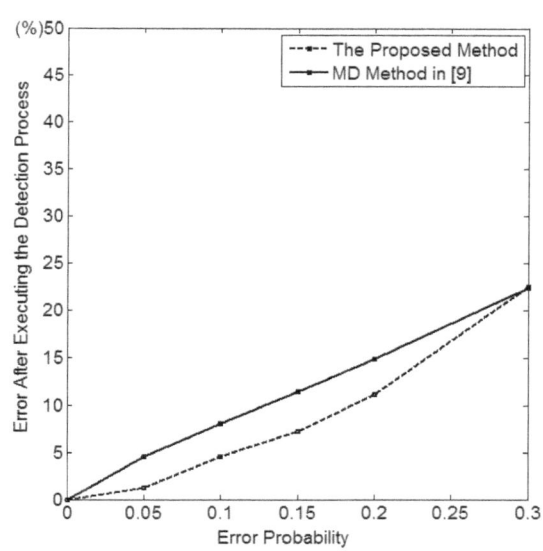

Figure 7. Comparison of Proposed Method and The Method in [9] On Probability of Errors after Executing the Detection Algorithm.

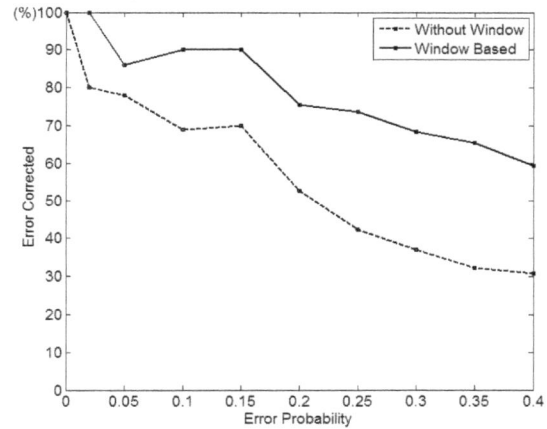

Figure 8. Comparison of With and Without Sliding Window On Probability of Errors Corrected.

when a new data is obtained, the Mahalanobis distance is calculated between the current arrival data and the mean of training data, then KDE is used to estimate the probability of this distance, if it is less than threshold , then the current arrival data is diagnosed as an outlier. From table IV, we can see that the performance of the proposed method is not better than the method in [9]. The reason is that the simulation performance of the outlier detection process depends on the sensor error rate, besides, the range in which the injected errors must lie is another factor influences the outlier detection performance, the closer the outlier to the ground true value, the harder it is to be detected.

In previous experiments, the range of injected errors is wider than the normal value of vital signs. Fig 5-7 give the comparison of the performance of the proposed method and the method in [9], in which range of injected errors is set as the same to the normal value range. From these figure, we can see that the proposed method has a better performance when the faults appear in the range of most normal vital sign data occurs.

4.2. Performance With Sliding Window

Fig. 8-10 show the comparison of the distance based method with and without sliding window. We can see

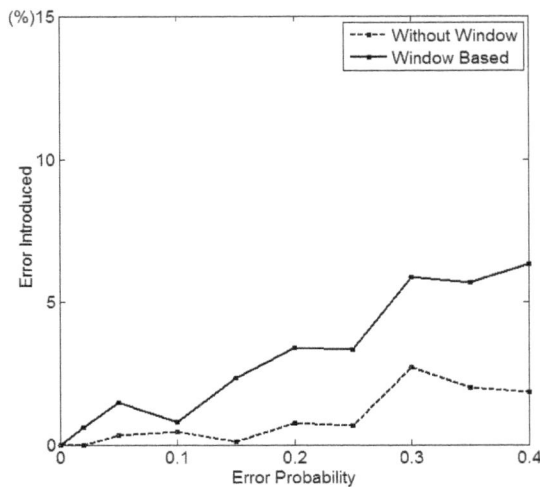

Figure 9. Comparison of With and Without Sliding Window On Probability of Errors Introduced.

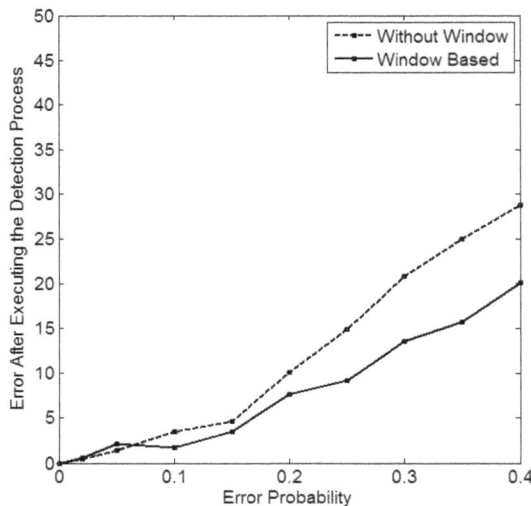

Figure 10. Comparison of With and Without Sliding Window On Probability of Errors after Executing the Detection Algorithm.

that, the method with sliding window has a high outlier detection rate. Although its false alarm rate is a little higher, but the error rate after executing the outlier detection algorithm is obvious lower.

5. Conclusion

Outlier detection is very important for BSNs to avoid false medical diagnosis and false alarms. In this paper, we formalize a distance based method for outlier detection of BSNs. This method consider both distance to k nearest neighbors and to the left and right neighbors. To deal with the condition like successive

errors, we formalize a sliding widow based method to improve the performance of the outlier detection method. To handle errors in the training data, we introduce a Hidden Markov Model based method to estimate the most probable ground truth values which have the maximum probability to produce the training data. Simulation results show that the proposed method possesses a good performance.

6. Acknowledgment

Part of this work has been supported by the National Natural Science Foundation of China (NSFC, No. 61373043, 61372073, 61003079), and the Fundamental Research Funds for the Central Universities (No. JB140316).

References

[1] B. Krishnamachari and S. Iyengar, Distributed Bayesian Algorithms for Fault-Tolerant Event Region Dectection in Wireless Sensor Networks. IEEE Transactions on Computers, 53(3):241-250, 2004.

[2] X. Luo, M. Dong, and Y. Huang. On distributed fault-tolerant detection in wireless sensor networks. IEEE Transactions on Computers, 55(1): 58-70, 2006.

[3] W. Wu, X. Cheng, M. Ding, K. Xing, F. Liu, and P. Deng. Localized Outlying and Boundary Data Detection in Sensor Networks, IEEE Transactions on Knowledge and Data Engineering, 19(8): 1145-1157, 2007.

[4] A. Annichini, E. Asarin, A. Bouajjani. Symbolic Techniques for Parametric Reasoning about Counter and Clock Systems. The Proceeding of 12th International Conference on Computer Aided Verification, LNCS 1855, pp. 419ÍC434, 2000.

[5] D.J. Kim and B. Prabhakaran. Motion fault detection and isolation in body sensor networks. IEEE International Conference on Pervasive Computing and Communications, pp. 147-155, 2011.

[6] Y. Chen, Y. Zi, H. Cao, Z. He, H. Sun. A data-driven threshold for wavelet sliding window denoising in mechanical fault detection. Science China (Technological Sciences),03: 589-597, 2014.

[7] Jie Ying, T. Kirubarajan, Krishna R. Pattipati. A Hidden Markov Model-Based Algorithm for Fault Diagnosis with Partial and Imperfect Tests. IEEE Transactions on Systems, Man and Cybernetics, 30(4): 463-473, 2000.

[8] O. Salem, A. Guerassimov, A. Mehaoua, A. Marcus and B. Furht. Sensor Fault and Patient Anomaly Detection and Classification in Medical Wireless Sensor Networks. IEEE International Conference on Communications, pp. 4373-4378, 2013.

[9] O. Salem, Y. Liu and A. Mehaoua. Anomaly Detection in Medical Wireless Sensor Networks. Journal of Computing Science and Engineering, 7(4): 272-284, 2013.

[10] Physionet, http://www.physionet.org/cgi-bin/atm/ATM.

[11] M.M. Breunig, H.P. Kriegel, R.T. Ng, et al. LOF: Identifying Density-Based Local Outliers. ACM Sigmod Record. 29(2): 93-104, 2000.

[12] K. Niu, F. Zhao and X. Qiao. An Outlier Detection Algorithm in Wireless Sensor Network Based on Clustering. Proceedings of 15th IEEE International Conference on Communication Technology, 2013

[13] S. Rajasegarar, C. Leckie, M. Palaniswami, and J.C. Bezdek, Distributed Anomaly Detection in Wireless Sensor Networks, IEEE ICCS, 2006.

17

Cognitive Relay Networks

Ayesha Naeem[1], Mubashir Husain Rehmani[2,*]

[1]Military College of Signals, NUST, Pakistan
[2]COMSATS Institute of Information Technology, Wah Cantt, Pakistan

Abstract

Cognitive radio is an emerging technology to deal with the scarcity and requirement of radio spectrum by dynamically assigning spectrum to unlicensed user. This revolutionary technology shifts the paradigm in the wireless system design by allowing unlicensed user the ability to sense, adapt and share the dynamic spectrum. Cognitive radio technology applied to different networks and applications ranging from wireless to public safety, smart grid, medical, relay and cellular applications to increase the throughput and spectrum efficiency of network. Among these applications, cognitive relay networks is one of the famous application where cognitive radio technology is applied. Cognitive relay network increases the throughput of network by reducing the complete path loss and also by ensuring cooperation among secondary users and cooperation among primary and secondary users. In this paper, our aim is to provide a survey on cognitive relay network. We also provide a detailed review on existing schemes in cognitive relay networks on the basis of relaying protocol, relay cooperation and channel model.

Keywords: Cognitive radio network, Cognitive relay network

1. Introduction

There is an increasing demand of spectrum resources in the past recent years due to emerging wireless technologies. Within the current framework of spectrum management, all the spectrum bands are allocated to some specific areas or services. This issue led the Federal Communications Commission (FCC) to move trend from static to dynamic spectrum allocation because large portion of licensed spectrum is underutilized in geographic and vast temporal regions. Spectrum utilization can be improved by allowing secondary user to use the licensed band opportunistically while causing no interference to primary users [1].

A novel technology proposed for effective spectrum utilization is Cognitive radio (CR) technology. In CR technology, secondary user senses the spectrum holes for effective utilization and then adapt the environment causing no interference to primary users. In order to detect the reappearance of primary user, cognitive user must continuously sense the available spectrum for effective communication and spectrum utilization [3].

Cognitive radio can be applied to different networks and applications in order to increase the network throughput and to ensure the effective spectrum utilization. Cognitive radio network plays a significant role in different applications like public safety, smart grid, medical, relay and cellular applications.

In cognitive radio network, secondary users dynamically senses to the local environment for available spectrum. Secondary devices in cognitive radio network experience diverse spectrum conditions due to dispersed geographical locations of other secondary users. Cognitive radio network can dynamically exploit the spectrum in order to support continuous transmission. In order to ensure spectrum opportunity and effective data transmission, a new concept is introduced that is Cognitive Relay Network.

Cognitive relay networks have been proposed to increase the throughput of network. In a cognitive relay network, a relay node is used to reduce the complete path loss. Cognitive relay network increases the throughput by cooperating among secondary users or cooperation among primary and secondary

---

*Corresponding author: Email: mshrehmani@gmail.com

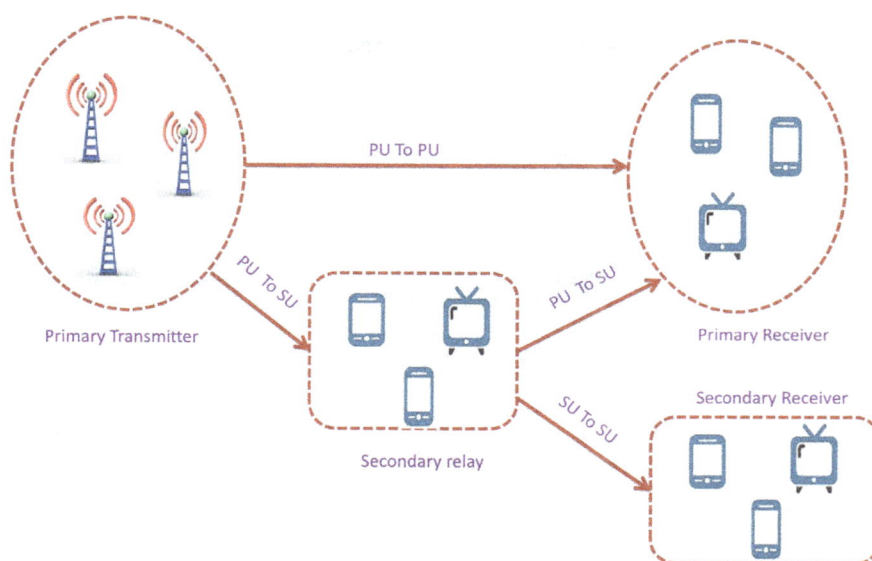

Figure 1. Basic architecture of relaying in cognitive relay networks [2]

users. Seamless data transmission can be realized and monitored in cognitive relay network because without cognitive relay, there is a direct link between source node and destination node. If primary user returns to that channel that is utilizing by primary user, then secondary nodes stop their transmission causing no interference to primary user. There are two phases in cognitive relay network. In the first phase, source node will broadcast all the information or data to all the intermediate nodes. In the second phase, depending on the protocol used, the message delivered to destination node via a relay node [4].

In this paper, we provide a survey on cognitive relay networks. Following are the contributions of this paper:

1. We provide the detail of relaying in cognitive radio network, its advantages and types of relaying in cognitive radio network

2. We discuss about existing schemes of relaying in cognitive radio network on the basis of relaying protocol, relay cooperation and channel model.

The organization of paper is as follow: In section 2, we discuss about relaying in cognitive radio network, advantages of relaying in cognitive radio network and types of relaying in cognitive radio network. Section 3 is about existing schemes of relaying in cognitive radio network. Last section 4 contains conclusion of paper.

2. Relaying in Cognitive Radio Networks

2.1. Advantages of Relaying in CRNs

One of the challenges faced by the cognitive radio networks is exploitation of transmission opportunity. Secondary users try to maximize their own spectrum utilization without interfering primary user resources. In order to maximize their transmission opportunity and efficient resource sharing, they might be cooperating or competing for the available resources. There are several advantages of relaying in cognitive radio network [5].In this section, we discuss the basics of relaying in cognitive relay networks. Figure 1 shows the basic architecture of relaying in cognitive relay networks. We classify the existing schemes of relaying in cognitive relay networks into three main classes, which we mention in Figure 2.

Cooperative transmission. In a wireless network by enabling different cooperative relay, spatial diversity can be improved. There are two types of cooperative transmission to increase the secondary throughput and also to increase the probability of transmission opportunity. One is transmission between secondary users in which secondary user act as a relay node for other secondary users transmission. The second one is transmission between primary and secondary users in which secondary user act as a relay for the primary user transmission [6] as shown in Figure 1.

Cooperative relay. Within a secondary network, handling of unbalanced spectrum is a great challenge. In order to resolve this challenge, cooperative relay concept is introduced. In this scheme, secondary users having low traffic demand act as relay nodes for other secondary users having high traffic demand, in order to improve the system or spectrum efficiency and performance[6].

Throughput maximization. In order to increase the throughput of whole network, there should be the best

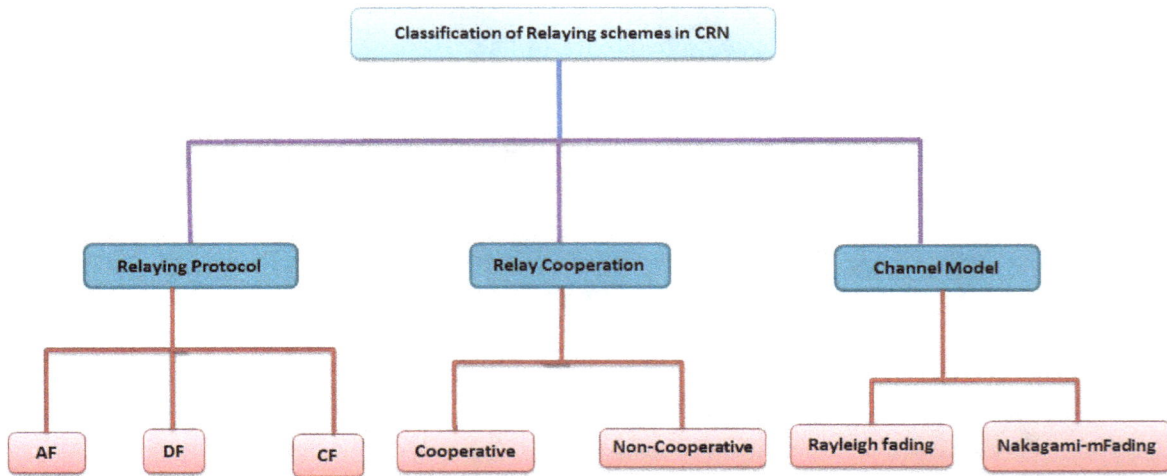

Figure 2. Classification of relaying in cognitive radio networks

selection of relay and destination node. Node having high traffic demand may not act as a relay node, as it may not exploit the throughput of whole network. In order to increase the throughput of network, there should be synchronization between nodes, so that they can negotiate the resource allocation and exchange information between both relays and channels control messages [7], [6].

Primary user detection and protection. In order to ensure the synchronization, there are MAC frames which consist of downlink and uplink transmission and control information. By making the fixed length of MAC frames, primary user protection ensures for dynamic spectrum sensing and also helps to reduce high transmission delays. Primary user detection becomes easy and efficient by ensuring frame synchronization [8].

Coordination. In a centralized CRN, nodes collect the information about spectrum availability and data demand in a frame format. This information is then broadcasted on a common control channel to ensure coordination among nodes and to ensure best decision about relay and destination node [9].

2.2. Types of Relaying in CRNs

Relaying protocols. After detecting the effective spectral holes in cognitive relay network, source node broadcast the signals and then these signals forward towards the destination node through relays. There are generally three different relaying protocols used in cognitive radio networks as shown in Figure 2, to determine what an individual relay should do after receiving a signal. These relaying protocols are Amplify-and-Forward

(AF), Decode-and-Forward (DF), and Compress-and-Forward (CF).

Amplify-and-Forward Relaying protocol (AF). In an AF protocol, signals received at relay station is first amplified and then retransmitted to different available bands simultaneously. Communication will be in two-hops if relay node shifted from primary and secondary users. Cognitive relay networks apply AF protocol as shown in Table 1 to achieve maximal throughput by exploiting the idle channels [10] [11] [12] [13] [14] [15]. The major advantage of implementing AF protocol is that it requires low cost implementation and is more flexible.

Decode-and-Forward Relaying Protocol (DF). In a decode-and-forward protocol, node acting as a relay node will decode the information or message, encode it again and then retransmit it towards the destination by selecting appropriate channel from spectrum pool. Different cognitive relay networks apply DF protocol for relaying scheme to enhance the network capacity and to ensure security [24] [23] [25] [26] [22] [19] [20]. DF relaying scheme takes large amount of time to decode and then to retransmit the message or information.

Compress-and-Forward Relaying protocol (CF). Received signal estimation is carried out by Compress-and-Forward protocol. In CF protocol, received signal is compressed first by the relay node, encoded and then retransmit towards the destination. This relaying protocol increases the signal redundancy received from the source node.

2.3. Relay Cooperation

Cooperative relaying in CRN. For traditional wireless technology, cooperative relay technique has been

Reference	Relay schemes	Relay Type	Channel Model	Cooperation	Simulator used	parameters evaluated
[11]	Efficient Multiple Relay	AF	-	Cooperative Relay	-	Average SNR
[16]	Efficient Multiple Relay	DF	Nakagami-m fading	Cooperative Relay	Monte Carlo	Selection diversity order
[10]	Power Allocation	AF	-	Cooperative Relay	-	Interference Power, System throughput
[17]	Cooperative Transmission	DF	-	Cooperative Relay	Monte Carlo	Transmission of PU and SU
[18]	Power Allocation	DF	Rayleigh fading	Cooperative Relay	Monte Carlo	Ergodic achievable rate
[19]	Outage Performance	DF	Rayleigh fading	Cooperative Relay	-	Outage probability
[20]	Outage Performance	DF	Rayleigh fading	Cooperative Relay	-	Outage probability
[21]	Outage Performance	DF	Rayleigh fading	Cooperative Relay	-	Outage probability
[14]	Outage Probability	AF	Rayleigh fading	Cooperative Relay	Monte Carlo	SER, outage probabilit
[22]	Outage Analysis	DF	Rayleigh fading	Cooperative Relay	Monte Carlo	Outage probability
[23]	Capacity Analysis	DF	Rayleigh fading	Cooperative Relay	-	Capacity loss
[24]	Capacity Analysis	DF	Rayleigh fading	Cooperative Relay	-	Capacity loss
[13]	Cognitive AF	AF	Nakagami-mFading	Cooperative Relay	-	Outage probability
[12]	Cognitive AF	AF	Rayleigh fading	Cooperative Relay	-	Outage probability
[25]	Cognitive Relay Networks	DF	Rayleigh fading	Cooperative Relay	Monte Carlo	Outage probability
[26]	Cognitive Relay Networks	DF	Rayleigh fading	Cooperative Relay	Monte Carlo	Outage probability
[27]	Power Allocation	AF	Rayleigh fading	Non-Cooperative Relay	Monte Carlo	Cumulative distribution
[28]	Geometric Approach	-	Rayleigh fading	Cooperative Relay	Monte Carlo	NNR and FNR
[2]	Distributed spectrum access	-	-	Cooperative Relay	-	Average total rate of PU
[29]	Cognitive Transmission	-	Rayleigh fading	Cooperative Relay	-	Spectrum hole utilization
[30]	Cellular cognitive relay network	AF	Rayleigh fading	Cooperative Relay	-	Outage capacity
[5]	Cognitive Transmission	-	-	Cooperative Relay	USRP-based testbed	Throughput Gain
[9]	Cognitive Transmission	DF	combat wireless fading	Cooperative Relay	-	Spectrum hole utilization
[6]	MAC protocol	-	combat wireless fading	Cooperative Relay	USRP-based testbed	Throughput gain
[31]	Network Utility Maximization	-	-	Cooperative Relay	-	Average throughput
[32]	underlay-based cognitive	-	Rayleigh fading	Cooperative Relay	-	Feasibility probability
[33]	CR Cellular Relay Networks	-	-	-	-	The normalized capacity , Path loss exponent
[34]	Interference and Delay Constrained	AF	Rayleigh fading	Cooperative Relay	Monte-Carlo	Normalized Effective Capacity
[35]	Power Allocation Strategy	AF	Rayleigh fading	Cooperative Relay	Monte Carlo	Power of cognitive transmitter
[36]	Outage probability	-	Rayleigh fading	-	-	Outage probability, maximum transmit power
[37]	Full Duplex	AF	-	Cooperative Relay	-	Resource scheduling
[38]	Full Duplex	AF	-	Non-Cooperative Relay	-	Self-interference, secondary downlink
[39]	Distributed Beamforming	-	Rayleigh fading	Cooperative Relay	-	Feasibility Probability
[40]	power loading strategy	AF	Rayleigh fading	Cooperative Relay	-	Channel capacity
[41]	Power allocation (PA) and multiple relay selection (MRS)	-	Rayleigh fading	-	-	Secondary system capacity
[42]	Multiuser CRN	DF	Rayleigh flat fading	Cooperative Relay	-	Outage probability
[43]	Spectrum access	-	-	Cooperative Relay	-	Network capacity
[44]	Optimal power allocation	AF	-	Cooperative Relay	-	Transmit rate
[45]	Multiple relay selection	-	Rayleigh fading	Cooperative Relay	-	Average SNR
[46]	Two relay selection schemes	-	Rayleigh fading	Cooperative Relay	-	Outage probability
[47]	Imperfect CSI	AF	Rayleigh fading	-	-	Interference probability
[48]	Spectrum access	-	-	Cooperative Relay	-	Average capacity
[49]	Spectrum sensing	-	Rayleigh fading	Cooperative relay	-	Sensing time
[21]	outage probability (OP)	DF	Rayleigh fading	Cooperative relay	-	Outage probability
[4]	Interference Constraints	-	-	-	-	Outage Probability of Secondary User
[50]	Power and Channel Allocation	-	Rayleigh fading	Cooperative relay	-	N spectrum bands
[51]	Half-Duplex Buffered CRN	DF	Rayleigh fading	Cooperative relay	Monte Carlo	Buffer gain
[52]	Resource Allocation Techniques	AF,DF,CF	Rayleigh fading	Cooperative relay	-	Spectrum management
[53]	Cognitive Relay Beamforming	-	Nakagami-m fading	Cooperative relay	-	Cumulative distribution of PU
[54]	Simplified Power Allocation	-	-	-	-	Average capacity
[8]	Interference of Primary User	-	Rayleigh fading	Cooperative relay	Monte Carlo	Outage probability

Table 1. Existing schemes of relaying in cognitive radio networks

* "-" =Not mentioned
AF= Amplify-and-forward
DF= Decode-and-forward
CF=Compress-and-forward

studied widely but cooperative relaying face some additional challenges in cognitive radio network. One of challenges faced in cognitive radio network is mutual interference between the secondary and primary users. Due to the generation of some false alarms in sensing phase, this may led towards the mutual interference between primary and secondary users. This mutual interference causes the outage probability for secondary users data transmission [9] [6].

Cooperative spectrum sensing may removes wireless fading and consists of two phases: First phase is, within the certain time period, detection of primary user existence. In the second phase, there is a fusion centre on which detection results are forwarded in the remaining time slot. Cooperative relay improves the secondary data transmission by having primary destination (PD) relays and cognitive destination (CD) relays. Primary and secondary users apply some relaying protocols (AF or DF or CF) and then send data or message to PD and CD.

Non-cooperative relaying in CRN. Non-cooperative relaying technique proposed to carry out spectrum occupancy analysis and measurement for individual users or radios to act autonomously and locally.

Spectrum sensing in non-cooperative technique consist of these categories: blind sensing and signal specific category. In blind sensing approach for cognitive radio network, there is a central entity known as fusion centre which collects all the sensing information and then decide which frequency can be used for data transmission. In signal specific approach, there is a requirement of primary user signal knowledge to send data over appropriate channel.

2.4. Channel model

Rayleigh Fading. In order to model multi path fading, Rayleigh fading is used widely having no line-of-sight path. Rayleigh fading is closely related to square distribution technique. For channel modelling in cognitive radio network, Rayleigh techniques is applied [45] [55] [47] [46] [49] [22] [19]. Rayleigh model is widely used for the propagation of refracted and reflected paths through on the radio link.

Nakagami-M Fading. In order to characterize the signal transmission from multipath fading channel, distribution that uses widely is Nakagami-M fading distribution. Cognitive relay networks use Nakagami-M fading distribution for in-door and land-mobile multipath radio link propagation [56] [13] [57] [16].

3. Existing Schemes in Cognitive Relay Network

Different cognitive relaying schemes proposed in order to ensure effectiveness of spectrum utilization and

to increase the throughput of whole network. These cognitive relaying schemes are shown in Table 1.

3.1. Cooperative Relaying Schemes in CRN

In cognitive radio network, cooperative relay technology is introduced in order to increase the network capacity of data transmission and to increase the throughput of network. Different cooperative relaying schemes proposed to allow communication or data transmission through relay nodes.

In [19], cooperative relay technique used to gain diversity in data transmission and also improves the secondary user spectrum sensing performance. In [20], cooperative relaying technique applied for reliable and effective communication among primary and secondary users to ensure the better communication range of secondary users. Author in [21] proposed a scheme that increases performance gain by applying cooperative relaying. Relaying technique in [14], ensures the coverage, reliability and throughput of whole cognitive radio network.

Spectrum utilization is the major concern for cognitive radio network. Author proposed combination of cooperative spectrum sensing and cognitive radio to increase the effective spectrum utilization [44] [10]. In order to maximize the achievable rate of cognitive user, cooperative relaying proves to be an effective approach [18]. To increase the capacity and the throughput of channel, cooperative communication used in cognitive relay network by making the intermediate node as a relay node [23], [24].

Cooperative relaying techniques used in cognitive radio networks can be used to improve spectrum sharing [23] [24] [13] [12] [34] [36], spectrum sensing [29] [30] [9], spectrum usage [38] [39] [33], spectrum efficiency [40] [44] and spectrum availability [42].

3.2. Amplify-and-Forward Relaying Schemes in CRN

AF relaying scheme applied in CRN for achieving maximum throughput. In [13], author deployed AF relaying scheme to gain maximal outage performance. In order to maximize the total power allocation rate in cognitive radio relay network, AF technique is used [10] [44]. In [27], total power constraint can be increased using AF relaying protocol, as secondary users communicate via a relay node. In order to enhance the spectrum usage in full-duplex cognitive relay network, AF relaying protocol are used to ensure spectrum efficiency [37] [38].

3.3. Decode-and-Forward Relaying Schemes in CRN

DF relaying techniques are proposed to ensure the confidentiality and integrity of data transmitted from source to destination via a relay node. In [24], DF relaying protocol is used to increase the cooperative diversity over the cognitive radio network. Two approaches of DF relaying scheme discussed in this paper: one is proactive DF approach and the other one is reactive DF approach. Scheme proposed in [58] analyse the primary user interference for the DF scheme applied in secondary relay network, resulting an effective outage performance.

3.4. Relaying Schemes on the basis of Channel Model

Rayleigh fading Relaying Schemes. In a cognitive relay network different channel modelling techniques used, one is Rayleigh fading. In [34] author applied Rayleigh fading to maximize point-to-point channel capacity and thus data transmission become efficient. For the effective throughput in a single and in a multiple relay, fading technique used in [35] is Rayleigh fading. Secondary users outage probability can be increased by using Rayleigh fading [36] [23].

Nakagami-mFading Relaying Schemes. Cognitive relay networks applied Nakagami-mFading for the outage analysis to maximize the transmit power over versatile fading channels [13]. Spectrum efficiency and spectrum sharing becomes more effective using Nakagami-mFading [56]. In [57], author investigates the outage probability and capacity analysis for multiple primary users.

4. Conclusion

In order to increase the throughput, to increase cooperation and to decrease the complete path loss, a novel technology is introduced which is cognitive relay networks. Cognitive relay network, increases the spectrum efficiency by cooperating among secondary and primary users and also among secondary users themselves. There are two phases in cognitive relay network. One phase is the broadcasting of all information from source to all intermediate nodes. Second phase is the transfer of information towards destination through relay node. This paper, provides classification of cognitive relay networks on the basis of relay type, cooperation in relay networks, and channel model used for relaying in cognitive radio network.This paper also provides detailed analysis of different relaying schemes applied in cognitive relay network.

References

[1] I. F. Akyildiz, W. Y. Lee, M. C. Vuran, S. Mohanty, Next generation/dynamic spectrum access/cognitive radio wireless networks: A survey, Computer Networks 50 (2006) 2127–2159.

[2] S. Bayat, R. H. Louie, Y. Li, B. Vucetic, Cognitive radio relay networks with multiple primary and secondary users: Distributed stable matching algorithms for spectrum access, in: Communications (ICC), 2011 IEEE International Conference on, IEEE, 2011, pp. 1–6.

[3] S. Haykin, Cognitive radio: brain-empowered wireless communications, IEEE Journal on Selected Areas in Communications 23 (2005) 201–220.

[4] J. Lee, H. Wang, J. G. Andrews, D. Hong, Outage probability of cognitive relay networks with interference constraints, Wireless Communications, IEEE Transactions on 10 (2) (2011) 390–395.

[5] J. Jia, J. Zhang, Q. Zhang, Cooperative relay for cognitive radio networks, in: INFOCOM 2009, IEEE, IEEE, 2009, pp. 2304–2312.

[6] Q. Zhang, J. Jia, J. Zhang, Cooperative relay to improve diversity in cognitive radio networks, Communications Magazine, IEEE 47 (2) (2009) 111–117.

[7] Y. Song, F. Zhang, S. Yubin, Energy efficiency and throughput optimization of cognitive relay networks, CIT. Journal of Computing and Information Technology 22 (3) (2014) 151–158.

[8] M. A. R. Gani, M. M. MR, Novel opportunistic cognitive relay network considering interference of primary user.

[9] Y. Zou, Y.-D. Yao, B. Zheng, Cooperative relay techniques for cognitive radio systems: spectrum sensing and secondary user transmissions, Communications Magazine, IEEE 50 (4) (2012) 98–103.

[10] Z. Liu, Y. Xu, D. Zhang, S. Guan, An efficient power allocation algorithm for relay assisted cognitive radio network, in: Wireless Communications and Signal Processing (WCSP), 2010 International Conference on, IEEE, 2010, pp. 1–5.

[11] M. Naeem, D. Lee, U. Pareek, An efficient multiple relay selection scheme for cognitive radio systems, in: Communications Workshops (ICC), 2010 IEEE International Conference on, IEEE, 2010, pp. 1–5.

[12] V. N. Q. Bao, T. Q. Duong, D. Benevides da Costa, G. C. Alexandropoulos, A. Nallanathan, Cognitive amplify-and-forward relaying with best relay selection in non-identical rayleigh fading, Communications Letters, IEEE 17 (3) (2013) 475–478.

[13] T. Q. Duong, D. B. d. Costa, M. Elkashlan, V. N. Q. Bao, Cognitive amplify-and-forward relay networks over nakagami-fading, Vehicular Technology, IEEE Transactions on 61 (5) (2012) 2368–2374.

[14] H. Yu, W. Tang, S. Li, Outage probability and ser of amplify-and-forward cognitive relay networks, Wireless Communications Letters, IEEE 2 (2) (2013) 219–222.

[15] E. E. Benitez Olivo, D. P. Moya Osorio, D. B. da Costa, S. Santos Filho, J. Candido, Outage performance of spectrally efficient schemes for multiuser cognitive relaying networks with underlay spectrum sharing, Wireless Communications, IEEE Transactions on 13 (12) (2014) 6629–6642.

[16] J. Bang, J. Lee, S. Kim, D. Hong, An efficient relay selection strategy for random cognitive relay networks, Wireless Communications, IEEE Transactions on 14 (3) (2015) 1555–1566.

[17] W. Jaafar, W. Ajib, D. Haccoun, A new cooperative transmission scheme with relay selection for cognitive

radio networks, in: Global Communications Conference (GLOBECOM), 2013 IEEE, 2013, pp. 949–954.

[18] Z. Shu, W. Chen, Optimal power allocation in cognitive relay networks under different power constraints, in: Wireless Communications, Networking and Information Security (WCNIS), 2010 IEEE International Conference on, IEEE, 2010, pp. 647–652.

[19] L. Luo, P. Zhang, G. Zhang, J. Qin, Outage performance for cognitive relay networks with underlay spectrum sharing, Communications Letters, IEEE 15 (7) (2011) 710–712.

[20] B. Prasad, S. D. Roy, S. Kundu, Outage performance of cognitive relay network with imperfect channel estimation under proactive df relaying, in: Communications (NCC), 2014 Twentieth National Conference on, IEEE, 2014, pp. 1–6.

[21] P. Yang, L. Luo, J. Qin, Outage performance of cognitive relay networks with interference from primary user, Communications Letters, IEEE 16 (10) (2012) 1695–1698.

[22] Q. Wu, Z. Zhang, J. Wang, Outage analysis of cognitive relay networks with relay selection under imperfect csi environment, Communications Letters, IEEE 17 (7) (2013) 1297–1300.

[23] J. Si, Z. Li, H. Huang, J. Chen, R. Gao, Capacity analysis of cognitive relay networks with the pu's interference, Communications Letters, IEEE 16 (12) (2012) 2020–2023.

[24] S. Sagong, J. Lee, D. Hong, Capacity of reactive df scheme in cognitive relay networks, Wireless Communications, IEEE Transactions on 10 (10) (2011) 3133–3138.

[25] D. Li, Cognitive relay networks: opportunistic or uncoded decode-and-forward relaying?, Vehicular Technology, IEEE Transactions on 63 (3) (2014) 1486–1491.

[26] T. Q. Duong, P. L. Yeoh, V. N. Q. Bao, M. Elkashlan, N. Yang, Cognitive relay networks with multiple primary transceivers under spectrum-sharing, Signal Processing Letters, IEEE 19 (11) (2012) 741–744.

[27] T. Wang, L. Song, Z. Han, X. Cheng, B. Jiao, Power allocation using vickrey auction and sequential first-price auction games for physical layer security in cognitive relay networks, in: Communications (ICC), 2012 IEEE International Conference on, IEEE, 2012, pp. 1683–1687.

[28] M. Xie, W. Zhang, K.-K. Wong, A geometric approach to improve spectrum efficiency for cognitive relay networks, Wireless Communications, IEEE Transactions on 9 (1) (2010) 268–281.

[29] Y. Zou, Y.-D. Yao, B. Zheng, Cognitive transmissions with multiple relays in cognitive radio networks, Wireless Communications, IEEE Transactions on 10 (2) (2011) 648–659.

[30] H. Cheng, Y.-D. Yao, Cognitive-relay-based intercell interference cancellation in cellular systems, Vehicular Technology, IEEE Transactions on 59 (4) (2010) 1901–1909.

[31] L. Ruan, V. K. Lau, Decentralized dynamic hop selection and power control in cognitive multi-hop relay systems, Wireless Communications, IEEE Transactions on 9 (10) (2010) 3024–3030.

[32] A. Piltan, S. Salari, Distributed beamforming in cognitive relay networks with partial channel state information, Communications, IET 6 (9) (2012) 1011–1018.

[33] S. Kim, W. Choi, Y. Choi, J. Lee, Y. Han, I. Lee, Downlink performance analysis of cognitive radio based cellular relay networks, in: Cognitive Radio Oriented Wireless Networks and Communications, 2008. CrownCom 2008. 3rd International Conference on, IEEE, 2008, pp. 1–6.

[34] L. Musavian, S. Aïssa, S. Lambotharan, Effective capacity for interference and delay constrained cognitive radio relay channels, Wireless Communications, IEEE Transactions on 9 (5) (2010) 1698–1707.

[35] Z. Zhang, Q. Wu, J. Wang, Energy-efficient power allocation strategy in cognitive relay networks, Radio engineering 21 (3).

[36] Z. Yan, X. Zhang, W. Wang, Exact outage performance of cognitive relay networks with maximum transmit power limits, Communications Letters, IEEE 15 (12) (2011) 1317–1319.

[37] L. Wang, F. Tian, T. Svensson, D. Feng, M. Song, S. Li, Exploiting full duplex for device-to-device communications in heterogeneous networks, Communications Magazine, IEEE 53 (5) (2015) 146–152.

[38] Y. Liao, L. Song, Z. Han, Y. Li, Full duplex cognitive radio: a new design paradigm for enhancing spectrum usage, Communications Magazine, IEEE 53 (5) (2015) 138–145.

[39] S. H. Safavi, R. A. S. Zadeh, V. Jamali, S. Salari, Interference minimization approach for distributed beamforming in cognitive two-way relay networks, in: Communications, Computers and Signal Processing (PacRim), 2011 IEEE Pacific Rim Conference on, IEEE, 2011, pp. 532–536.

[40] T. Nadkar, V. Thumar, U. Desai, S. Merchant, Judicious power loading for a cognitive relay scenario, in: Intelligent Signal Processing and Communication Systems, 2009. ISPACS 2009. International Symposium on, IEEE, 2009, pp. 327–330.

[41] M. Choi, J. Park, S. Choi, Low complexity multiple relay selection scheme for cognitive relay networks, in: Vehicular Technology Conference (VTC Fall), 2011 IEEE, IEEE, 2011, pp. 1–5.

[42] L. Fan, X. Lei, T. Q. Duong, R. Hu, M. Elkashlan, Multiuser cognitive relay networks: Joint impact of direct and relay communications, Wireless Communications, IEEE Transactions on 13 (9) (2014) 5043–5055.

[43] C.-H. Huang, Y.-C. Lai, K.-C. Chen, Network capacity of cognitive radio relay network, Physical Communication 1 (2) (2008) 112–120.

[44] L. LU, W. JIANG, H. XIANG, W. LUO, New optimal power allocation for bidirectional communications in cognitive relay network using analog network coding, China Communications 7 (4) (2010) 144–148.

[45] J. Xu, H. Zhang, D. Yuan, Q. Jin, C.-X. Wang, Novel multiple relay selection schemes in two-hop cognitive relay networks, in: Communications and Mobile Computing (CMC), 2011 Third International Conference on, IEEE, 2011, pp. 307–310.

[46] J. Si, Z. Li, X. Chen, B. Hao, Z. Liu, On the performance of cognitive relay networks under primary user's outage

constraint, Communications Letters, IEEE 15 (4) (2011) 422–424.

[47] J. Chen, J. Si, Z. Li, H. Huang, On the performance of spectrum sharing cognitive relay networks with imperfect csi, Communications Letters, IEEE 16 (7) (2012) 1002–1005.

[48] Q. Li, Q. Zhang, R. Feng, L. Luo, J. Qin, Optimal relay selection and beamforming in mimo cognitive multi-relay networks, Communications Letters, IEEE 17 (6) (2013) 1188–1191.

[49] L. Zhang, J. Yang, H. Zhou, X. Jian, Optimization of relay-based cooperative spectrum sensing in cognitive radio networks, in: Wireless Communications, Networking and Mobile Computing (WiCOM), 2011 7th International Conference on, IEEE, 2011, pp. 1–4.

[50] G. Zhao, C. Yang, G. Y. Li, D. Li, A. C. Soong, Power and channel allocation for cooperative relay in cognitive radio networks, Selected Topics in Signal Processing, IEEE Journal of 5 (1) (2011) 151–159.

[51] Y. Chen, V. K. Lau, S. Zhang, P. Qiu, Protocol design and delay analysis of half-duplex buffered cognitive relay systems, Wireless Communications, IEEE Transactions on 9 (3) (2010) 898–902.

[52] M. Naeem, A. Anpalagan, M. Jaseemuddin, D. C. Lee, Resource allocation techniques in cooperative cognitive radio networks, Communications Surveys & Tutorials, IEEE 16 (2) (2014) 729–744.

[53] P. Ubaidulla, S. Aïssa, Robust distributed cognitive relay beamforming, in: Vehicular Technology Conference (VTC Spring), 2012 IEEE 75th, IEEE, 2012, pp. 1–5.

[54] M. Choi, J. Park, S. Choi, Simplified power allocation scheme for cognitive multi-node relay networks, Wireless Communications, IEEE Transactions on 11 (6) (2012) 2008–2012.

[55] X. Zhang, Z. Yan, Y. Gao, W. Wang, On the study of outage performance for cognitive relay networks (crn) with the nth best-relay selection in rayleigh-fading channels, Wireless Communications Letters, IEEE 2 (1) (2013) 110–113.

[56] H. Kim, S. Lim, H. Wang, D. Hong, Optimal power allocation and outage analysis for cognitive full duplex relay systems, Wireless Communications, IEEE Transactions on 11 (10) (2012) 3754–3765.

[57] H. Tran, T. Q. Duong, H.-J. Zepernick, Performance analysis of cognitive relay networks under power constraint of multiple primary users, in: Global Telecommunications Conference (GLOBECOM 2011), 2011 IEEE, 2011, pp. 1–6.

[58] W. Xu, J. Zhang, P. Zhang, C. Tellambura, Outage probability of decode-and-forward cognitive relay in presence of primary user's interference, Communications Letters, IEEE 16 (8) (2012) 1252–1255.

Permissions

List of Contributors

Isabel Montes, Romel Parmis and Roel Ocampo
Computer Networks Laboratory, Electrical and Electronics Engineering Institute, University of the Philipines Diliman

Cedric Festin
Networks and Distributed Systems Group, Department of Computer Science, University of the Philippines Diliman

Tian Zhang
School of Information Science and Engineering, Shandong University, Jinan 250100, China

Wei Chen and Zhigang Cao
State Key Laboratory on Microwave and Digital Communications, Tsinghua National Laboratory for Information Science and Technology (TNList), Department of Electronic Engineering, Tsinghua University, Beijing 100084, China

Wei Zhang
School of Electrical Engineering and Telecommunications, The University of New South Wales, Sydney, NSW2052, Australia

Shree Krishna Sharma and Symeon Chatzinotas
SnT - securityandtrust.lu, University of Luxembourg, Luxembourg

Mohammad Patwary
FCES, Staffordshire University, United Kingdom

Dongliang Duan
Department of Electrical and Computer Engineering, University of Wyoming, Laramie, WY

Liuqing Yang
Department of Electrical and Computer Engineering, Colorado State University, Fort Collins, CO

Shuguang Cui
Department of Electrical and Computer Engineering, Texas A&M University, College Station, TX

Mahdi Azarafrooz and R. Chandramouli
Department of Electrical and Computer Engineering, Stevens Institute of Technology, Hoboken, NJ

Donghai Zhu, Xinyu Yang and Peng Zhao
Department of Computer Science and Technology, Xi'an Jiaotong University, Xi'an, China

Wei Yu
Department of Computer and Information Sciences, Towson University, Towson, MD, USA

Tianyu Wang and Lingyang Song
School of Electrical Engineering and Computer Science, Peking University, Beijing, China

Zhu Han
Electrical and Computer Engineering Department, University of Houston, Houston, USA

Yichuan Wang and Xin Liu
Department of Computer Science, University of California, Davis, CA 95616, USA

Xiaoyan Wang, Yusheng Ji and Hao Zhou
Information Systems Architecture Science Research Division, National Institute of Informatics, Tokyo, Japan

Zhi Liu
Global Information and Telecommunication Institute, Waseda University, Tokyo, Japan

Jie Li
Faculty of Engineering, Information and Systems, University of Tsukuba, Tsukuba Science City, Japan

Cong Xiong, Geoffrey Ye Li and Lu Lu
School of Electrical and Computer Engineering, Georgia Institute of Technology, Atlanta, GA, USA

Daquan Feng
School of Electrical and Computer Engineering, Georgia Institute of Technology, Atlanta, GA, USA

National Key Lab on Commun., University of Electronic Science and Technology of China, Chengdu, China

Zhi Ding
School of Electrical and Computer Engineering, University of California, Davis, CA, USA

Helena Mitchell
Center for Advanced Communications Policy, Georgia Institute of Technology, Atlanta, GA, USA

Huu Tam Tran, Harun Baraki and Kurt Geihs
University of Kassel, Distributed Systems Group, Wilhelmshöher Allee 73, Germany

Praveen K. Muthuswamy and Koushik Kar
Department of Electrical, Computer, and Systems Engineering, Rensselaer Polytechnic Institute

Aparna Gupta
Lally School of Management and Technology, Rensselaer Polytechnic Institute

Saswati Sarkar
Deptartment of Electrical and Systems Engg, University of Pennsylvania

Gaurav Kasbekar
Deptartment of Electrical Engg, Indian Institute of Technology, Bombay, India

Lanchao Liu and Zhu Han
Department of Electrical and Computer Engineering, University of Houston, Houston, TX, USA

Zhiqiang Wu
Department of Electrical Engineering, Wright State University, Dayton, Ohio, USA

Lijun Qian
Department of Electrical and Computer Engineering, Prairie View A&M University, Prairie View, TX, USA

V.Balaji, S.Anand and C.R.Hota
Department of Computer science and Information Systems, BITS Pilani, Hyderabad Campus, 500078, India

G.Raghurama
Department of Electrical and Electronics Engineering, BITS Pilani, Goa Campus, 403726, India

Nima Namvar
Electrical and Computer Engineering Department, University of Tehran,Tehran, Iran

Walid Saad
Wireless@VT, Bradley Department of Electrical and Computer Engineering, Virginia Tech, Blacksburg, VA

Behrouz Maham
Electrical and Computer Engineering Department, University of Tehran,Tehran, Iran

Haibin Zhang, Jiajia Liu and Cheng Zhao
School of Cyber Engineering,hbzhang@mail.xidian.edu.cn, Xidian University, Xi'an 710071, P.R. China

Ayesha Naeem
Military College of Signals, NUST, Pakistan

Mubashir Husain Rehmani
COMSATS Institute of Information Technology, Wah Cantt, Pakistan

Index

A

Adaptive Services, 114
Alleviate Cellular Congestion, 82
Architecture, 29, 57-58, 60, 69-70, 91, 114-115, 117, 122, 177, 185

B

Bandwidth Contracting, 124
Bandwidth Scavenging, 1-3, 5-7, 9
Bernstein Convex Approximation Technique, 124
Body Sensor Networks, 176, 182

C

Cell Selection, 165, 175
Cellular Networks, 26, 35-36, 82, 89, 98, 108-109, 111, 113, 171, 174-175
Cellular Operation, 82
Coalition Formation Algorithm, 76-78
Coalitional Games, 72-73, 81, 105
Cognitive Radio, 31, 37-40, 45-46, 55, 72-73, 75, 81, 91-92, 97-98, 107, 110-113, 142-143, 145, 155, 157-159, 164, 184-190
Cognitive Radio (cr), 46, 107, 143, 157, 184
Cognitive Relay Network, 184-186, 188-190
Compressive Sensing, 143, 147, 155-156
Compressive Spectrum Sensing, 143-144, 146, 154
Congestion Advertisement, 46-47, 49, 52-53
Congestion Control, 1-3, 5-6, 8-10
Context Information, 165-166, 168-171
Context-aware Matching Game, 165
Cooperative Multiple-input Multiple-output (mimo), 11
Cooperative Packet Forwarding, 1
Cooperative Retransmission, 91-92, 94-97
Cooperative Spectrum Sensing, 38-39, 45, 74, 143, 157, 159, 164, 188, 191
Coordinating Agents, 114

D

Decentralized Spectrum Sensing, 143
Df-af Selection, 11-14, 20
Distance Based Method, 176, 182
Diversity, 11-14, 20, 24-27, 30-33, 36-38, 42, 45, 56, 83, 86-87, 98, 102, 110, 145, 152, 158, 165, 185, 187-189
Dynamic Spectrum Access, 46, 70, 81, 91, 110-112, 157, 164, 189

E

Effective Intra-flow Network Coding, 56, 70
Efficiency and Reliability, 38
Evolution Analytics, 117-118

G

Game Theoretic Approach, 91
Game Theory, 46, 73, 81, 92, 98, 103, 112
Graphical Anti-coordination Game, 46

H

Hidden Markov Model (hmm), 176-177

I

Infrastructure Coverage, 1
Internet of Things, 1, 8-9, 114, 122-123
Intra-flow Network Coding, 56-57, 61, 69-70

K

Kernel Density Estimation (kde), 176

L

Learning, 46-47, 49-51, 53-55, 83, 90, 111, 113, 155, 157-159, 164
Ledbat Lbe Algorithms, 1
Less-than-best-effort, 1
Localization, 143-146, 149, 152-156
Logit-response Dynamic, 49

M

Machine Learning, 157-159, 164
Mimo Relaying, 11-14, 20, 24
Mp-lbe Design, 3, 8
Multipath Flows, 1

N

Network Model, 73, 96

O

Opportunistic Scheduling, 29-30, 36, 82-85, 90
Orthogonal Multiple Access, 26
Outage Probability, 11-14, 20, 24, 165, 187-189
Outlier Detection, 176-177, 181-183

P

Performance Metrics, 39-40, 82

Portfolio Optimization, 124-128, 142
Price of Anarchy, 46, 49, 53
Pricing Mechanism, 101-102

R
Rayleigh Fading, 12-13, 24, 36, 171, 187-189
Revolutionary Technology, 184

S
Secondary Subflows, 3, 5, 8
Service Co-evolution, 114-116, 120-122
Sliding Window, 176-177, 181-182
Small Cell Networks, 165-166, 173-174
Software Defined Wireless Mesh Networks, 56
Space Division Multiple Access (sdma), 26-27
Spectrum Leasing, 91-92, 97-98, 100-101, 108, 112-113
Spectrum Markets, 100, 109-110, 124-126
Spectrum Sharing, 46-47, 49-55, 91, 99-100, 107, 109-113, 126, 142, 189, 191

Spectrum Trading, 99-105, 107-112, 125-126, 142
Stable Matching, 107, 165-166, 170-171, 173-174, 189
Stackelberg Game, 91-92, 94-95, 98, 102, 105, 108
Synthetic Trace, 86, 88
System Model, 13, 39, 73, 83, 92, 144-145, 166

T
Tcp Congestion Control, 1, 5
Trading Model, 91-92, 94, 104
Traffic Engineering, 82

U
User-cell Association, 165

V
Vehicular Ad Hoc Networks (vanets), 72

W
Wireless Mesh Networks, 56-58, 69-71
Wireless Two-tier Networks, 165